GW01311794

Aero Press Books
14445 Northridge Drive
Charlotte, N.C. 28269

Rev. 1-3-15

Acknowledgments

Einstein once stated that genius is 1% inspiration and 99% perspiration. Rest assured designing and building a helicopter requires a lot of the aforementioned perspiration. Despite the many frustrations and time-consuming aspects of the project, I enjoyed every phase - the technical, the engineering, design layouts, part detailing, Bill-of-Material generation, part fabrication, assembly and finally culminating with flight testing. As with any self-absorbing project the time spent is time subtracted from one's social life. It is with sincere gratitude I acknowledge my wife's support from start to finish. I could not have completed this project without her.

Forward

The Helicopter simply put is an aircraft that uses rotating blades to provide lift, propulsion and control. They vary in size and shape but the majorities have the same major components such as a main rotor, tail rotor, power plant, transmission, cabin and landing gear all supported by a structural airframe. In flight, the main and tail rotor blades rotating at high speed generate lift and translation of the craft. Due to the fact that the speed of the rotor blades is independent of the actual helicopter speed, a helicopter can hover over a spot on the ground, rise and descend vertically, fly backward, sideways left or right, as well as forward. The inherent benefit of these flight characteristics is that it does not require a formal runway to takeoff or land meaning even an area the size of a tennis court is adequate. Another benefit of the high rotational speed of the rotor blades relative to the surrounding air is that helicopters are not as prone to the destabilizing effects of wind shear and gusts as compared to fixed wing aircraft.

This book is a compilation of the research, engineering and physical experiences I had over the last few years in my quest to design and build an ultralight helicopter. It all started when I purchased plans for one off the internet only to find that they were not truly feasible to build nor would they have been airworthy. Intrigued I began to investigate what was truly required and then I was hooked. This book is for those who would like to know more about helicopters and their function as well as those who might wish to go so far as to continue to develop an ultralight of their own design. I've enjoyed the experience and wish to share what I've learned from it.

Table of Contents:

Introduction

As you can see from the Table of Contents this book covers a lot of territory. Subsections 0.1 thru 0.5 will familiarize the reader with the design / build portions of the book while subsections 0.6 thru 0.8 will provide interesting facts regarding the history of helicopter development. Each section was clearly noted so that the reader may easily find those topics in which they have the most interest and skip over or save for later reading the remaining sections.

0.1 Inside This Book

This book was compiled from the knowledge acquired during the research, engineering, design, part fabrication and assembly of my full-function ultralight helicopter. Prior to tackling this task my only experience with helicopters was being airlifted from the Atlantic Ocean during an Air Force Sea Survival course held off the coast of Virginia.

Every phase of the development process I experienced to build this helicopter is divulged in this book. It presents the information assuming the reader has no prior helicopter background. In addition it does offer in-depth detailed calculations for those interested in the analytical and the theoretical of this helicopter's development. All the calculations presented were performed using Mathcad, an engineering software program.

Additional reference materials are provided for the reader seeking further knowledge in areas such as advanced rotor aerodynamics, rotor planform theory, etc. or training instruction on the various pieces of equipment required to fabricate parts.

The Appendix includes various fabrication reference charts, a Federal Aviation advisory circular which answers frequently asked questions concerning ultralights, a military specification on helicopter control requirements, my patent on laminar airflow and "supplemental information". In the supplemental information section there is a further clarification of relevant technical terms as well as interesting analogies to explain them further.

0.2 Topics Included

In Part 1 "THE BASICS", after a brief discussion of what I consider as prerequisites for "The Helicopter Builder (Chapter 1)" and discuss the properties of "Atmospheric Air & Airfoil Theory (Chapter 2)". It begins with getting the reader familiar with the helicopter as if planning to pass an FAA flight exam. All terminology in this book is identical to that used by the FAA for the key components. Subjects the FAA requires to be known such as the subsystems of "The Helicopter (Chapter 3)", "Basic Helicopter Aerodynamics (Chapter 4)", "Basic Flight Aerodynamics (Chapter 5)" and operation of the "Flight Controls (Chapter 6)" are covered.

The only difference here is that I am presenting the information as it applies to my specific ultralight helicopter. Note: If you want to know more than what is presented I suggest downloading both the FAA's "Pilots Handbook of Aeronautical Knowledge, FAA-H-8083-25" and the "Helicopter Flying Handbook, FAA-H-8083-21A" which at the time of this writing are available on the internet.

Part 2 "ENGINEERING" focuses on the engineering involved starting with "Helicopter Design Relationships (Chapter 7)" and "The Calculations (Chapter 8)".

In the "Helicopter Design Relationships" chapter, I start with a simple comparison of some very successful helicopters where respective weight, horsepower and rotor blade diameter are tabulated. This leads into six generally accepted helicopter relationships, which basically define the overall design envelope for my helicopter. Chapter 8 "The Calculations" mathematically finalizes those parameters for this ultralight helicopter as per the rotor blade shape and size, the type rotor assembly and the required engine power.

When I stated that I started from scratch on this project I didn't mean to imply that I was going to "reinvent the wheel". Although I personally started with a blank piece of paper I made every effort to utilize in my work all the latest information of theory and design that was available including using experimental data when current formula did not predict repeatable results.

After a cursory look at other helicopters and the subsequent correlation of functional relationships, the next subchapters become more detailed. Simply knowing the relationship between the power required and helicopter weight doesn't explain the mechanics of how this power is transferred into lift. In the airfoil section (Subchapter 8.2.3) the rotor blade shape is selected. Once this is established the performance plots for the airfoil are obtained from the reference book "Theory of Wing Sections". These plots provide blade lift and blade drag as the blade angle is varied (Subchapter 8.2.5). Although over simplified here this information is later used in "The Rotor" section

(Subchapter 8.4) to determine the rotor diameter. Once the rotor diameter is defined and the type rotor established detailed step-by-step calculations (using Mathcad engineering software) are presented for the defining rotor rpm and rotor blade width (chord). This information defining the overall rotor such as NACA0012 drag equation, rotor blade radius, cord and rpm are used in the "Energy Balance Equations (Subchapter 8.5.1)" to generate specific rotor shaft power during hover and flight at sea level.

Subsequent calculations include finding the Range and Endurance (Subchapter 8.6.2), "Maximum Forward Speed" (Subchapter 8.6.3), "Rate of Climb" (Subchapter 8.6.4), "Horsepower required versus Angle-of-Attack (Subchapter 8.6.5)", "Maximum Altitude (Subchapter 8.6.6)", "Ambient Air Temperature impact on Angle-of-Attack setting (Subchapter 8.6.7)". In addition the "Optimal Engine Horsepower calculation was derived from an energy balance equation and from general formula, including the theoretical hovering pitch angle calculation for the helicopter. This work more or less defines the basic helicopter from the "power" perspective. The additional calculations required to support structural integrity, gearbox strength, rotor blade stress, etc will continue in Part 4 "DESIGN DETAIL" where those subsystems are analyzed.

In Part 3 "THE BUILDER'S TOOLBOX", I take a step back to describe the "Raw Materials (Chapter 9)" and "Commercial Parts (Chapter 10)" available for the helicopter build. These descriptions vary depending on the subject material but in general include the basis for selection of these items. For example when choosing a one-inch diameter rod, one made of aluminum weighs one third as much as a steel one but also deflects under load three times as much. By increasing the diameter of the aluminum piece and manipulating the wall thickness you could maintain the weight saving and be equal in deflection to the steel piece. For weight saving structures this is useful knowledge. In some instances I state why a material is rejected when it appears to be the logical choice. For example composites may be the current trend today in aircraft design but being aware that the tail of Flight 587 (Airbus A-300) fell off in flight because of a composite bracket failure, I decided not to incorporate any composites in my design other than as filler material.

In addition to my comments I have included a significant number of charts to enhance the readers understanding of the subject. This is especially helpful with the power transmission selection of belts, gears and sprockets required to deliver the engine power to the main and tail rotor shafts. These charts enable one to rather quickly grasp the capabilities and limitations for instance of a belt drive. If you know how fast the mating shafts must rotate then the charts will state what the diameters of the pulleys should be and how many belts are required. This selection process continues until that belt pulley combination to be used not only satisfies the power transfer requirements but also weighs the least.

Also included is a section "Fabrication Machinery (Chapter 11)" devoted to the machinery required to make the helicopter. The initial cost for the equipment represents around 25% of the total cost of the helicopter, provided you elect to do all the work "in house" with your own staff or by yourself. If you subcontract the work to outside vendors the helicopter will basically cost approximately three times as much based upon current estimates, as compared to doing the work with owned equipment. I do not attempt to give in depth knowledge since to be a machinist technically requires a formal six-year training program but I do provide instruction based upon my own learning experience. This self training enabled me to become proficient enough to fabricate parts to within an industry accepted part dimensional size variation of within five thousands of an inch (0.005"). Those parts that initially gave me grief such as putting a vertical slot in a hardened spherical bearing come with their own step-by-step process sheet. Off hand the two most important things I learned regarding the operation of the lathe and the milling machine was to always use cooling water while the material was being formed and to make sure the material was totally secured before attempting to cut it. The cooling water minimizes thermal expansion of the part while it is being cut, thereby what you measure while the part is still in the machine will measure the same after the part is removed and reaches room temperature. This insures good fabrication dimensional control. Having seen a part totally destroyed in a split second because the material wasn't secured definitely made me a believer in the importance of making sure the part is anchored down before doing anything. This part concludes with a section "Suppliers (Chapter 12)" noting suppliers for the parts, tools, raw materials etc, as well as possible sources for new or used engines and gearboxes necessary for the project.

Part 4 "DESIGN DETAIL", "Visualizing the Design (Chapter 13)" and "The Subsystems Analyzed (Chapter 14)" really go hand-in-hand. Before you start formulating a CAD (computer aided drafting) full-scale three-dimensional model you must have a good idea of what you want to include. Chapter 13 is where everything comes together. All the parts selected during the engineering phase have their physical dimensions transferred into subassembly drawings for each module (such as the rotor, swashplate, drive, etc.) which in turn forms the basis of the overall helicopter layout. The engineering work includes accommodating government technical specification MIL-H-8501A, which dictates how the controls should work and establishes parameters for flying and ground handling qualities required for military helicopters (the USAF spec is MIL-H-83300). As the helicopter layout takes shape based upon the above fixed engineering data (referred to as hard points) it is noted that some of the inputs are subjective. An example would be the determination of your seating position. In contrast to the calculations required to get to this point, it turned out to be one of the most time consuming tasks. It took many variations to finally arrive at the best ergonomic seating position allowing the easy reach of the foot pedals, the joystick, the collective, and the throttle while still in position to effectively monitor the instruments.

As to the detailing process itself, in the old days the layout was completed on an "E" size piece of paper (36" x 48") positioned squarely on a horizontal drafting table three feet wide by six feet long. Lines were drawn by pencil with the aid of T-squares and plastic triangles, while circles were made with various sizes compasses. At the end of the day your neck was stiff from leaning over your board for hours at a time. Just before the advent of computer-aided-drafting the newer model drafting tables could rotate the entire board upward which helped your neck but old habits die hard and many a time putting a pencil down meant it wound up on the floor subsequently breaking the lead tip off. I mention this because if you think using paper and pencil is the easier way to layout your helicopter you are going down the wrong path. Covered in this chapter, I discuss the benefits of computer aided drafting "General Comment (Subchapter 13.1)", the modeling of the helicopter in three-dimensional space "The Three-Dimensional Model (Subchapter 13.2)", generating the details for the parts "Details (Subchapter 13.3)", generating a bill-of-materials "Bill of Materials (Subchapter 13.4)" and also explained is the significance of proper revision documentation "Revision Documentation (Subchapter 13.5)" as it applies to your personal project including its relevance in the international manufacturing community as dictated by those ISO 9000 series numbers you see on the manufactured items you buy today. That does not mean it can't be done without software but it would definitely be twice as time consuming and tedious.

Chapter 14 is where the engineering is completed and all the parts to be used in the subassemblies are selected. The calculations discussed in this section are comprehensive as would be expected for such a complex aircraft as a helicopter. For instance stress analysis is performed on the structure and the rotor blade; gearbox transmission strength evaluated to insure proper performance over the flight operational range, cockpit layout and instrument selection finalized, etc. Although some preliminary sketching is performed the actual layout into dimensionally accurate subassemblies is completed as stated in chapter 13.

Part 5 "THE HELICOPTER TAKES SHAPE" addresses the transformation of raw material into physical parts "Fabrication (Chapter 15)" and their subsequent placement into physical subassemblies to form the completed helicopter "Assembly (Chapter 16)". "Programming (Chapter 17)" is discussed as it relates to engine speed control via a PLC (Programmable Logic Controller) as rotor shaft flight loads fluctuate.

Part 6 "THE FINAL PHASE" covers helicopter testing "Testing (Chapter 18)" relating some first hand experiences as to what to expect. Something seemingly as simple as finding an airfield to use for testing is not as straightforward as one might expect. Also included are items that should be learned from these flight test sessions. Included are the startup checklist, data sheets to record critical run-up information after startup, and suggested procedures for tethering the helicopter for restricted flight while getting experienced with the controls. Next up is the "The FAA (Chapter 19)" where Subchapter 19.1 states FAA regulation FAR 103 which covers Ultralight restrictions and Subchapter 19.2 deals with a discussion of the flight manual. As an example included is the flight manual generated for this helicopter, the Airsport 254.

"Legal Paperwork (Chapter 20)" covers the steps necessary to register the aircraft with the FAA. Chapter 21, which focuses on cost, is divided into two areas. The first part is centered on the estimated cost "today" to build your helicopter as defined in the Airsport 254 Bill of Materials, while the second addresses the cost to keep it maintained, stored and flying. The final topic (Chapter 22) discusses how to transport the helicopter and modifications that can be made to an existing tag-a-long trailer that would allow compliance to local traffic laws. Chapter 23 titled "The End" is a personal account of the overall project and what direction it should take from here.

The book ends with a comprehensive Appendix.

0.3 How to Maximize Your Benefits from this Book

This book was developed to achieve two objectives. The first objective was to enable readers to understand how to design a rigorously engineered inexpensive ultralight helicopter with a low operating cost. The second objective was to give you a manufacturing overview of what it takes to transform your design from raw materials to physical parts to assemble into your helicopter.

This book enables the reader to not only build a helicopter but also provides the opportunity to expand on the knowledge I have compiled to take this design to the next level. The reader determines the ultimate outcome. If you actually intend to build a helicopter from the design process on up, every portion of this book must be totally absorbed, every detail understood, and all raw materials and component parts sourced (by "sourced" I mean being able to buy the raw material or commercial part). Just to drive home the "sourced part" further, when I was just about finished assembling the framework of the helicopter the engine supplier I intended to use "Two Stroke International" went bankrupt. This resulted in a substantial amount of rework to the CAD model in order to incorporate a replacement engine and dictated the fabrication of altered support framework. In addition, the peak torque of the new engine occurred at a different RPM that resulted in the need to spend additional money for new pulleys and belts to accommodate the new optimal drive train ratio. The point is that if you achieve a good working knowledge of the helicopter as defined in this handbook you will be able make revisions to the design if necessary to finish your project if things beyond your control change.

Obviously, to get the most out of this book you must actually build the helicopter. If this is not feasible you could still learn a substantial amount from this handbook by simply building your own pseudo-helicopter in the form of a computer 3-D model from the subsystem design parameters I used and have included in Part 4 titled "Design Detail". Once this CAD model is complete you can view it from every angle, generate pictures of the various views and certainly feel very proud of yourself for creating this helicopter part by part as your own personal 3-D helicopter. The neat aspect about this approach is, it costs you absolutely nothing to create your 3-D parts, yet you can rotate them, move them around, change their color, and even modify them just like real machined parts only they reside in digital space. The other benefit is the extent of helicopter knowledge you acquired by actually building the various subsystems piece by piece. For example just reading about the function of a swashplate you can't grasp the reality of its true function. When you create and assemble one, part by part as a model, the whole concept of how a swashplate functions becomes real. Creating this 3-D model should be considered a test of your resolve. If you find the completion of the model to be unrealistic then obviously there is no reason to spend the money to try and build it! Rather use that money to attend helicopter flight school and use this handbook as a study guide.

To build an actual helicopter, the first step would be to check out Chapter 21 titled "Costs". If these estimated dollar numbers seem acceptable, your next step would be to extract from the 3-D model every part and make a fully dimensioned detail of it, number the detail and list it in a BOM (Bill of Materials) document. Then obtain a quote for all the commercial parts and raw materials detailed in your helicopter Bill of Materials and my Machinery list found in Chapter 11. Once you obtain the estimate of this helicopter and machinery cost in today's dollars, I recommend adding an addition 20% to that amount as contingency for escalating costs due to inflation. If this estimated cost is still acceptable and you want to proceed then the next step would be to buy the Lathe and Milling machines and proceed from there to start making your subassemblies.

If these costs are way beyond your budget but you would still like to have a physical model of your helicopter I would suggest investigating the possibility of buying one of those new 3-D printers. Then you could scale down your computer model and have the 3-D printer make all the parts for you to assemble.

The options are numerous and how far you take the project is up to you. I spent years on my helicopter and this book is the culmination of that effort. I have accomplished what I set out to do - design, build and operate my own helicopter. Hopefully this handbook will help someone to complete their own version adding to or improving on what I have presented or perhaps it will simply be a source of information for those wanting to expand their knowledge of ultralight helicopters.

0.4 Background Experience Required:

Although this topic is discussed more completely in Chapter 1 one important prerequisite to building a helicopter, whether from a kit or from plans, is having an above average hands-on mechanical aptitude. For example, as a kid you should have been able to build LEGO models from their instruction sheets and even better yet having built and programmed a LEGO "MINDSTORM" robot, as a pre-teen in addition to owning either a bike, dirt bike, go-kart or jet ski should have been able to maintain it, as a teen have a working knowledge of what makes your car go, and finally as an adult being in possession of your own set of English and metric wrenches including an electric drill, worn drill bits and a dull hacksaw.

Notice how I assume this is an adult project. The reasoning is that every age group referenced before adult is just too preoccupied growing up to tackle such a project. I remember when I was fourteen how gung-ho I was to customize a 1935 Chevy pickup I had acquired. I got so far as to take the flat panel dash out and replace it with one from a 1955 Oldsmobile. That restoration effort was interrupted by such things as school, service, college, and married life. Finally, after some concentrated effort the Chevy was finally completed, thirty years later! You need the aptitude and the desire but also the time.

0.5 Standard Conventions Employed

The standard conventions used are as follows:
   Helicopter nomenclature established by the FAA
   Drafting guidelines from " Modern Drafting Practices and Standards" published by General Electric
   Welding symbols from "Symbols for Welding & Nondestructive Testing" published by the American Welding Society
   Bill of Materials format in accordance with ISO 9000 guidelines

0.6 Brief History of the Helicopter

The time span for the helicopter to truly invent itself covers almost 2400 years. It starts with the Chinese flying top first used around 400 B.C. and culminates in the early 1950's. Subsequently helicopter improvement focused more on research and development of existing methodologies than on invention. That's not to say everything dealing with helicopter evolution is over, its just that at the present time no one has come up with something new to try that hasn't already been suggested. Driving home the point further I have included the history of those people involved over the years, which in some aspect further advanced the knowledge or functionality of the helicopter. This becomes very educational as you scroll down this list and realize how slow this inventive process progressed.

Actually, you could parallel this same slow progress as it applied to the development of rocketry, which started with the Chinese rocket first used in the 13th century to the United States development of the Saturn V, which put a man on the moon in 1969. No rocket built to date has surpassed the shear lifting power of that rocket. Since 1969 for the rocket, as with the early 1950's for the helicopter no paradigm shift from existing theory has occurred to radically change anything. That's not to say improvements aren't continually being made. It's just that both are at a very advanced state of sophistication already.

Reading this chronological developmental time line is important because it forms a basis for your future work in this area. Of course I said the thought of vertical flight has been around for over 2000 years but in reality once the stage was set by having the right accessible metallurgical materials to build with, along with a suitable power plant and a certain engineering knowledge of the physical world it took only 44 years to go from the Wright Brothers first flight, December 17,1903 to having a helicopter, the Bell Model-47 receive a United State's certification for airworthiness in 1947.

The fact the Bell Model-47 was granted this certificate indicated to me the first place to start designing my ultralight helicopter was to understand every facet of this helicopter's theory, so that's where I started. If you follow the chronology thru you will understand the process of how the Bell-47 came into being.

The Beginning:

2500 B.C. a story states that the Chinese emperor Shun supposedly built himself an air chariot.

400 B.C. the Chinese invented a flying top, a stick with a propeller on top, which flew by manually generating a rotational spin to the assembly then releasing it.

Aristotle (Greek) proposed the concept that air has weight and Archimedes (Greek) developed the "Law of Floating Bodies" which later was used as the basis of lighter-than-air craft.

During the thirteenth century Galileo (Italian), Roger Bacon (England), and Pascal (French) added to the understanding of air as a gas. Defining properties such as the fact that it's pressure decreases with altitude and subsequently hot air rises and cold air descends. This work provided the beginnings of aerodynamics, the study of bodies thru air and other gaseous fluids.

DaVinci (Italy, late 15th century) generated sketches for a vertical flying machine utilizing a screw shaped propeller he simply called the airscrew. His actual model utilized bird feathers for the rotor blades and a mounting hub to secure them. The main difference between it and the Chinese flying top was the way energy was generated to rotate the propeller/ blades. The flying top was manually spun prior to flight, while with DaVinci's model the blades increased rotational speed as the craft descended due to the air movement. DaVinci is credited with being the first to understand that the faster air flows over a wing like surface the more lift is produced. His later flying machines attempted to simulate the flapping motion of birds to support manned flight. Unfortunately none of these ornithopters flew and the complexity of wing design was left to another time.

Sketch 0.6-1

Lomonosov (Russia, 1754) designed a vertical flight model powered by springs.

The first manned flight took place in 1783 by French nobleman Francois Pilatre de Rozier in a hot-air balloon built by Frenchmen Jacques and Joseph

Montgolfier.

Launoy and Bienvenu (France, 1784) their spring powered model had two four bladed rotors rotating counterclockwise with respect to one another.

Sketch 0.6-2

Sir Cayley (England, 1790's) developed models that used elastic components for power. In 1810 he wrote a dissertation on the aerodynamic fundamentals for both rotary and fixed-wing aircraft. Later in his career he built gliders and a helicopter, which had two steam powered counter-rotating rotors laterally spaced.

Sketch 0.6-3

Phillips (England, 1842) his 10 kg model was steam powered.

Heri Giffard, (France) added steam engine driven propellers to a dirigible allowing him to maneuver the craft in the air, which initiated the era of aerial navigation proving the functionality of mechanical power. In 1852 he flew a pilot directed flight over Paris at an impressive speed of 6 mph.

In 1860 Francis Wenham, an English scientist, was one of the first to conduct studies in wing design and concluded a flat straight wing surface provided far less lift than a curved one. Furthering his work Horatio Phillips, a Canadian, utilized a novel approach to study airflow by building a wind tunnel to collect data. This research enabled him to patent an "under cambered" airfoil which provided substantially more lift than prevailing designs. Due to this detailed work Phillips is considered to be the father of the science of aerodynamics. Wenham is also credited with pioneering the biplane or double winged aircraft increasing wing surface area without resorting to widening the span. He is also credited with suggesting the adaptation of a gasoline engine to power the airplane.

Gustave de Ponton d'Amecourt (France, 1863) built a steam-powered model but more significantly he was responsible for the actual word "helicopter" for this type of aircraft.

Sketch 0.6-4

1863
Gustave de Ponton d' Amecourt

Penaud (France, 1870's) built models.

Achenbach, (Germany, 1874) His model was the first to use a tail rotor to cancel main rotor torque.

Sketch 0.6-5

Achenbach
1874

Nicholas Otto (German) invents the internal-combustion engine in 1876. This event was a major paradigm shift in power delivery. Within four years Gottlieb Daimler perfects the gasoline-fueled engine.

Forlanini (Italy, 1878) built a 3.5 kg steam powered flying model.

Sketch 0.6-6

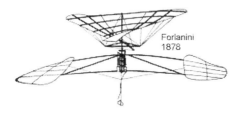

Forlanini
1878

Thomas Edison (United States, 1880's) worked with models but concluded helicopters would not be capable of sustained flight until engines developed with a weight-to-power ratio below 1 to 2 kg per horsepower.

The century ends with German brothers Otto and Gustave Lilienthal building a one-man, non-powered, hang glider in which Otto pilots to become the first person to fly a heavier-than-air craft. From their direct observations of bird wing manipulations in flight, to either increase stability or lift, they formulated a theory of flight, which laid the groundwork for current day aerodynamic studies.

The Twentieth Century:

On December 17, 1903 the first controlled sustained flight of an engine powered aircraft takes place at Kitty Hawk, N.C. USA by Orville Wright. The aircraft named the "Wright Flyer" was airborne for 12 seconds traveling 120 feet. The most significant contribution of the Wright brothers, Orville and Wilbur, was the development of the control system for turning and maneuvering in the air. Lilienthal's hang glider maneuvered by shifting the pilot's weight whereas in the Wright's plane the pilot operated controls which altered specific aerodynamic airflows to change the aircraft's direction. They were also the first to realize keeping the craft straight during these maneuvers required a rudder.

Soon after their success the study of aerodynamics spread worldwide with technical research facilities being built everywhere to address the technical and physical demands of aircraft flight. The three most noted facilities were France's Eiffel Laboratory, Russia's Aerodynamic Institute of Kotchino and Germany's University of Gottingen. Within a short period of time the major aircraft stability problems were solved such as the importance of proper center of gravity location, use of a decalage angle (the differential angle between the wing and tail, wing positive, tail negative) to dampen the effects of pitching motion, and use of a dihedral angle to handle lateral stability.

Renard (France, 1904) was the first to use a two-cylinder engine to power a helicopter with side-by-side rotors. He also developed the first rotor hub with a hinge so the rotor blade could pivot, commonly referred to as the "flapping hinge".

Sketch 0.6-7

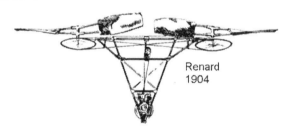

Renard 1904

Breguet-Richet (France 1907) Gyroplane #1 made a tethered flight one meter (3 ft) off the ground for about one minute carrying an observer. The craft had four bladed rotors, weighed 580 kg and was powered by a 45 HP engine.

Sketch 0.6-8

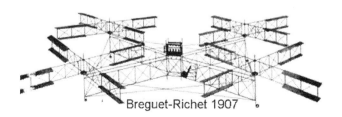

Breguet-Richet 1907

Cornu (France, 1907) is credited with the first helicopter flight with a pilot. The flight of the tandem counter-rotating two bladed rotor machine lasted 20 seconds and managed to raise above the ground about 0.3 meters (1 ft). His concept of controlling the aircraft was to utilize vanes in the rotor downwash slipstream. It did not work well and the overall flight was very unstable.

Sketch 0.6-9

Emile and Henry Berliner (Untied States, 1909) developed a two-engine coaxial machine that briefly lifted off the ground with a pilot without being tethered.

Sikorsky (Russia, 1910) built a three bladed dual coaxial rotor machine which could fly but not with the added weight of a pilot.

In 1909 aircraft racing begins with the "International Air Meet" held in Riems, New York. The owner of the New York Herald, James Gordon Bennett, sponsored the event. Glen Curtiss flying his "Golden Flyer" biplane by attaining an average speed of 46 mph won the event. In 1911, Germany's Count Zeppelin designs and builds a large rigid frame dirigible called the "Schwaben" which had the capacity to carry 32 people. It was the world's first air passenger service and by the start of World War I had carried 35000 ticket-purchasing passengers to various places on the continent. In England, the Royal Flying Service is founded and in 1913 the U.S. Army Signal Corps forms an aviation branch. The world's largest air force was the French Air Service having more than 200 planes by 1912.

Yuriev (Russia, 1912) was the first inventor of the helicopter with a single main rotor and a small anti-torque tail rotor. Although conceptually the first of this design the craft made no successful flights.

At the start of World War I in the summer of 1914 only Russia had a multi-engine airplane, the "Russly Vitiaz" (Russian Knight) powered by four 100 hp German engines. The first one was build in 1913 by future helicopter builder Igor Sikorsky and in 1915 a larger version produced called the "Ilia Mourometz" (The Giant) which had over a 100 foot wingspan, could carry 1000 pounds of bombs and had five machinegun positions. When the Armistice was signed, ending the war, over 60,000 military airplanes were built. The airplane, now less than fifteen years old, was no longer just a novel toy and substantial manufacturing plants replaced the garage-shop operation of the past with the likes of companies such as Boeing, Curtis and Dayton-Wright to name a few. Technological advancements to the plane itself during this period include all metal structure, more powerful supercharged engines for higher altitude operation, cantilevered wings for better aerodynamic characteristics, a plethora of instrumentation, and many advanced concepts for future exploitation. I mention these facts to point out how far behind the development curve the helicopter was in comparison to the airplanes technological sophistication at the time.

Petroczy and von Karman (Austria, 1916) first tethered helicopter with payload to attain an altitude of 50 meters.

In 1918 the U.S. Post Office starts airmail service between New York and Philadelphia to determine the practicality of such a service. In 1919 a U.S. Navy flying boat, the NC-4 piloted by Lt. Comdr Albert Reid, flew from Long Island to the Azores, the first Atlantic crossing by airplane. In 1920 Army Major R. Schroeder, flying a turbo-supercharged observation plane, reached an altitude of 31,115 feet setting the World Record.

De Bothezat (United States, 1922) Utilizing four six bladed rotors at right angles to one another, this craft incorporated a mechanism to alter the mean blade pitch angle which enabled the amount of upward thrust of each rotor to be varied during flight. The current term for this capability is called differential collective or simply collective. This addition resulted in a helicopter that was very stable. In addition to being able to carry passengers vertically up to an altitude of 4 to 6 m it was the first to be ordered by the Army.

Sketch 0.6-10

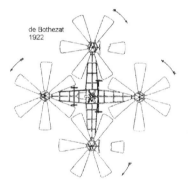

de Bothezat
1922

Juan de la Cierva (Spanish, 1920's-1930's) develops the autogiro, the first practical utilization of a rotary wing for direct lift. This rotary mechanism could not hover or initiate vertical flight since it was not powered directly; however, due to its capability to windmill as it forced forward lift was achieved. The early autogiro simply was designed into a wingless airplane. In 1923 Cierva flew his craft over the Cuatro Vientes airport at an altitude of 82 feet covering a 3-mile circuit to become the first officially documented flight of a roto-wing configured aircraft. Cierva is credited with being the first inventor to use a flap hinge in a viable rotary wing aircraft. The craft was so successful that about 500 C6 autogiros were manufactured and sold. Advances made in rotor design as a result of his work included the better understanding of high in-plane blade stress due to flapping, the addition of a lag hinge, and direct aircraft control via rotor disc manipulation. This also led to the development of the fully articulated rotor hub. Most importantly during this time of the 1920's and 1930's, the substantial amount of rotary wing analytical and experimental work generated because of his autogiro enabled the groundwork to be set for advancing helicopter design.

Sketch 0.6-11

Jaun de la Cierva 1920-1930

Oemichen (France, 1924) this craft traveling 360 m set the first helicopter distance record. Not only did this 120 horse-powered machine have four main two-bladed rotors it also incorporated eight driven propellers to maintain control. Five were used for maintaining attitude, two for propulsion, and one for Yaw stabilization.

Sketch 0.6-12

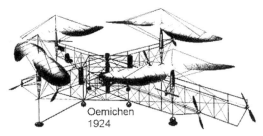

Oemichen
1924

Pescara (Spain, 1924) his helicopter had two coaxial four bladed rotors and set a new distance record at 736 m. He was the first inventor to use effective cyclic control of the main rotors to change the tilt of the rotor disc. Although innovative in design the helicopter stability was still an issue.

Sketch 0.6-13

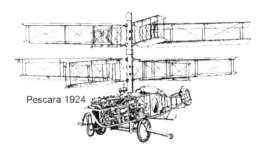

Pescara 1924

In 1925 production of Ford's all-metal corrugated-skin tri-motor 12-passenger transport begins and soon becomes the cornerstone of U.S. airlines for years. Then in 1927, Lindbergh completes the first trans-Atlantic nonstop solo flight from Roosevelt field, Long Island to Le Bouret in Paris. His plane named "The Spirit of St. Louis" utilized the "Earth Inductor Compass" for navigation. It was one of the many new devices being developed that helped eliminate a pilot "flying blind". Elmer Sperry's addition of a gyroscope to instrumentation like the " Artificial Horizon gage" established a fixed base point that allowed accurate flying through all types of inclement weather such as rain, sleet and snow and the darkness of night by instantly showing any deviation in direction. The "Artificial Horizon" instrument also indicated aircraft altitude. These devices were designed such that vibration, the earth's magnetic fields or the metal of the aircraft did not affect their functionality and in the case of the cockpit compass did not "hunt" like land/sea magnetic compasses.

Emile and Henry Berliner (United States, 1920-1925) their radical design placed a tilt-able rotor with wood propellers at each wing tip on a biplane. Control was attained as a function of rotor tilt.

Brennan (England, 1920's) this helicopter, powered by propellers mounted on the rotor blades themselves eliminated the need to counter balance main rotor torque. Cyclic control was handled by inducing blade warp via blade mounted mechanical tabs. Mechanically, the machine was extremely complex.

Baumhauer (Holland, 1924 - 1929) this 1300 kg helicopter consisted of a single 15 m diameter two-bladed main rotor and a tail rotor for counteracting torque. Each rotor was driven independently with its own engine - 200 hp for the main and 80 hp for the tail. The rotor blades could flap, the rotor teeter, and mechanical cyclic control was handled via a swashplate. Ultimately, controlling two independent engines at the same time created directional control issues and further development discontinued after crashing.

D' Ascanio (Italy, 1930) This 95 hp helicopter had two coaxial two bladed rotors each 13 m in diameter both incorporating flap and feathering hinges. Servo tabs integrated into the blade enabled cyclic and pitch changes. This craft held the records for altitude (18 m), distance (1078 m) and endurance (8 min. 45 sec) for several years. Unfortunately, controlling the craft was difficult.

Blecker (United States, 1930) His machine used four wing shaped blades each with its own propeller driven from a fuselage-mounted engine. Control was maintained manipulating surface deflectors on the blades and the tail.

Yuriev (Soviet Union, 1931) This 1100 kg, 120 hp, helicopter utilized an 11 m diameter four-bladed main rotor, which incorporated cyclic and collective control. Two small counter-rotating tail rotors were used to handle main rotor torque.

Hafner (England, 1935) He improved on the direct rotor hub tilt employed by Cierva by developing a "spider" mechanism for controlling cyclic pitch.

Wilford (United States, 1930's) develops a cyclic controllable hinge-less autogiro rotor.

In 1931 Germany build a twelve-engine transport plane capable of carrying 169 passengers. It was called the Dorier DO-X. In England a S6B Supermarine racing seaplane attained a speed of 407.5 mph. In the USA an Army tri-motor Fokker transport stayed aloft for 150 hours and 40 minutes by air-to-air refueling from an aerial tanker.

In 1932 Sperry invented the automatic pilot for aircraft. Its function allowed hands free navigation once the heading and altimeter are set. Also during that time the variable-pitch propeller was introduced enabling increased engine efficiencies over a wider rpm range. Flush head rivets replaced conventional bullet-head rivets in the aircraft skin assembly process reducing aerodynamic drag during flight. In 1934 wing flaps are first used on Boeing's 180 mph twin-engine Model 247 to eliminate wing stall at slow speed during landing. This 10-passenger plane is considered the first modern airliner.

1935: by this time in both America and Europe the autogiro was quite advanced. It was more successful than the helicopter because it required less power and employed a structurally simpler rotor but it could not accomplish vertical flight.

Dorand (France, 1935) This helicopter had a 450 hp engine and a gross weight of 2000 kg. It employed coaxial 16.5 meter diameter two bladed rotors each having fully articulated hubs and was controlled via a cyclic mechanism. Directional control was handled by differential applied rotor torque. The craft held the forward speed record at the time (44.7 kph), altitude (158 m), distance (44 km), and duration (1 hr 2 min). Technical data: gross weight per hp = 9.7; disc loading = 1.9 lbs / sq. ft.

Focke (Germany, 1936) This helicopter had a 160 hp engine and a gross weight of 950 kg. It utilized two side-by-side 7 m diameter three-bladed rotors that were angled slightly towards each other for increased stability. Each rotor had tapered blades and an articulated hub. The cyclic controlled direction and longitude, a differential collective the roll. For control in forward flight the tail had both horizontal and vertical surfaces that aided stability and trim. The craft was stable and set the following records: altitude (2440 m), speed (122.5 kph), endurance (1hr 21min). Technical data: gross weight per hp = 13; disc loading = 2.5 lbs / sq. ft.

Sketch 0.6-14

Focke 1936
D-EKRA

1935: Boeing introduced the "Stratoliner" which was the first passenger plane in the world to be pressurized so it could cruise at 20,000 feet to fly above any weather induced turbulence. The inside cabin pressure (typically set to 8000 ft) was maintained such that passengers did not have to wear oxygen masks which would have been a necessity had the plane not been pressurized. During this period retractable landing gear appears on advanced aircraft such as America's P-35 fighter.

Flettner (Germany, 1938-1940) His helicopter used a 140 hp engine and had a gross weight of 1000 kg. It had two closely spaced 12 m diameter two-bladed rotors mounted 0.6 m from each other so that when operational the rotor's blades were intermeshed. This craft was called the synchropter. Technical data: gross weight per hp = 15.7; disc loading = ~approximately 1.2 lbs / sq. ft.

Sketch 0.6-15

Flettner 1938

Pullin (Britain, 1938) this helicopter called the W6 was also a side-by-side configuration but utilized 7.6 m diameter three-bladed rotors, a 205 hp engine and had a gross weight of 1070 kg. Technical data: gross weight per hp = 11.4; disc loading = 2.4 lbs / sq. ft.

Bratukhin (USSR, 1939-1940) Named the Omega 1 this helicopter powered by two 350 hp engines had a gross weight of 2300 kg and utilized a side by side three-bladed rotor configuration each having a diameter of 7 m. Technical data: gross weight per hp = 7.2; disc loading = 6.1 lbs / sq. ft.

Sketch 0.6-16

Bratukhin 1939

Germany (1941) The Focke-Achgelis Fa-223 had a three-bladed 12 m diameter side-by-side rotor configuration. Its gross weigh was 4300 kg. It used a 1000 hp engine and could reach an altitude of 5000 m, travel 300 km, and had a 120 kph cruise speed with six passengers plus a 900kg payload. Although many consider this to be the world's first successful helicopter design, its development was not pursued. Technical data: gross weight per hp = 9.4; disc loading = 3.8 lbs / sq. ft.

Sketch 0.6-17

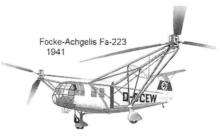

Focke-Achgelis Fa-223
1941

D-CEW

Sikorsky (United States, 1939-1941) This three bladed main rotor with a small torque canceling tail rotor further enhanced with cyclic stick, pedals, and collective stick with twist grip throttle, is considered the first truly practical operational helicopter. His R-4 model powered by a 185 hp engine had a gross weight of 1100kg, utilized a single main three blade rotor having a 11.6 m diameter and a single tail rotor. The main rotor cyclic handled longitudinal and lateral control while the tail rotor directional control. The actual pilot operated control and locations established on this craft were so successful they became standardized and are still in use today. This was a simple straightforward concept in comparison to past designs and revolutionized the helicopter industry. *It was the first helicopter to go into actual production.* The Sikorsky R-4 helicopter was first to be ordered in quantity by the Army.   Technical data: gross weight per hp = 13; disc loading = 2.13 lb / sq. ft.

Sketch 0.6-18

Bell (United States, 1943) This 10.7 m diameter two-bladed helicopter also with a small torque canceling tail rotor, used a teetering type hub and rotating stabilizer bar for the main rotor. It was powered by a 178 hp engine and had a gross weight of 950 kg. This design was so reliable that the Bell Model 47 was the first helicopter to receive an American Certificate for Airworthiness. Technical data: gross weight per hp = 11.7; disc loading = 2.16 lbs / sq. ft.

Sketch 0.6-19

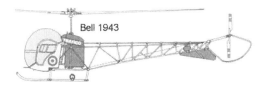

U.S. military production of warplanes in World War II reached 96,315 units in 1944. Significant planes of the war include USA's Mustang fighter, which featured a "laminar flow" wing having half the drag of the conventional airfoils in use at the time. This wing design was a direct result of the scientific aerodynamic research performed by NACA (National Advisory Committee for Aeronautics). This plane was one of the fastest propeller-driven fighters of the war attaining speeds up to 475 mph.

Germany's Me.163 was the world's first liquid-fueled rocket powered fighter which had a 5,500 fpm climb rate and a top speed of 560 mph., Germany also produced the first true jet engine fighter the Me.262 with a top speed of 500 mph.

Piasecki (United States, 1945) His Company, which focused on the design and production of tandem rotor helicopters eventually, became the Boeing Vertol Company. The successful PV-3 had two 12.5 m diameter three-bladed rotors mounted in tandem fashion one behind the other. Power came from a single 600 hp engine, and had a gross weight of 3100 kg. Technical data: gross weight per hp = 11.3; disc loading = 2.5 lbs / sq. ft.

Sketch 0.6-20

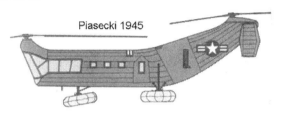

Piasecki 1945

Breguet (France, 1946) This 240 horsepower helicopter, the G-11, had two coaxial 8.5 m diameter three-bladed fully articulated counter rotating rotors. The rotors incorporated flap and lag dampers. It had a gross weight of 1300 kg. Technical data: gross weight per hp = 11.9; disc loading = 2.3 lbs / sq. ft.

Hiller (United States, 1946-1948) This 178 hp helicopter, the Model 360, had a gross weight of 950 kg, and used a single 10.7 m diameter main rotor and a small torque canceling tail rotor. Control of the main rotor was accomplished with a gyro stabilizer bar that had built in controllable tabs that the pilot could manipulate to effect rotor orientation. Technical data: gross weight per hp = 11.7; disc loading = 2.1 lbs / sq. ft.

Sketch 0.6-21

Hiller 1946

U.S. ARMY 554062

Kaman (1946-1948) His contribution to the helicopter rotor design was the introduction of the servo-tab to control rotor pitch by twisting versus rotating the rotor blade at the pitch bearing.

Mil (USSR, 1949) He developed various size helicopters using the single main rotor, small torque canceling tail rotor configuration, one of which was the 570 hp Mi-1 which had a 14 m diameter three-bladed main rotor and a gross weight of 2250 kg. Technical data: gross weight per hp = 8.6; disc loading = 2.99 lbs / sq. ft.

Kamov (USSR, 1952) He developed various size coaxial configured helicopters. His 225 hp Ka-15 had 10 m diameter three-bladed rotors and a gross weight of 1370 kg. Technical data: gross weight per hp = 13.4; disc loading = 3.567 lbs / sq. ft.

Yakolev (USSR / 1952) He designed the tandem four- bladed rotor Yak-24 helicopter.

Karman (United States, 1951-1954) He designed and built the first helicopter to be powered by a gas turbine which he called the K-225. Since then this type of power plant has become the standard in the industry because of its excellent power to weight ratio. In 1954 Karman introduced the first twin turbine helicopter the HTK-1 synchropter.

By 1952 the last of the propeller-driven strategic bombers (B-36) were being replaced by the jet engine B-52. Jet engines more or less updated military aircraft across the board by this time. The USA dominated the world aircraft record category and in 1953 had flown to the highest altitude, 42,000 feet and flew the farthest on a tank of fuel 535 miles. In the fifty years following Orville Wright's historic first flight the progress made in the development of aircraft was simply astounding.

This history of the evolution of the present day helicopter was more or less complete by the early 1950's. Research and development has continued to the present day but it would seem that all the novel approaches to rotary winged flight have been tried. For all practical purposes this chronological timeline of significant helicopter events leading to its invention is complete.

0.7 Brief History of Ultralights

When Orville and Wilbur Wright took turns flying their plane back in December 17, 1903 to establish once and for all that controlled flight of a powered heavier-than air craft was possible, many enthusiasts from all over the world wanted to get involved with flying. These amateur aircraft builders were helped along by some of the "do-it-yourself " magazines that published plans. In 1910 Popular Mechanics gave build instructions for hang gliders and in the 1912 edition of "Practical Flying" plans were given for the Curtis biplane and the Bleriot monoplane. Before World War I aircraft builders were more or less grouped together with little distinction made between the garage-shop "do-it-yourselfers", the research driven experimenters, and those interested in mass production. All that would change after World War I.

The aircraft industry was well established at the war's end. In the United States literally thousands of surplus trainers and scouts were available for hire at low cost to those that wanted to fly. In spite of the fact production aircraft were available, there still existed a mind-set among some of the population to explore the possibility of building an aircraft more as a hobby than to save time or money.

The "amateur" builder movement had its first notable event in 1924 at Dayton, Ohio. The organizers scheduled two National Air Races for planes using engines of 80 cubic inches or less and a cargo capacity of 150 pounds. This specification restriction enabled amateur builders to get into the air-racing scene, which had been an exclusive event for more powerful factory-built pure racers or surplus military planes. The engine sizes being manufactured for aircraft use at the time were much larger than the allowable size for these two particular races. This meant that new lightweight aircraft designs had to be developed by amateurs to enable the use of smaller power plants.

By the mid-1920's most of the war-surplus planes reached their useful life and were replaced by new aircraft for those that could afford them. Those that couldn't, attempted to build their own using modified motorcycle engines or Ford model T (late 1920's) and model "A" (early 1930's) engines. Some of the more successful homebuilt aircraft of the time were: J.A. Roche's "Aeronca C-2, (1929)" and his subsequent "C-3, (1931-1936)", the "Parasol" designed by Ed Heath, the "Lincoln Sport" a biplane developed by S. S. Swanson, and the "Air Campers" developed by B. H. Pietenpol. Supporting this effort, magazines such as "Modern Mechanix" annually printed a compendium of plans and data specifically for the homebuilders and the "Flying and Glider Manual", which started publication in 1929, gave implicit instruction to walk a homebuilder thru an entire aircraft build.

Photo 0.7-1

The next major impetus of this grass roots home-build movement following Orville and Wilbur Wright's success was probably Lindbergh's solo 33 and a half hour transcontinental flight in May of 1927 from Roosevelt Field in New York to Paris, France. He became one of the greatest American heroes because his fame was worldwide. In fact I doubt if any American since has had a French parade to honor them. In New York, on his return in June 14, 1927, an estimated 4.3 million people lined the streets from the Battery to Central Park just to see him. He also had a lucrative flying contract offered to him. If Lindberg's instant fame didn't entice the aircraft homebuilder to press forward I'm not sure if any other incentive could have done more to spur them on. As the years passed this infatuation by dedicated individuals to build homebuilt aircraft continued to grow. The momentum of this movement peaked during the years of 1930 thru 1933.

It should be understood that during this time period regulations didn't exist for either the flier or the aircraft. Those that did have formal flight training and engineering expertise were the exception. The majority were just accidents waiting to happen. Those daring individuals seeking flight experience typically would take the "slow and low" approach toward proficiency, which was originally used by the French to train aviators during World War I and nicknamed "grass cutting". This basically meant that you found a long field to use for take off and repeatedly flew in a straight line at takeoff speed literally hopping across the field performing what is called in today's flight terms as a "touch and go". Then as pilot skill increased you would attempt to fly over fences prior to actually attempting normal flight maneuvers. Unfortunately, as you may have already experienced on commercial airlines some landings can be rather abrupt and on these home built planes the learning curve for takeoff and landing of the "touch and go" phase of the flight training left many a plane crumpled in the field due to the inability to absorb the structural stress of the repeated impacts. The positive part of the "slow and low" process was that the number of fatalities was relatively low although the true extent of the number of homebuilt adventures that ended up this way will never be known due to the backyard nature of the event and the fact that mandatory aircraft accident reports were not required.

Parallel to homebuilt powered aircraft flight were homebuilt gliders, a sport introduced in 1929. Here again formal pilot training was virtually non-existent but future pilots did fare a little better because enthusiasts typically formed clubs for mutual support and training thru out the States. The first national flying club to be organized was in fact a glider club called the "Soaring Society of America" founded in 1932. At the time Germany produced some of the exceptional gliders and some good craft were made in the US as well. As momentum built for the sport so did the competition for manufacturers to sell their kits (usually $100 for plans, $500 for kits) and many claimed how easy their gliders were to fly. The result was increased crashes, as many a young man launched his craft from hillsides and cliffs, with no more than book training. As these instances increased more and more communities started to spread the word of the need to stop the carnage. It was the community-club nature of the glider organization that appears to have started the movement calling for regulation rather than individual aircraft builders.

As for the homebuilt planes two major problems presented the greatest challenge for builders - arrogance and ignorance. The arrogant were determined to show the world what a proper plane should look like. Keep in mind the Wright brothers' aircraft was a glider they designed with an engine strapped on. It was, after all, a "homebuilt" put together by bicycle shop mechanics. During this time there were mono wing, bi-wing and tri-wing aircraft all painting the horizon and all having their advantages. The airplane was still a work in progress and what it should look like was undefined. Homebuilders thought they could still take the "clean piece of paper" approach to the layout. Don't get me wrong I welcome "design arrogance" or to put it in a more acceptable way "thinking outside the box". Such thinking can at times create a paradigm shift. One such thinker of the time, Steve Wittman introduced the spring steel landing gear that became a universal choice for light aircraft. Later there was the thinker who solved the astronaut space walk drifting problem during the Gemini space program by suggesting the use of simple Velcro pads on the side of the capsule to enable the astronaut to hold on. Unfortunately design arrogance without hard engineering to back it up rarely succeeds in the field. Over the years I have heard many a "blue sky idea (an idea that just happens to come off the top of the head)" in various engineering sessions and virtually all were eventually discounted usually for two basic reasons. Either the proposed idea wouldn't package in the space provided or the numbers didn't add up to match the design requirements. Needless to say the majority of those truly unique aircraft never survived the rigors of aerodynamic loads. One of the most memorable of the arrogant plane designs was a popular French kit plane called the "Flying Flea" popular during the years 1936 thru 1937. It was notable due to the high number of fatalities resulting from poor engineering of the earlier models.

As for the "ignorance" part, the majority of homebuilders, just didn't know enough about material science, aerodynamics, fatigue limits, vibration management, structural design or basic engineering to tackle such an aggressive project as the design and build of an aircraft or glider. It was not unheard of to see critical aircraft joints being nailed together or cheap pine being substituted in the plans where aircraft grade spruce was specified. Engines used were simply lifted from motorcycles or Ford cars without any regard for aircraft upgrades for reliability. As a result the engine became the primary headache for aircraft owners. Basically the general consensus at the time was that structural integrity was second only to pilot ineptitude as a source of accidents.

Co-incident with Lindberg's transcontinental flight in 1927 the government that year instituted rules for interstate flying such as airworthiness criteria and licensing, enforced through the Bureau of Air Commerce (now the FAA, Federal Aviation Agency). The standards set forth under the airworthiness requirements read like a military specification where design standards were extremely high, rigorous structural stress analysis had to be performed on all critical assemblies, only aircraft grade materials could be used, etc. Conformance to the above was considered as a bare minimum before any new aircraft design could be licensed. At the time the Bureau of Air Commerce introduced the rules for interstate flying individual states did not have to comply and could self regulate. This left a loop hole for those that still felt they could produce a better "mousetrap" but it limited builders because fewer could finance what was now being demanded for government compliance for a unique homebuilt design.

Eventually, due to the appalling record now becoming more and more evident across the US, all of the States one by one adopted the federal regulations and virtually overnight the casual amateur builder was eliminated. By the start of World War II only Oregon had not signed on to the Bureau of Air Commerce (BAC) Regulations. If you wanted to pursue your hobby you had to move there, which some did. By the end of World War II even Oregon joined the federal ranks and the amateur builder was left high and dry. It should be noted that the BAC did grant a few exceptions to some non-standard aircraft via an "experimental" certificate for craft with specific purpose such as crop dusting, racing, and those aircraft under development for certification.

Several events occurred to bring back the amateur builder. In 1947 there were the National Air Races sponsored by the Goodyear Tire company and in that same year George Beaugardus's completed a round-trip (August thru October) from Oregon to Washington D.C. for the sole purpose of demonstrating the reliability of amateur built aircraft. Flying under a 90-day experimental license his flight so impressed government officials that the Government regulations restricting recreational aircraft were amended. In 1948 full official status was given to the homebuilder and the class was established as the "amateur-built" aircraft category. Technically a "homebuilt" is an airplane used specifically for the purpose of recreation and education and is licensed by the Federal Aviation Administration (FAA) in the "amateur-built" subcategory of the "experimental" licensing category. For an aircraft to be considered amateur-built the builder must complete at least 51% of the project himself with the remainder being commercial prefabricated items or components used from factory built aircraft if necessary.

Once the FAA established the "amateur-build" aircraft category what remained was federal guidelines for those individuals who wanted to fly but did not possess a formal pilot's license. As I mentioned earlier glider enthusiasts had a national organization as early as 1932, homebuilt aircraft enthusiasts were too individualistic and did not require the ground support effort that gliders did therefore club organization did not take place until 1953 with the formulation of the "Experimental Aircraft Association (EAA)". The objective of the EAA was to promote the dissemination of knowledge to the amateur builder community thru out the nation via publications sent out from their main office in Oshkosh, Wisconsin. Soon local chapters of the organization spread throughout the US and the organization became a powerful voice in the formulation of future governmental regulation concerning amateur built aircraft. Although in took time, thru EAA's efforts in the early 1990's the FAA generated a specific regulation labeled "FAA part 103" strictly for ultralights. This regulation, which is the most lenient in the world, covers the parameters for single person ultralight flight. (See section 19.1 to review that document)

0.8 Brief History of Helicopter Theory

Before I attempt to discuss the evolution of helicopter theory by the geniuses that developed the concepts, let's start with stating some of the prior mathematical knowledge these men drew upon starting with the natural laws more formally known as the conservation laws. These basically state that with any particular physical system no matter what actions may occur in the system certain measurable quantities remain constant. The classic three are: the Law of the Conservation of Momentum, the Law of the

Conservation of Angular Momentum, and the Law of the Conservation of Energy. Those laws are stated as follows:

1. Law of Conservation of Momentum: this law states that the linear momentum of a particle is unchanged if no unbalanced forces act on the particle.

2. Law of Conservation of Angular Momentum: if no resultant external moment acts on the system the angular momentum of a system of bodies about a fixed axis is unchanged.

3. Law of Conservation of Energy: states that the total mechanical energy of a system remains constant when subjected solely to forces that depend on configuration or position.

These laws are self-evident if you think about it, and you have been exposed to them in various forms since you were a child. The first to come to mind is the one that states, "For every action there is an equal and opposite reaction". These natural laws may be self-evident but it is the work of the engineer, mathematician or scientist to work out the detailed physics to understand why it is so. For example Einstein with his famous relationship $E=mc^2$ (E being energy, m the mass, and c the speed of light) generated the link between Conservation of Mass with the Conservation of Energy. Prior to this revelation both were thought to be independent self-evident relationships. The reason these laws are so important is that they allow predictions about how a system behaves without knowing the exact analytical details during the course of the reaction. The laws provide a method to make a direct connection between the "before and after" state of the system when undergoing a reaction. The first investigations into helicopter theory started with these natural laws in mind, and as the physics progressed the analytical development would explore every element acting on the system between the "before and after".

Additional significant background knowledge used by the helicopter theorists included Newton's Laws (first published 1687):

• Newton's Laws stated that the following facts apply to a particle in a balanced force system:

1. A particle at rest will stay at rest, and a particle in motion will remain in motion in a straight line and without acceleration unless acted upon by an unbalancing force.

2. If an unbalancing force acts on the particle, it will move in the direction of the resultant force at acceleration proportional to the magnitude.

3. When two particles direct forces on each other, these forces will be collinear, equal in magnitude but opposite in direction (or as stated earlier "for every action there is an equal and opposite reaction").

Also important is the "Law of Conservation of Mass" which states the mass of a body remains unaffected when subjected to any ordinary physical or chemical change.

Additionally the concepts of Circulation and the Vortex are extremely significant. This progression of thought sought to define airflow circulation about a point. These concepts require elaboration on the terminology because later in the chapter it becomes the mathematical foundation explaining the generation of lift at each point on a forward traveling airfoil.

To understand these complex subjects some "basic" terms require definition. I say "basic" because I define these terms only to the extent necessary to present the material forthcoming. A more rigorous study would be required to completely delve into the mathematics of these complex subjects.

The terms we need are as follows:

(1) Inviscid Flow: The properties of an ideal fluid are that it has no viscosity. It does not resist a shear force. This is an extremely important concept to understand in our future discussion of aerodynamic lift. It is incompressible and has uniform velocity distributions when flowing. This flow is referred to as "inviscid flow" and does not support eddy currents or turbulence.

(2) Euclidean Space: In ancient Greece, Euclid of Alexandria first defined the description of a point's position in space. Since then updated as follows: "points" in the three dimensional space are defined by real numbers in each of the three defining co-ordinates (x, y, z) and "geometric shapes" as equations and / or inequalities. In one dimension we get the line, in two dimensions the Cartesian plane and in higher dimensions it is a coordinate space.

(3) A Function: This is a relation between a set of inputs and some set of allowed outputs with the property that each input generates exactly one output. Lets examine how this relates to "X" marks-the-spot. Lets say to get to "X" we must travel 100 miles (d) and we want to find out how long it will take us to get there traveling at various set speeds (s) between 10 to 60 mph. To determine "travel time" (T) the exact relationship between distance and speed must be established in order to generate a single "travel time" output for each speed input. Obviously to get "travel time" we must divide the distance by the speed (d / s), "s" being the variable because we are varying it from 10 to 60 mph. The "travel time" (T) we want calculated depends on (or is a function of) what specific input of speed we put into the equation "d / s". The expression for this is $T(s) = d/s$. You would refer to this expression as "T of s" or "T as a function of s" is equal to d / s. If you were to place this function into a math program such as Mathcad you would first define the range of the variable "s" to go from 10 mph to 60 mph. Then you would set the value of the speed incremental, for example to calculate every 10 mph increase. The output would be a tabulation of "travel time versus speed" over the entire speed range i.e. 10, 20, ...60 mph.

Of course this is a very simple form of a function. The more complex types could be represented as $f(x, y, z)$ or stated as "f" as a function x, y, and z. This would mean there were three varying inputs to determine in the equation before an output was obtained for each, for example $f(x, y, z) = (x + y + z)^2$. The Mathcad tabulation for this would pick off the first value of y and z then calculate the formula varying all the x inputs. The next pass it would keep the z variable the same, increment the y variable one step then recalculate the formula varying all the x inputs again. This process would continue until every combination was calculated.

(4) The Vector: A vector is something that has "magnitude" and "direction". If it starts at the origin it is referred to as an origin-based or zero-based vector. In the "X" marks the spot example; the magnitude would indicate how fast they wanted us to get to the "X". The direction would be determined by the orientation of the straight-line arrow going from the starting point (origin) to that point in three-dimensional space, point X (x, y, z). For example traveling at "50 mph in the direction towards point X" is a vector quantity. Just aimlessly traveling at a speed 50 mph is referred to as a scalar quantity, as is mass, enthalpy and density. Vectors can be added, multiplied, generate a scalar product (dot product) or a vector product (cross product) in accordance with the rules of vector calculus to obtain specific results.

(5) Vector Field: The name says it all. To have a field of vectors there must be a multitude of vectors involved, or as stated in vector calculus, it is an assignment of a vector to each point in a subset of Euclidean space. Vector fields can usefully be thought of as representing the velocity of a moving flow in space.

(6) Flux: This is defined as the amount of something passing thru a surface. It is a total and not "per unit area" or "per unit volume". The total Flux depends on the size and strength of the Vector Field. In summation "flux" depends on the magnitude of the source, the angle between, and the size of the surface.

(7) Divergence: This is "flux density" or stated another way it is "the amount of flux entering or leaving a point". Divergence = Flux / Volume. Flux (a fixed amount of something) divided by the volume involved is similar to density of a substance where mass is divided by the volume (it takes up). Divergence represents the rate of change of volume of a flow.

(8) Gradient: When I initially heard this word in class the first thing I thought of was the temperature gradient that occurs across a metal plate when you put an oxy-acetylene torch to one end of it (4130 steel starts to melt at around 2610 degrees F and the torch temperature 5580 degrees F). Soon after the blue flame contacts the metal plate you start to note how the steel changes color the longer you apply the heat. As this color / surface temperature continues to rise at that point the plate turns cherry red (at around 2600 degrees F). The color diminishes in intensity the further you move out radially. Based upon this knowledge my instincts told me the math definition had something to do with a rate of change of something (heat?) going on with respect to all points on that plate.

The math definition of a "gradient" is the rate of change of a function. It's a vector that points in the direction of the greatest increase of a function. My intuition was right with the torch example. At any three dimensional point on that plate we get the temperature plus when we evaluate the heat transfer equations to get the "rate of change" of that function (its derivative), we obtain the gradient giving us the direction to move to get to a hotter point. Here's the twist. The vector direction calculated at that point doesn't necessarily mean if you travel in that direction you will get to the torch. For example if there are holes in the plate the gradient vector direction will change as the heat transfer taking place in the plate negotiates the steel path around the hole.

Stepping back a bit look at Microsoft Word's synonyms for "gradient". They are: incline, slope, rise, pitch, grade, and hill. The first five words imply a fixed "rate of change". We are not interested in something that simple, lets move on to the hill reference where the "rate of change" of the function (equation defining the hill shape) changes as a function of your three-dimensional position on the surface of the hill. As a college prank we are given the equation (F) for the hill and told to give the co-ordinates of its highest point but we are not given a starting point. If we are smart and realize that the top of the hill represents the maximum increase of the function, any movement in any direction from that point means we are going back down the hill. The gradient at that point is zero because there is no direction to move that improves on height of where you are. This means to solve this problem we have to find the point where the Gradient F $(x, y, z) =$ (dF/dx, dF/dy, dF/dz) = 0. This means the "rate of change" of F with respect to the x direction is 0; the "rate of change" in the y direction is 0, as with the "rate of change" in the z direction. Our solution will take some effort because every time we plug in some x, y, and z coordinates we get a new direction to move. After moving in that direction a little bit a new gradient at that point must be calculated for our next move. Eventually, we will reach a point on the hill such that the Gradient F is zero. That point is the top of the hill and we get to leave.

(9) Circulation: This is the amount of force that pushes something along a closed boundary or path. A good example of this is a leaf in a whirlpool. The circulation would be the amount of force required to push it as it traveled around the circle. The more push the more circulation.

(10) Curl: This is circulation per unit area, rate of circulation, amount of twisting at a single point or better yet, the rotation of a flow. To get a better grasp of this concept lets explain the right-hand rule. I first came across this while studying electromagnetics. In the case of current-induced magnetic fields along a straight wire if you curled your fingers and pointed your thumb in the direction of the electrical current, the curl of your fingers would give you the magnetic field direction. Using the right-hand rule if your thumb is in the direction of flow and the flow is starting to circulate in the direction my curled fingers are pointing then this is called a right-hand curl.

The "vorticity" term is used to denote the extent a fluid will rotate about itself (its "curl"). The curl F of a vector field can be interpreted as the vorticity per unit area of flux (i.e. a flowing substance) in a small area (i.e. at a point). One of the uses of the curl is to determine whether flow (represented in direction and magnitude by F) is rotational. Flow is irrotational if curl F=0.

(11) Vortex: Meteorology defines a "vortex" as a mass of fluid in which the flow is circulatory and its "vortex line" is the locus of the centers of circulation (centers of two-dimensional vortices). In the field of Fluid

Dynamics the solid core of the vortex is called the vortex tube and the centerline of this circulating mass of particles the vortex filament. A vortex tube is a tube that is the locus of vortex lines drawn thru every point of a closed curve. The maximum rate of change of this function is called the gradient vector function. A vortex is a rather dynamic physical happening. Consider water flowing down the drain. You would think the path of least resistance would be to go straight down the sinkhole without all the swirling motion but it doesn't flow that way.

Mathematically rotational fluids are extremely difficult to analyze because an infinite set of vortex lines exist. In contrast in an irrotational fluid the fluid flow reduces to a single vortex line. Guess which type of flow the fluid dynamic guys decided to analyze? Yes, it was irrotational flow where the vorticity equals zero and there is no curl. How did they get that to happen? Helmholtz's back in 1859 simplified the tornado vortex type problem by replacing the fluid medium of atmospheric air with an ideal gas for the calculations. This flow is referred to as "inviscid flow" and does not support eddy currents or turbulence. His second simplification was that the inviscid flow had no vorticity which means the flow did not rotate (or "curl") about itself (flow remained irrotational). Resulting from these assumptions he concluded that the vortex strength is constant along the vortex filament. The vortex cannot end in the fluid therefore it either follows a closed path, extends to infinity or start/end at solid boundaries and if initially irrotational remains irrotational.

In this model the tornado vortex has been reduced to a vortex tube of uniform cross-section and constant strength along its vortex line whose flow vorticity is zero. In the real world where the viscous fluid is ambient air and the start of the vortex tube of the tornado is on boundary earth, its vorticity is diffused with its interaction with air's viscosity such that the width of the vortex filament becomes so large it is no longer recognizable as a vortex line. At that point in the sky the tornado's vortex diffuses over a large area.

With these points in mind the helicopter theoretical journey can begin.

If you would like to obtain a better understanding and some interesting reading of these topics please see the section titled Supplemental information in the Appendix.

Now let's look at the timeline and the players involved in the evolution of aerodynamics.

Da Vinci (Italy, late 15th century) is credited with being the first to understand that the faster air flows over a wing like surface the more lift is produced.

Sir Cayley (England, 1790's) in 1810 wrote a dissertation on the aerodynamic fundamentals for both rotary and fixed-wing aircraft.

Helmholtz's (German, 1800's) theorems published 1859 dealt with three-dimensional motion of a fluid in the region outside the laminar flow layer where the influence of frictional forces are small and can be ignored.

Francis Wenham (England, 1860) was one of the first scientists to conduct studies in wing design and concluded a flat straight wing surface provided far less lift than a curved one.

Horatio Phillips (Canada, 1860's) utilized a novel approach to study airflow by building a wind tunnel to collect data. This research enabled him to patent an "under cambered" airfoil which provided substantially more lift than prevailing designs. Due to this detailed work Phillips is considered to be the father of the science of aerodynamics.

"Momentum theory" originally developed for the understanding of marine propeller motion by W. J. Rankine in 1865 and furthered by R. E. Froude in 1885. In 1920 A. Betz extended it still further to include rotation of the slipstream. When applied to a helicopter rotor, the model is that of an actuator disc of zero thickness. This circular surface is defined to be able to support a pressure differential and accelerate air through it.

Diagram 0.8-1

Momentum theory flow model for hover

Momentum conservation equates the rate of change of momentum between upstream and downstream positions to the rotor thrust

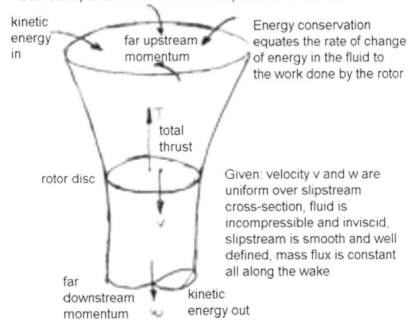

kinetic energy in

far upstream momentum

Energy conservation equates the rate of change of energy in the fluid to the work done by the rotor

total thrust

rotor disc

Given: velocity v and w are uniform over slipstream cross-section, fluid is incompressible and inviscid, slipstream is smooth and well defined, mass flux is constant all along the wake

far downstream momentum

kinetic energy out

v is rotor induced velocity at rotor disc
w is rotor induced velocity in far wake

PSB

Ships moved forward with solid shaft driven propellers and so did aircraft but helicopters having rotor blades solidly mounted to the rotor shaft had a tendency to fall over sideways in forward flight. Propeller theory at the time still had a long way to go to explain all the analytical intricacies involved in helicopter rotor dynamics.

In 1923 Juan de la Cierva used his "seat-of-the-pants" experience to formally address the outstanding issues of rotor flight control (which by the way was years in advance of the theorist efforts to model it). To duplicate the flight characteristics of his flexible palm wood model he added a flap hinge at the connection point of the rotor blade to the rotor hub on his autogiro. This rotor design arrangement allowed individual differential blade lift to be balanced out, thereby providing the capability for successful forward flight. This rotor was used on his C-4 autogiro built in 1923, which was later demonstrated at the Royal Aircraft Establishment in 1925. At that show Cierva's rotor so impressed Englishmen Glauert and Locke they initiated a formal analysis into rotary wing theory.

Extended Momentum theory – Glauert in 1935 broadened the scope of momentum theory, as it applied to marine propellers, to address the higher axial speeds of airplane propellers. Later these basic fluid flow fundamentals were utilized to understand the macro nature of helicopter rotor airflow. In this case the global analysis of the rotor and the total air flow as a whole was performed using the basic laws of conservation of energy, momentum or mass.

The following theories start to investigate the physics involved "in between" the "start and end" process described by the pure momentum analysis, for example what happens force-wise at the actual rotor blade.

Blade Element theory,

Diagram 0.8-2

Blade Element Theory

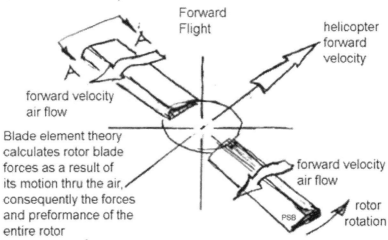

Forward Flight

helicopter forward velocity

forward velocity air flow

Blade element theory calculates rotor blade forces as a result of its motion thru the air, consequently the forces and preformance of the entire rotor

forward velocity air flow

rotor rotation

PSB

Initially brainstormed by William Froude in 1878, however Stefan Drzewiecki accomplished the major analytical work during 1892 thru 1920. Although momentum theory could predict what would happen if a certain volume of air mass flowed through a certain slipstream diameter how that happened due to the actual interaction between the rotor blade and the air mass was not factored in. Blade Element theory attempted to do that by predicting the induced drag on the rotor blade when generating lift. This induced drag over the blade length generates the induced power loss that ultimately determines the power required to provide a desired amount of lift.

Combined Blade Element and Momentum theory - During the years of 1915 thru 1919 various attempts were made to merge these two theories. In 1915 A. Betz's studies led him to believe the ideal aspect ratio (the ratio of the square of the span (length) of an airfoil to the total airfoil area), or the ratio of its span to its mean chord (width) value was a function of blade shape. G. De Bothezat, 1918, worked to establish airfoil characteristics of various blade planforms (shapes). A. Fage and H.E. Collins, 1917, tried to resolve the inconsistency between theoretical and actual numbers generated for induced velocity (a velocity increase over that of the free-air caused by the presence of a airfoil in the stream or reworded as the velocity of the induced flow through a rotor) of the blade as a function of aspect ratio by introducing a correction factor.

Although the result of this approach was more comprehensive than any analysis preceding it, it still didn't generate the perfect solution. Something was still missing in the final analysis to generate numbers that matched the observed empirical data. The airflow discussed in this theory is local to the blade surface. The revolving rotor blades in the process of producing lift generate a resultant pulsed airflow as the rotor blade interacts with the impacted air. This resultant airflow is non-uniform with respect to the rotor blade sweep area. In contrast pure Momentum theory has that airflow being uniform. In this combined theory when the differential form of the Momentum theory equation is used one can obtain a non-uniform air inflow distribution for hover and vertical flight. Using this approach rotor thrust and power can be calculated for a given blade pitch, plan form, twist and cord as a function of rotor blade radius.

Airfoil Bound Circulation Analysis and Vortex theory – N. E. Joukowski (1912 thru 1929) laid foundations for Vortex theory using bound circulation (the solution used fluid dynamic laws covering the influence of vorticity (angular rotation of the flow within the slipstream) and the principles of the Biot-Savart law, Kelvin's theorem, and Helmholtz's laws. Lanchester, Prandtl and others came to the conclusion that the lift on a three dimensional airfoil is a result of a bound circulation about it. The driving force for the induced velocity of this bound circulation results from trailed vorticity in the wake. Blade lift is generally produced at the tip and the resultant wake vorticity is generated by concentrated tip vortices that form in helices trailing below the rotor disc. Vortex theory is an analysis that calculates this flow field of the rotor wake, including the rotor disc induced velocity.

Lifting-line theory, shown next is the simplified three-dimensional diagram of wing theory as it applies to an aircraft wing during flight. In this model a straight line replaces the actual "wing-lifting surface" and it's associated circulating lift component substituted by what is referred to as a vortex filament. At each span-wise location the strength of the vortex is proportional to the local lift intensity. Those vortices surrounding the span are called "bound vortices" while the remainder extending downstream are "trailing vortices." According to Helmholtz's theorem the vortex filament cannot terminate in the slipstream. The superposition of the individual trailing vortices generates the vortex strength variation along the wing's representative straight line, and the induced downward velocity vector (downwash) at and behind the wing.

Diagram 0.8-3

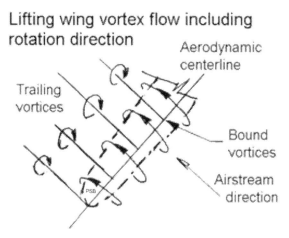

Lifting wing vortex flow including rotation direction

Aerodynamic centerline

Trailing vortices

Bound vortices

Airstream direction

Obviously once this vortex flow (shown above) is converted from this simple aircraft wing's translating surface in ambient air, to that of a helicopters rotating lift system of individual blades, the resulting pictorial becomes complicated, as you will see in the diagrams that follow.

Diagram 0.8-4

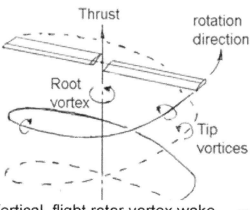

Thrust

rotation direction

Root vortex

Tip vortices

Vertical flight rotor vortex wake

As shown in the preceding diagrams at this time in the development of helicopter theory, a substantial effort was made to go beyond momentum theory to include all reactions occurring at, on or near the rotor blade during its rotation about it's centerline. Additionally all secondary effects were included such as the impact of residual ambient air wake disturbance and potential energy forced flow on the periphery of rotor motion.

One should keep in mind that during this period, the field of physics was deeply involved with deriving the mathematical formulae to systematically explain how everything in the universe functioned, from the basic atomic building block to the vast universe itself. Resulting from this comprehensive effort in the scientific world, new highly specialized fields of study emerged such as nuclear physics, aerodynamics, thermodynamics, etc. I point this out because when you read this chronological listing of helicopter theory development terms such as vorticity, solenoidal, bound circulation etc. one must keep in mind there was a great collaboration going on between researchers directly or indirectly through published scientific papers to progress the overall knowledge of the workings of nature. At the rate those mysteries were being unraveled by the minds of these great men, it seemed inevitable that eventually a single formula would be generated to cover every aspect of the workings of the physical world. Unfortunately for us and our project that never happened.

Diagram 0.8-5

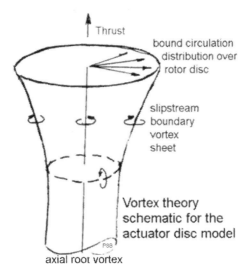

As for the furthering of helicopter theory, Prandtl and Goldstein definitely made their contribution. They both sought solutions using the concept of finite-blade vortex theory to analyze rotor power. Their approaches were specifically directed to address the loading at the blade tip. In this regard Prandtl is credited with presenting the concept of the wake-induced velocity at the wing to explain how disturbances in the aerodynamic environment must be included in the overall analysis of rotor power.

Diagram 0.8-6

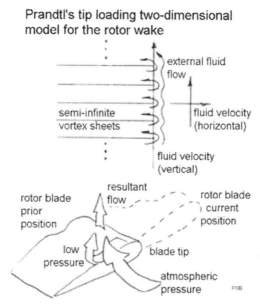

Prandtl's tip loading two-dimensional model for the rotor wake

This was significant since it produced results closer to experimental data than those generated by Momentum theory alone. Although Prandtl and Goldstein both sought solutions for high inflow rotor wake structure, their individual approach differed. Prandtl used a simple two-dimensional model to derive a blade tip-loss factor due to the vorticity in the far wake, whereas Goldstein worked with three-dimensional helical vortex sheets. What wasn't accounted for in their work was the inclusion of the overlapping wake generated by low inflow rotors although this blade-wake interaction did impact tip losses.

Diagram 0.8-7

Geometry of the tip vortex of the rotor wake in forward flight

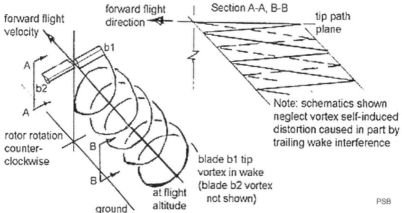

In an effort to compile the entire overlapping theoretical work-taking place at the time I have included below some of the other highlights of the period:

• Airscrew Theory (1920) this general theory utilized vortex theory and Prandtl's wing theory in its development.

• Betz (1929) generates a detailed analysis of propeller wake vortices as a system.

• S. Goldstein (1929) develops the Vortex theory described above, to determine the velocity induced by the wake vorticity, the Biot-Savart law is followed.

• Lock (1930) critiques Goldstein's Vortex theory to the actuator disc model, with reference to propeller design, to establish the solution limits.

• Lock and Yeatman (1934) using Goldstein's vortex theory generate tables of Goldstein's theoretical results and explains their use with Prandtl's approximation.

• Glauert (1935) determines from his "Wake model" that the optimum rotor produces a wake with a constant pitch helix.

• Prandtl generates an expression for blade tip loss based upon his two-dimensional model of rotor wake generated by a low inflow rotor. His blade tip loss factor ranges from 0.96 to 0.98 and in 1941 Sissingh indorses these numbers to be used in calculations since they correlated well with experimental data.

• Kaman (1943) elaborates on Goldstein's vortex theory with respect to a hovering or vertical climb helicopter rotor.

• Root Cutout, that section of the rotor comprised of the rotor hub, pitch bearings, lead, lag and flap hinges or basically all surfaces that do not produce lift and have a high drag coefficient (typically between 10 to 30% of the rotor diameter), are factored into the overall mathematical computation.

• Gessow (1948) standardizes NACA's terminology, symbols and reference planes used for helicopter analysis. He also investigated the effects of rotor blade twist and taper on rotor blade efficiency in 1948, and with Myers in 1952.

• Potential theory looks at the airflow as a fluid dynamics problem and solves the equations derived to generate the stream function or velocity potential.

• Numerical Vortex theory introduced with the advent of high-speed computers, probably in the early 1960's with the advent of IBM's model 360 computers. With this approach the simplified models discussed up to this point, now include detail previously omitted, such as discrete tip vortices, distorted wake geometry, blade elasticity, wind gusts etc. just to name a few. In 1968 Jenny, Olson and Landgrebe came to the conclusion hover power was underestimated by both classical and numerical methods if the wake was uncontracted. This error increased as a function of blade tip Mach number, blade solidity, twist, and loading. By 1970 Clark and Leiper realize that the numerical analysis solution must include "exact four dimensional (x, y, z, plus real time position) wake geometry" when the oncoming blade tip vortex encounters this following blade vortex flow. It is then that they introduce the concept of distorted tip vortex geometry for hover analysis.

Work continues to this day to fine-tune the numerical analysis to perfection. I'm not up to date on the current status of this subject but undoubtedly the computational algorithms used in the programs are being improved. Even something as common as a wind gust across one section of functioning helicopter rotor plays havoc with a numerical simulation. Just defining a wind gust would be a nightmare for the programmer to include all the boundary conditions. Mathematically it's right up there with generating the formula that defines how long the milk you pour in your coffee will take to disperse the swirl if you don't stir it. Complex, you can bet on it. Even if the capability today to predict helicopter performance is 99.999% possible, it is still mind- bending that back in 1925 Juan de la Cierva had already realized that simply enabling the rotor blade to pivot up as a function of lift / airflow was the key to getting his autogiro to fly straight. That was brilliant!

PART 1-THE BASICS
Chapter 1- The Helicopter Builder

1.1 The Modular Design Build Approach

The simplest way to stay interested in completing your helicopter whether via plans or a kit is to approach the project in small doses or what is commonly referred to as modules. Those modules are already defined by the FAA in their subsystem definition of the helicopter as it applies to obtaining flight certification, for example the rotor, swashplate, controls, etc. By using the FAA defined modules for our build two goals are met at the same time. The work on the module is completed and the knowledge needed for the written part of the flight certification is acquired. In my experience concentrating on one module at a time enabled me to get a deeper understanding of the subsystem from both the engineering prospective and at the fabrication level. Staying focused on the particular module at hand gives you a better understanding of how all the components work together versus utilizing some shotgun approach where every time the work gets complicated you move to the next section. The reward for tackling this project on a module-by-module basis is that once that first module is completed, any time you need an incentive to keep going just looking at something completely finished is extremely rewarding. In addition time is not wasted going back and forth trying to assess where you left off. Always keeping a forward momentum with a specific goal in mind is the key to a sense of accomplishment and also to not being overwhelmed by what remains to be done.

1.2 Basic Capabilities

1.2.1 Electronics / Programming

If you have never worked with 12-volt engine electronics including coils, solenoids, relays, alternators, etc. then the best place for you to gain some basic knowledge is to purchase one of the several Radio Shack's electronics experimenter's kits and work through the step-by-step instructions to build various electronic devices. This is necessary because even though most engines include the necessary wiring and components to run "out of the box" so to speak, you will definitely need to add electrical components to your helicopter. Whether the addition is something as simple as installing an engine starter button to the joystick or to the more complex task of adding engine instruments to a cockpit instrument panel, you will save yourself time and money if you are familiar with how circuits work. Something as basic as omitting a fuse in a tachometer circuit could cost you hundreds of dollars when an unexpected surge destroys it.

In addition once an engine is selected, maintaining its' peak performance requires being able to periodically check or reset the timed spark plug firing to the stated specification range. Learning how circuits work can be fun and educational at the same time and once you learn the basics you can utilize that knowledge for virtually any electrical device you acquire such as a dirt bike, car, lawn mower, etc.

Programming knowledge is an optional skill requirement for this helicopter project. Its main function is to control main rotor RPM. Most PLCs (programmable logic controllers), which would be utilized to handle speed control, have their input based upon a standard "ladder logic" electrical schematic. Once the PLC is programmed, RPM inputs from the main rotor sensors are compared to the set value in the PLC. If that value is outside the set range by being either too high or too low, a servo actuator manipulates the carburetor linkage according to a predefined algorithm so the desired rotor rpm is obtained. Learning how to generate ladder logic schematics or developing control algorithms is beyond the scope of this manual. If you are interested in understanding how to program inputs and output for control the best place to start is by purchasing a LEGO Mindstorm set. This is a fantastic way to learn about servo motor control, ultrasonic, touch and light sensors, etc., all while programming in a machine language variation used for the United States Martian explorer SOJOURNER. In Chapter 17, I go over the system used for main rotor speed control.

Being able to program a PLC is not as critical to this project as the circuit knowledge. This helicopter will fly in a manual mode without a PLC based main rotor speed control. In the mechanical mode the pilot simply regulates rotor rpm via a hand throttle located on the collective lever which in effect is no different than regulating car speed using a gas pedal.

1.2.2 Mechanical Design

For someone designing a helicopter from scratch a basic requirement in my opinion would be to have a mechanical engineering degree from an accredited college. To design you need to have a working knowledge regarding the strength of materials, how stresses cause failure, what impact vibration has on the endurance of a part, the limiting conditions for fluid flow, pump efficiency, and last but not least an understanding of thermodynamics as it applies to engine heat transfer.

If you do not have a mechanical engineering degree yet possess good mechanical aptitude and a strong desire to move forward with this project, there are two suggestions I can make to help you on your way. The first is to take some mechanical engineering courses on-line or through correspondence (actually, while I was in the Air Force they offered excellent college level engineering correspondent courses which could be transferred to an accredited college program). The second possibility for those on a limited budget would be to buy the following books:

Engineer-in-Training Reference Manual, Mechanical Engineering Manual, and 1001 Solved Engineering Fundamentals Problems all written by Michael R. Lindeburg P.E. These books are the only ones endorsed by the National Society of Professional Engineers for the purpose of passing your P.E. (professional engineer) exam. I would take the time necessary to read and understand every page dealing directly with engineering and work out all the reference fundamental problems. If you get stuck go to either a college for help or try online. Let me be perfectly clear do not attempt to design or redesign anything without this rudimentary background mechanical knowledge. If you can't acquire the needed knowledge base then stick with a pre-manufactured kit or buy an FAA certified used helicopter.

1.2.3 Fabrication Background

Actually this subject brings back a unique memory from my past. At the time I was working for a company that designed and built "state of the art" blown film co-extrusion lines. This equipment processed raw plastic pellets manufactured by companies such as Dow, Exxon, Dupont, etc. and transformed them into plastic film continuously wound on cardboard cores. Then when the predetermined width and diameter of the film was reached it would be cut-off and the free end transferred to new empty cores. This process would continue on specialized high speed indexing winders until the production run was completed. The end product of this equipment could be something as simple as single layer garbage bags, grocery bags, and garment bags or as complex as five layer co-extrusions commonly used for packaging cereal. Needless to say the list of what could be produced from the equipment was extensive. On this particular day I was in the machine shop with the foreman when one of the machinists interrupted our conversation to ask a question about a part he was working on. Much to my surprise the foreman, who was normally a measured and controlled individual snapped back and said " JUST MAKE IT LIKE THE PICTURE!" The picture he was referring to of course was the detail print the machinist was working from. "MAKING IT LIKE THE PICTURE" is ultimately your fabrication skill goal as well.

I began this project with no lathe or milling machine experience. The lathe generates round parts and the milling machine produces flat surfaces and both are used for about 90% of the machining operations required for the fabrication of the helicopter parts. In spite of lack of training within a relatively short time I was producing high quality parts in direct conformance to the dimensions and tolerances stated on the detail drawings. The word "tolerance" in the manufacturing field is the allowable finished part dimensional variation from the dimension stated on the detail print. Typically for most parts it is plus or minus 0.005".

Even if you have no experience on these two pieces of machinery let me assure you, you can obtain the skill you need if you work at it. As an aid in this endeavor I have included a section in this book devoted to helping you get started based upon my learning experience on not only these two pieces of equipment but on all the machinery and tools you will be working with. This tutor guide of mine is not a substitute for actual machinist training but for the limited range of skill required to generate these particular aluminum parts, it is a good way to get up to speed quickly with the equipment.

1.2.4 Assembly Experience

If you were the one who fabricated all the parts from the detail prints for this project, then this is the fun part and you will have no difficulty assembling your helicopter. If you were assembling this from a kit then I would venture to say that if you already maintain your own car doing such things as changing the plugs, rotating the tires, replacing air filters etc. you meet the basic criteria.

Those of you who do not possess the above prerequisites and do not own any wrenches and never put a LEGO set together will have some work to do prior to touching any of the parts. In the programming section 1.2.1, I suggested the purchase of a Lego's Mindstorm set as a simple way to understand how PLC's, Programmable Logic Controllers, operate. This set allows you to not only build a functional robot but also gives you the chance to control it's motion either thru your computer link to it, or remotely via a Wi-Fi connection. The point here is building something from assembly drawings like the LEGO robot is a great way to start. The next step would be to take things apart that have been discarded such as old stoves, computers, printers, vacuum cleaners etc. The point here is to get some skill and understanding how parts go together and maybe an insight as to why it was designed to fit a certain way. Those individuals I said met the basic criteria have been doing this sort of stuff since they were kids.

Before moving on, here are TWO IMPORTANT ASSEMBLY TIPS:

1. NEVER tighten the first bolt on mating parts having a multiple bolt pattern until all bolts are inserted first. This simple procedure will prevent you from doing one of the following; pulling your hair out, getting extremely irritated, or grabbing your drill and drilling thru the interference (drilling thru parts to open up hole diameters should be avoided because it compromises the original designed part strength). Make it a habit to secure all bolts in position before tightening them. For those parts, which have a right and left side, such as the tail section, it is a helpful to have all bolts for both sides in place before beginning the tightening sequence. Trust me you'll be surprised how interferences typically disappear if you work at getting all

bolts in place before tightening them.

2. USE A TORQUE WRENCH when tightening all bolts. The recommended bolt torque chart is found in section 10.1.1. The inherent strength of this helicopter comes from the clamping action of the bolted connections. These connections must develop the proper force to keep them secured. This only occurs when the bolts are preloaded to a set value by tightening the nut to the proper torque. Tightening nut / bolts without a torque wrench is unacceptable for the integrity of this project. More detail information about proper bolting design and the complete bolting process is included in Chapter 16.

1.2.5 The Most Important Skills

Patience is an important requirement in all areas of this project. I spent over one year on the development of the rotor blade design before I produced the first set of blades. Mind you that this was not working eight hours a day forty hours a week on this project but I must admit a lot of time was expended before I was satisfied with both the design of the blades and the equipment subsequently built to fabricate them. That year I spent learning and experimenting is shortened into one chapter in this book (Chapter 14.2). The information contained there can enable the serious builder to be in position to build a high quality set of balanced rotor blades in less than a month. The point here is to have the patience to complete what you started perfectly. For example, for the CAD design work the 3-D model should be faultless, the parts fabricated should replicate the detail print exactly, and the final assembly performed in strict adherence to assembly drawings. At any point if something doesn't conform you should stop, toss it and start over even to the point of going back to the CAD model and revising it if something can't be made or assembled as intended.

Remember the old saying "the job is not complete until the paperwork is done" well this is really where the patience part comes into play. It is highly improbable that you are going to finish this project with the original design 3-D model and the design detail drawings generated from that model. Somewhere during the build something is going to change, whether it is forced upon you by having a required bearing discontinued or your own decision to incorporate something better.

After field-testing the completed helicopter, you may want additional features and as a result, more changes occur to the model. The patience here comes into play when you force yourself to update your 3-D model. When I finished my first build-to-print helicopter I did in fact force myself to update the 3-D model, all the detail drawings and even the Bill of Materials (the parts list). However during field testing I noticed some things could still be improved such as the use of a larger rotor rpm meter (couldn't read the existing one due to the engine vibration), a laser pointer incorporated into the joystick (the laser was calibrated to a bull's-eye target mounted on the cockpit windshield to indicate proper joystick position for hovering), a collective angle positioning indicator (this allowed accurate setting of the angle of attack for hovering), etc.

All these additional changes were rather unique in that I wasn't sure which ones were truly final. As a result I decided to leave the original 3-D model and all associated documentation as they were and generate a second model that I called the "AS BUILT MODEL". This in a sense gave me peace of mind knowing I still had a complete reproducible finished helicopter set of documents in my hands based upon the original model while I edited the " AS BUILT MODEL". This fact took the pressure off the documentation issue until further field-testing verified the worthiness of any change.

1.3 Ultralight versus Experimental Helicopter

1.3.1 The Ultralight

These aircraft must meet the requirements and conditions outlined in FAA FAR part 103. That document in its entirety (along with a very comprehensive "FAA Advisory Circular" in the Appendix) can be found in Chapter 19 titled "The FAA". I highly recommend you read those two documents at this time if you have any future thought about building your own ultralight helicopter.

Summarized below are the basic restrictions governing an ultralight helicopter:

Maximum empty weight of 254 pounds

One person only

Maximum fuel capacity of 5 gallons

Maximum level-flight calibrated airspeed of 55 knots at full power

A maximum power-off stall speed of 24 knots calibrated airspeed

Other conditions and limitations apply for this aircraft classification but the above line items highlight the key requirements.

1.3.2 The Experimental

Part 103 of the FAA regulation states as follows, an "ultralight vehicle is a vehicle that is used or intended to be used for manned operation in the air by a single occupant." The exception to this rule is the Experimental aircraft that is basically a two-seat trainer version of the single seat ultralight. The FAA recognizes the importance of flight instruction to those individuals who desire to fly an ultralight by granting an exemption to this federal aviation regulation to select fight training schools. Under this exemption a two-seat ultralight (Experimental) may be used for training purposes as long as an authorized flight instructor flies it. The experimental craft due to the nature of its two-seat design is allowed the following additional capabilities:

Maximum empty weight of 496 pounds

Maximum fuel capacity of 10 gallons

Maximum level-flight calibrated airspeed of 75 knots at full power

A maximum power-off stall speed of 35 knots calibrated airspeed

Other conditions and limitations apply for this aircraft classification but the above line items highlight the key requirements.

1.4 Packaged Kits, Build from Plans, Buy Used

1.4.1 Packaged Kits:

At the time I began this project there were several reliable helicopter kit manufacturers – Robinson and Rotorway International for full size helicopters and UltraSport for an ultralight. As of this writing only two remain. The Robinson company, around since 1980 sells a two place helicopter, the R22 Beta II which has an rotor diameter of 25ft 2inches, uses a 124 horsepower engine, has an empty weight of 865 pounds and a base price of $276,000. The other company is Rotorway International (founded 1961), which sells its model A600 Talon, costs $97,700 and has a 25ft diameter rotor.

The UltraSport company used to sell three models of helicopters with the lightest being within the ultralight limit of 254 pounds. Although at that weight you were not required to have a FAA helicopter license to operate the helicopter, the company would not sell you a kit unless you had one, probably because of liability issues. Unfortunately, I can no longer find them on the Internet so I believe they no longer operate in the US.

Recently I came across a company called Mosquito Aviation, which sells an ultralight Mosquito XEL model using 18 ft diameter rotor blades, a 60 hp engine and has an unassembled cost of $35000. Apparently, this rather new company has good following and should be investigated further if you are interested in a prepackaged kit.

1.4.2 Build from Plans

In spite of these few available kits it seems there are many who will sell you plans for building a helicopter. I actually purchased two sets before realizing that such plans were not only flawed from an engineering standpoint but also downright dangerous for anyone to attempt to pilot. For example no one in his or her right mind would sit in front of an unguarded high-speed chain drive with only one-foot separation between you and it. There was also the issue of obtaining the rotor blades. There were no plans for building them but after investigating many avenues the reality was that no one will sell rotor blades to you "the home builder" on a one-up buy.

After this futile exercise to find a set of properly engineered helicopter plans I came to realize that no one selling these ultralight design packages actually had any shown flying in real time even on UTUBE. I came to the conclusion in order to build a good reliable ultralight helicopter I would have to design the helicopter myself. The plans, bill of materials and instruction set I have compiled along with the development of this book resulted in the helicopter shown on the cover of this book.

1.4.3 Buy Used

I recently came across the website called "Barnstormers.com" which offers free classified ads for experimental, homebuilt and ultralight aircraft. This would be a good place to start if you go the used route. Problem is getting parts for any used ultralight you buy. I purchased a used Kawasaki jet ski a while back and can no longer buy an engine for it from the manufacturer because it is discontinued. UltraSport seemed like a major player several years ago now they don't even exist. The Mosquito ultralight, who knows? If I did buy a used helicopter I would definitely want a certified set of spare main and tail rotor blades to go with it just in case.

1.5 The Mental Discipline Demanded

I have already talked about completing tasks as modules but the main point here is to work towards a schedule but not be dominated by it. In the real world the schedule is what every company works from, it is the driving force, which makes all managers, work overtime, all bosses insane, all raises ultimately based. If you don't believe this just ask anyone who worked for Apple founder Steve Jobs. I read that he could bring grown men to tears by imposing impossible schedules down employee's throats. Ultimately that type of pressure is why people quit their jobs. Building your helicopter should be a fun and a rewarding adventure. You do not need the self-imposed added pressure to work with a rigid schedule or you to will ultimately quit. A schedule is necessary to outline the sequence of events necessary to complete the project, but your schedule should be flexible. If your computer goes offline and you lose an hour's worth of CAD design work you're not going to get fired when you have to redo the lost work. So when things go wrong, which they will, just take a deep breath and chill-out. This is where the patience part comes in.

Along those same lines, I can remember when I purchased my Kawasaki 650 standup jet ski. After reading countless "Warning" statements which if not correctly followed could result in personal injury or loss of life. Then even more "Caution" statements which if not strictly observed could result in damage to, or destruction of the Jet Ski. The subsequent paragraph made the point that you would definitely take a spill on your first attempts to ride. It emphasized that this was part of the fun of a watercraft because of the challenge it provided and knowing the best way to fall, as in Judo, was essential. The point here is that accepting and coping with setbacks is very important on any project of this magnitude requiring a learning process.

1.6 ISO 9000 Standards

If you decide to go the whole route, like I did, and do the design- build thing then somewhere along the line you may believe your design will change the world by overcoming several major problems with light weight helicopters such as "main rotor rapid RPM degradation under certain flight conditions due to low blade inertia" or "inherent control issues due to cyclic stick sensitivity" (see section 18.3.2) or in general attempting to reduce complexity. Based on your confidence in your design you might decide to manufacture it. I realize from where you are in this book the point of going into a manufacturing situation is way out there, but just let's suppose you do reach that point.

If you are going to sell these magnificent helicopters you must go beyond the FAA certification process (see Internet) and become knowledgeable with the International Organization of Standardization ISO, founded in 1947 in Geneva, Switzerland.

ISO 9000 is a European Standard with the following objectives - to promote development of standardization to facilitate international exchange of goods and services and to promote cooperation in intellectual, scientific, technological, and economic activity. These ISO 9000 standards are "carved-in-stone" things you must comply with in order to obtain ISO certification by this International Organization. ISO 9001, 9002 and 9003 are the categories in which you can apply for certification, ISO 9001 is for companies which are involved in manufacturing or the creation and delivery of a service, ISO 9002 for companies involved in other functions outside of those covered in ISO9001, and ISO 9003 for distributors.

If interested in the manufacturing end, ISO 9001 is the applicable standard to seek certification. Obtaining it will give an edge over competition in International sales. It is a difficult and costly process to go through the certification process in the hope of achieving foreign market share. Here's what happens next. You must gather all the information possible about what the inspectors will be looking at and then work to be sure that your design will pass on EVERY point. This is like getting your car inspected. If you fail you have to bring the car back only in this case it will cost you thousands of dollars for each rerun.

Back in 1995 when our company got our ISO 9001 certification it cost us $30,000 and that was with every department including purchasing, manufacturing, engineering, quality control and sales all working together and passing on the first inspection.

What do the inspectors actually look for? Basically, it all boils down to traceability. In my engineering department the first thing they asked for would be the Bill of Materials for any machine currently being built by the shop. I admit I am an organizational fanatic so I simply went to the library shelf containing current machine builds and handed them the relevant binder for each machine. They perused the contents and asked for blueprints of several of the parts listed. I then provided them with the requested prints and with blueprints in hand; all of us went to the shop. There they compared the revision date of each print they had with the print revision available in the shop where building of the part took place. This type of cross checking for discrepancy or error went on for several days.

The point is this is an inherent weakness in any manufacturing organization. First of all the shop never wants only one print to work from. The foreman wants one, the machinist wants two because one will always get too dirty to read during the manufacturing process or the first blueprint sent was not clear enough. The number of prints the shop has is not the problem. The problem occurs when there is a design change that has to be made. It is engineering's responsibility to update the drawing indicating what was changed in the drawing revision section then issuing the drawing to the shop. Then the revised blueprint gets made and put in company mail to be delivered to the shop foreman. Here comes the problem, "SOMEWHERE ALONG THE LINE THE MACHINIST WILL NOT BE WORKING WITH THE REVISED PRINT" and you know when that will happen? Either when the ISO inspectors are there or when the part happens to be a critical one that is delaying shipment. In my experience this is how to solve the problem. First of all I sealed off the print room to any outside department. If anyone needed a print a formal request had to be made. Prints released to the shop followed normal channels thru office mail and were sent to the shop foreman however all drawing revisions went thru me. I personally brought the print to the shop foreman and retrieved the old ones. Sounds brutally archaic in these times but I never had time and money wasted due to the old saying of "one hand not knowing what the other was doing".

The other area the inspectors dwelled on was how reworked parts were handled. In the course of a week's manufacturing some parts are going to fail inspection. Some are salvageable and some scrapped. That determination took place at a weekly review board with engineering, manufacturing and quality control represented. In our case it was held by the departmental managers to determine what was still functional. If a part was found useable the part history found its way into the machine "AS Built" folder retained in the Engineering Records Library. If the review board found that the part could be used for a machine the question was where do you store it until used in the assembly process. The answer to that question was a fenced in "Almost OK Holding Area" located in the shop.

The last main probe by the ISO inspectors was how major revisions to a model were handled. For example this included product changes by the supplier of say transmissions, pumps or motors. Our company handled it by not only retaining "As Built" folders for every machine but also a model history folder indicating major changes by serial number or build date with explanatory notations. For example a notation might state: All machines built before 1990 had Baldor motors - see revision XXX for new motor replacement.

The emphasis here is make sure you take the time to document everything you do. Keep track of your parts, raw materials and changes. Initially, you may say sure that sounds easy. If at first you used 3/4 inch diameter piece of aluminum tubing you bought from a hardware store then later decided to use a 3/4 inch aircraft grade 6061-T6 tube instead that you purchased from a specialty supplier, you will not be able to see the difference and yet the two are worlds apart from a strength standpoint. If you don't note that change of material you have made a costly error. It is inevitable that on your first build you will make design changes. If you don't consistently keep your documentation current you will have to tear down your helicopter after it's first flight to figure out what exactly you built. What a time consuming project that would be! That is exactly what ISO standards were written for to avoid a waste of time or the possibility of error, either for you now or in the future after many of your craft are out there flying.

Chapter 2- Atmospheric Air & Airfoil Theory

When I was nineteen I was sent to Elgin AFB in Florida to undergo testing in an altitude chamber as part of crewmember training. The objective was to give you first hand experience of what it was like to encounter a sudden loss of aircraft cabin pressure at a normal cruising altitude of 30,000 feet. Up until that time I never really gave too much thought to the subject of atmospheric pressure. Of course it wasn't too much of a stretch to figure out that if the pressure a scuba diver experiences doubles for every additional 33 feet in the water, it wasn't unrealistic to think the greater the distance from earth to the vacuum of space the atmospheric pressure must decrease. In fact both air pressure and density decrease exponentially with altitude for all practical purposes.

There is no substitute for actual physical experience and this training was geared to do just that! You can read all you want about a subject but until you actually do it, everything is just your mental concept of the event. They put twenty of us in a sealed metal cylinder and before the simulation began the instructor pointed to a inflated 6 " diameter balloon and a beaker full of water. He said when the pressure in this chamber was equivalent to 30,000' the balloon would have a diameter of over 9" (provided it didn't burst first) and the water would boil. Both events happened. At that altitude the oxygen concentration is approximately 70% less and the effects quickly impact your judgment. Let me assure you, something as simple as writing numbers on a pad backward from one hundred to zero started to get difficult on the second pass. Just before I put my oxygen mask on I could feel my particular reaction to the oxygen deprivation, it was a mild stinging sensation in my ankles.

Water boiling, balloon bursting, mind going blank, ankles tingling, all as a result of atmospheric change, wow what a learning experience. Before I continue on the subject of helicopter design and function, I really think getting a better appreciation of the working environment of a rotor blade and the engine should be briefly discussed. In the following sub-chapters this will be elaborated on so you can get a better handle on what atmospheric air is and how it's mass, viscosity, density, water content and oxygen content impact helicopter performance. The objective is to get a heavier-than-air helicopter to use air as the fluid medium, having a sea-level density of 0.0752 pounds per cubic foot, to overcome the forces of gravity.

2.1 The Properties of Atmospheric Air
General Information:

Before any rigorous discussion of helicopter function can begin you have to know some basic facts about the operational environment the helicopter will be working in, that being atmospheric air. Atmospheric air is designated as standard up to about 65,000 feet and can be considered to be a uniform gas for aerodynamic purposes. Here are some characteristics of a gas such as atmospheric air:

Diagram 2.1-1

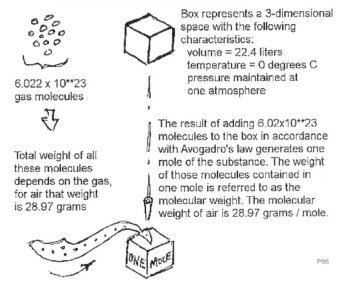

DEFINITION- MOLECULAR WEIGHT

6.022 x 10**23 gas molecules

Box represents a 3-dimensional space with the following characteristics:
volume = 22.4 liters
temperature = 0 degrees C
pressure maintained at one atmosphere

Total weight of all these molecules depends on the gas, for air that weight is 28.97 grams

The result of adding 6.02x10**23 molecules to the box in accordance with Avogadro's law generates one mole of the substance. The weight of those molecules contained in one mole is referred to as the molecular weight. The molecular weight of air is 28.97 grams / mole.

ONE MOLE

PSB

- Atmospheric air has a molecular weight of 28.97 g/mole (that means according to Avogadro's law the number of molecules, regardless of the gas being checked, contained in a volume of 22.4 liters (0.791 ft^3) at a temperature of ^{0}C (centigrade) (32 ^{0}F) and at 1 atmosphere (14.7 psia) equals 6.022 x 10**23 molecules, this defines a mole of the substance.

- In the case of the gas mix in atmospheric air the sum of the weight of those molecules (the molecular weight) is 28.97 g/mole or (0.064lbm/mole) or more simply stated a mole of any gas is equal to it's molecular weight. The point here is that air has mass due to the molecules in it and that mass will physically respond to any mass moving thru it such as a rotor blade.

- It has a specific gas constant R of 53.3ft-lbf/lbm-^{0}R, as the name implies this specific gas constant is unique to air. It is derived by dividing the universal gas constant by the molecular weight of air. It is used in the ideal gas equation (p x V = m x R x T) based upon Avogadro's law to develop a relationship between air's pressure p, volume V, mass m, and temperature T.

- It has a critical pressure of 547 psia and a critical temperature of 235.8 ^{0}R. These critical properties of air state the pressure and temperature where the liquid phase and gas phases are indistinguishable.

- Without getting too far into the subject of thermodynamics as it gets very complicated very quickly, lets review the First Law of Thermodynamics. This law states energy cannot be created or destroyed. As a result in any defined system in which something is done all energy must be accounted for. There are two types of systems: the closed type a process which works on the contents of a sealed space such as a "bomb calorimeter" and the open type which is applicable to flow over an airfoil surface. Bernoulli's conservation energy equation not only applies to closed systems but also for open systems if modified for non-adiabatic processes (in non-adiabatic processes heat or other energy can cross the system boundary).

• Next up is the topic of enthalpy. This is the value given to the total useful energy of a substance, in this case air, which can do work. Enthalpy is also divided into two parts - internal energy and flow energy. This is an important fact to know since those brilliant men who initially researched rotor performance overlooked it. Let me give you a simple example to demonstrate. Take your bicycle tire pump and start pumping air into one of the already inflated tires. After several pump cycles, feel the temperature at the base of the pump- warm isn't it? The temperature rise is due to you doing work on the air and the heat generated as a result of the air's molecular reaction to that input and for the purist reading this it is also true that some of the heat is generated by the friction of the plunger seals against the air pump cylinder walls during the pump cycle.

Right around now it would be a good idea to see a diagram so here it is: Diagram 2.1-2

FIRST LAW of THERMODYNAMICS-open system

(Bernoulli's energy conservation law extended to non-adiabatic process)

Q = delta U + W(flow) + delta E(kinetic) + delta E(potential) + W(piston)

U1, E1(kinetic)
E1(potential)

INPUT

Piston Work

Cyclic piston travel

z1

Q

z2

W(flow)
U2, E2(kinetic)
E2(potential)

OUTPUT

Definitions:
m = mass of a substance (kg)
v = velocity of the flow (m/sec)
vs = specific volume = the volume occupied by one unit
    of mass of a substance
p = pressure (kpa)
z = input or output height of flow (m)
Q {kJ} = heat flow out of system (q = kJ /kg)
delta U {kj /kmole} = U2 - U1 = change in internal energy
    of the flow internal energy includes all the kinetic and
    potential energies of the atoms or molecules of the
    system (u = kJ /kg)
W{flow} {kJ} = (p2 x vs2) - (p1 x vs1) = this is the
    pressure work done to creat the flow, for example it
    takes flow work to blow up a ballon. Specific
    flow work W = kJ /kg
delta E(kinetic) {kJ} = 1/2(mv**2) = mechanical energy
    used to move or rotate a substance
delta E(potential) {kJ} = mgz = mechanical energy
    possessed by a substance due to its relative
    position in a gravitational field
W(piston){kJ} = work that the steady-flow device does to
    the substance. Specific piston work W = kJ /kg
h{kJ /kg } = enthalphy = u + (p x vs) = represent total
    useful energy of the substance.
    Molar enthalpy H = kJ /kmole

Using enthaphy to represent internal energy and flow work the above equation becomes: q = (h2 - h1) + (v2**2 - v1**2) / 2 + (z2 -z1)g + W(shaft)

PSB

This fact was missed in the early research of rotor performance since no account was made of atmospheric air's total enthalpy change due to the work done at the molecular level after impacting the rotor blade, specifically its tip. The initial analysis focused only on the local event of airflow at the rotor blade and not from the macro view of the overall atmospheric air flight environment. In the real world the enthalpy of this air in certain flight circumstances can result in such a loss of rotor performance that a flying helicopter in this zone can cause it to literally drop out of the sky. This condition is called "settling-in-ones-downwash". The condition occurs when rotor blades continually impact the same air volume. There is more detail about this later on in the book but what is important here is that the total atmospheric air working environment and its enthalpy must be taken into account in order for an accurate prediction of helicopter power and lift. Of course today this is understood and computer simulations attempt to take all this into account in their analysis but it is important to be aware that air continues to play a part after the initial hit by the rotor blade during its angular path thru the air.

To explain some more thermodynamics this time we will look at air's capability to perform a "thermal process." What do I mean by "a thermal process"? Let's take a brief moment to look at thermodynamics from a human physiological view. If you walk outside your house and see a man covered in sweat, mentally you might wonder how he got that way. One possible process was excessive jogging, or running (internal heat generated due to the work performed by the body), another could be too much sun (external heat to the body provided by solar radiation). Either way his body's response to its thermal energy rise was to evaporate water from the skin (sweat) to cool the body down in order to maintain it's 98.6 F normal operating temperature. In this example we noted two different processes, which could cause this man's sweaty "thermo-elevated (dynamic)" condition to occur. In gaseous systems such as atmospheric air the resultant thermodynamic properties of the system depend on the process type. The names given to the specific processes are - adiabatic (energy constant within the system boundary), isobaric (constant pressure), isothermal (constant temperature), isometric (constant volume), isentropic (reversible adiabatic), and throttling (constant enthalpy adiabatic). For another example, more germane to the subject at hand, lets go back to the bicycle pump operation described earlier to define a thermodynamic process. In this case the process is to raise the pressure of the ambient air so it can be transferred to the tire. Sounds simple enough and using a variation of the ideal gas law where $(p1 \times V1) / T1 = (p2 \times V2) / T2$, if we know P1 (ambient air pressure entering the pump), V1 (the initial volume of the pump equal to the inside pump circumferential area multiplied by the distance the pump handle could travel) and T1 (outside air temperature) you can obtain the temperature rise T2, where P2 is the pressure noted at the gage and V2 the volume of the pump (equal to the inside pump circumferential area multiplied by the distance remaining that the pump handle could not travel because of the pressure). Obtaining the final temperature of this process is straightforward and the change in enthalpy, internal energy (includes all potential and kinetic energies of the molecules in a substance) and entropy (energy no longer available to perform work) can be calculated just knowing the beginning (pump handle up) and the end point (pump handle down). The values obtained are independent on the process path taken. For example how fast or slow the pump handle was pushed down does not matter. The formulae for calculating enthalpy, internal energy, and entropy are not going to be given because I do not want to delve too deep into the subject other than to say to calculate these values specific heats of the gas must be known and for air the specific heat for a constant-pressure process, c_P is 0.240 BTU/lbm-^{0}R and a specific heat for a constant-volume process, c_V is 0.1714 BTU/lbm-^{0}R (the specific heats stated here are related to R in the relationship R = c_P - c_V); and k, which is the ratio of specific heats = c_P / c_V = 1.4. Now, just to give more depth to the number of possible processes ideal gases can be analyzed here's a quick snapshot: Constant-Pressure-Closed System; Constant-Volume-Closed System; Constant-Temperature-Closed System; Isentropic-Closed system (reversible adiabatic); Polytropic-Closed system; Isentropic-Steady-Flow systems, Polytropic-Steady-Flow systems; Throttling-Steady-Flow systems.

Unfortunately, we're not done with the bicycle pump example because even though enthalpy, internal energy and entropy do not depend on the path or process taken by the pump handle travel the calculation of the amount of heat generated does. Of course it's not hard to comprehend that the force applied to the pump handle (large, small, fast, slow or any combination) will dictate the amount of heat generated. With that said thermodynamics has a special field of study called combustion power cycles, which keeps things simple. The theorists tossed out the combustion part dealing with the actual burning involved and used what is called an air-standard cycle to simulate what was occurring in the real device, such as a gas or diesel engine. Formally, three such air-standard cycles come to mind the Otto, Diesel and Carnot. The main point here is from a thermodynamic point of view, simple atmospheric air, is not so simple and I never even touched on the subject of steam tables!

• The air in the balloon (a flexible container), as described in the previous section when I was in the altitude chamber, will change its volume (balloon diameter) as a function of temperature and external pressure in accordance to the ideal gas law. The pressure exerted by air on its container (the balloon for example) is always normal to that surface.

• Air is compressible since it is just a gas, but the atmospheric air volume change (compressibility) due to impact with the rotor blade only becomes a factor as the tip speed exceeds 50% of the speed of sound. At this point the effects of air being compressed causes flow / lift / drag issues and wing-section performance calculations "based upon incompressible flow theory" do not agree with observed data. Putting some numbers to the above statement, given an ambient static pressure of 2116 pounds/square ft. the impact pressure (at the wing leading edge) is 34 pounds/square foot when the wing speed is 115 mph, 321 pounds/square ft. @ 345 mph, 1,494 pounds/square ft. @ 690 mph. At a leading edge, forward speed at the speed of sound the pressure is so great that the air pulse (that's the pulse which allows sound to be transmitted thru the air) can no longer move forward of the leading edge. With both sonic and supersonic wing speeds the air dynamics are so complex that the analysis is a distinct science and way beyond the scope of this book. For that type of detail you must consult the book "Theory of Wing Sections". The impact of this is, not to design a helicopter where it's rotor blade tip speeds enter the region introducing compressibility factors thereby falsifying results obtained from the formula generated from theory for performance calculations. Additional reasons to stay away from this region will be discussed as other characteristics are mentioned.

• Air resists shear force but deforms continuously in order to minimize that force thus the air will take the shape and volume of its container. The term used to measure air's resistance to flow when acted on by gravity or a pressure differential is viscosity. In general, for gases, the viscous forces take place at the molecular level, noting that it is the "Van der Waal's" forces that keep gas molecules attracted to each other. During the operation of our rotor blade, atmospheric air can be treated as a continuous fluid. Under these conditions Newton's law of fluid friction is obeyed whereby in simple shearing motion the shear stress is proportional to the rate of shear. The name given to the proportionality constant which makes the relationship an equality is called the "absolute viscosity". Kinematic viscosity is the name given to the term comparing the ratio of absolute viscosity to mass density. Since mass density changes with temperature and pressure it stands to reason kinematic viscosity is a function of those variables also. Values for absolute viscosity and kinematic viscosity as a function of temperature are stated in Table 3.

• It should be noted that when a gas is below it's critical temperature (stated above as 235.8 °R for air) the forces of molecular attraction (Van der Waal) dictate corrections to the ideal gas equation represented by $P \times V = n \times R^* \times T$ to reflect actual real world behavior. The reason for this is because at low temperatures the actual pressure exerted by a real gas versus an ideal gas is reduced by the degree of molecular attraction. The corrections to the above formula to account for Van der Waal forces are: $(p + a / V^2)(V-b) = n \times R^* \times T$; where the a / V^2 term corrects for pressure and the b term for molecular volume in a dense state.

• Forward motion of an object thru air will be resisted due to its viscosity. When the motion stops so does air's resistance against it. The molecular spacing of the molecules comprising atmospheric air are far apart as compared to those of a liquid. Their kinetic energy is greater and the distance traveled between collisions is much larger. In our helicopter performance calculations the atmospheric air utilized is considered a Newtonian real gas, which by definition means it is compressible, will experience friction and turbulence in flow, has a non-uniform velocity distribution, and shows finite viscosities. Just for the record ideal type fluids exhibit none of these characteristics and are more suited for theoretical comparison type analysis, a typical example is using the ideal gas law to find the density of an ideal gas.

• In our calculations we will set the rotor blade tip speed and angle-of-attack so the air flow over the airfoil surface is laminar versus turbulent. The definition of laminar flow (from the Latin word "laminae" or layers) is that all of the fluid particle paths are parallel to the overall flow direction. Typically this condition exists when the Reynolds number is approximately less than 2100. In this Reynolds number range the viscous forces are strong enough to hold the flow together. For example the section of smoke rising virtually straight up from a cigarette would be considered laminar. Then when that same smoke starts to swirl this would be the turbulent region. Turbulent flow is characterized like the smoke visual where there is an aberrant three-dimensional flow deviating from the overall direction of flow. Typically the Reynolds number in this case is greater than 4000. One other point, which I will elaborate in greater detail in the airfoil selection section, is that when an airfoil enters conditions that promote turbulence, airfoil lift simply drops off and strong vibrations are generated in its place. Both of those conditions being detrimental to smooth helicopter control operation.

• The thermal conductivity of air at 32 degrees F is 0.014 BTU/ hr-ft-^{0}F. Knowing the thermal conductivity of air enables one to design the cooling system required to keep our internal combustion engine operating at the optimum water temperature. That heat transfer takes place at the fins of the radiator. The rate at which heat can be removed from the fins of the radiator by air flowing thru them is governed by air's thermal conductivity. If the radiator fins are made of copper the radiator has thermal conductivity of 224 BTU/ hr-ft-^{0}F if it is made from aluminum the thermal conductivity is 117 BTU/ hr-ft-^{0}F. Obviously, the copper or aluminum fins are far more capable of transferring heat than air by a factor of over a 8000:1 so it stands to reason in order to transfer heat from the radiator by air flow a fan must be employed to force a large magnitude of air over the fin surface.

• As for gas dynamics in this case the "Net Transport Theory" is employed. This theory is based upon the fact that hot gas (or air) molecules have a higher velocity than cold molecules and have a greater probability of contacting cold areas more frequently than cold molecules contacting hot areas. Additionally, thermal conductivity is virtually independent of pressure but does exhibit a linear value increase with increasing temperature.

• The Psychrometric Chart. One can easily determine enthalpy, specific volume and other mathematical relationships for air directly from this chart. I find it's most interesting use is to explain the range of personal comfort in both outdoor and environmentally controlled indoor places. Now, I realize how comfortable I am in a room has nothing to do with designing a helicopter, but I thought since this chart is basically the corner-stone of all HVAC design you should know it exists. The chart can be found in the Appendix.

• The SDP (standard datum plane) is a theoretical altitude where the pressure of the atmosphere is 29.92 " Hg and the weight of air is 14.7 psi.
• Atmospheric air is never 100% dry and typically contains a certain percentage of water. Its humidity refers to the amount of water contained in the air as a function of temperature. A stated 50% humidity on a 90-degree day has more water content than on a 50 degrees day. At 100% humidity at any temperature that's the maximum amount of water the air can support. Furthermore as air's water content increases the air becomes less dense and aircraft performance decreases. Those instruments that rely on air's density to determine altitude will be affected by humidity variations.

Table 1. AIR COMPOSITION
Air composition is as follows (average of dry air at sea level and not including typical concentrations of pollutants)
(1) As a percentage of total weight: oxygen (O2) 23.15, nitrogen (N2) 75.54, ozone (O3) 1.7×10^{-6}, carbon dioxide (CO2) 0.05, argon (Ar) 1.26, neon (Ne) 0.0012, krypton (Kr) 0.0003, helium (He) 0.00007, xenon (Xe) 5.6×10^{-5}, hydrogen (H2) 0.000004, methane (CH2) trace, nitrous oxide (N2O) trace. Note: the ratio of nitrogen to oxygen being 3.32;

(2) As a percentage of total volume- oxygen (O2) 20.95, nitrogen (N2) 78.08, ozone (O3) 0.00005, carbon dioxide (CO2) 0.03, argon (Ar) 0.93, neon (Ne) 0.0018, krypton (Kr) 0.0001, helium (He) 0.0005, xenon (Xe) 8.0×10^{-6}, hydrogen (H2) 0.00005, methane (CH2) trace, nitrous oxide (N2O) trace. Note: the ratio of nitrogen to oxygen being 3.78

Table 2. U.S. STANDARD ATMOSPHERE (1962)
Data below is based upon the following sea-level baseline: pressure p_o=29.91 in; mass density md= 0.002378 slugs per ft^3 (0.0765 lbm/ ft^3); temperature T_o=59 F. Tabulation key, Altitude = h (ft), pressure ratio = p/p_o, density ratio md/ md_o, speed of sound = V_s (ft per sec.) Note: This is a truncated tabulation, the original table is stated in 1000-foot increments starting at sea level and continues to 100,000 ft.

h ~	T	~ p/p_o (actual pressure)	~ md/md_o	~ V_s
0 ~	59	~1.0000 (14 696 psia)	~ 1.00	~ 1,116
5,000 ~	41.17	~ 0.8320 (12.227 psia)	~ 0.8617	~ 1,097
10,000 ~	23.34	~ 0.6877 (10.106 psia)	~ 0.7385	~ 1,077
15,000 ~	5.51	~ 0.5643 (8.293 psia)	~ 0.6292	~ 1,057
20,000 ~	-24.62	~ 0.4595 (6.753 psia)	~ 0.5328	~ 1,036
40,000 ~	-69.70	~ 0.1851 (2.720 psia)	~ 0.2462	~ 968
60,000 ~	-69.70	~ 0.0708 (1.040 psia)	~ 0.0941	~ 968
80,000 ~	-61.81	~ 0.0273 (0.401 psia)	~ 0.0355	~ 977
100,000 ~	-50.84	~ 0.0108 (0.159 psia)	~ 0.0137	~ 991

Table 3. PROPERTIES of AIR (at sea level)
Where temperature "T" is F°; density "d" (lbm/ft^3); kinematic viscosity
is "kv" (ft^2/sec); absolute viscosity "av" (lbf-sec/ft^2).

```
  T ~ d      ~ kv              ~ av
  0 ~ 0.0862 ~ 12.6 x10^-5 ~ 3.28 x10^-7
 20 ~ 0.0827 ~ 13.6 x10^-5 ~ 3.50 x10^-7
 40 ~ 0.0794 ~ 14.6 x10^-5 ~ 3.62 x10^-7
 60 ~ 0.0763 ~ 15.8 x10^-5 ~ 3.74 x10^-7
 68 ~ 0.0752 ~ 16.0 x10^-5 ~ 3.75 x10^-7
 80 ~ 0.0735 ~ 16.9 x10^-5 ~ 3.85 x10^-7
100 ~ 0.0709 ~ 18.0 x10^-5 ~ 3.96 x10^-7
120 ~ 0.0684 ~ 18.9 x10^-5 ~ 4.07 x10^-7
250 ~ 0.0559 ~ 27.3 x10^-5 ~ 4.74 x10^-7
```

Table 4. SATURATED AIR WATER CONTENT AS A FUNCTION OF
TEMPERATURE
Fahrenheit / Water content (pounds in 1 pound of air)
40 / 0.00520; 50 / 0.00765; 60 / 0.01105; 70 / 0.01578; 80 / 0.02226
90 / 0.03108; 100 / 0.04305

Conclusions:
Based upon the above the following generalizations can be made:
(1) Air density is greater when the temperature is colder (in fact air is a
solid at – 273 degrees C) and at lower altitudes. Increased density means
more mass per unit volume, which implies more molecules in the space.
Based upon Newton's 3rd law of motion stating for every action there is an
equal and opposite reaction, then with more air molecules in the way of an
object moving thru them more drag on the object is generated. On the
positive side according to Bernoulli's principle, the greater the differential
velocity between two flowing air masses over an object the lower the
pressure, therefore if the air mass is denser, the pressure differential over an
airfoil is greater, and more lift produced.
(2) Atmospheric air as stated in Table 2 (at sea level and a density of
0.0765 lbm / ft^3) is a mixture of gases. For the sake of simplicity if oxygen
were isolated from the other constituents it would take (by weight) 4.32
pounds of air to obtain one pound of oxygen. At 10,000 feet altitude, from
Table 2, atmospheric density drops to 0.0565 lbm / fT^3 (md/md_o x md = 0.
7385 x 0.0765 lbm/ ft^3) correspondingly the amount of oxygen available is
reduced by 26.15%. Following along those same lines, if you are hovering at
sea level but the ambient temperature is 100 degrees F instead of 59 degrees
then from Table 3 the atmospheric air density is 0.0709 lbm / ft^3 and this
amounts to a 7.3 % oxygen reduction.
Since everyone knows oxygen is required for combustion it stands to
reason that if less oxygen is available per unit volume entering during the
engine's intake stroke, less power will be generated. Based upon this
knowledge those performance curves (horsepower / rpm and torque / rpm)
stated by the engine manufacturer for a specific engine (in this case a

reciprocating piston-type operating on either the Otto, Diesel, or dual cycle) must be de-rated for the density of the operating environment. Noting that these engine manufacturers typically base these performance data sheets upon a SCFM (standardized) airflow, which is taken as air at a standard density of 0.075 lbm/ ft^3. Using the SCFM density of 0.075 (which equates approximately to an ambient temperature of 68 degrees F at sea level) instead of the 0.0765 lbm/ fT^3 (based upon 59 degrees From Table 2) we obtain an oxygen reduction of 24.7% when operating at an altitude of 10,000 ft and a 5.5% reduction when the ambient temperature is 100 degree F. If the engine manufacturer states his engine produces 50 hp @ 5500 rpm then at an altitude of 10,000 ft the output power drops to 37.7 hp @ 5500 rpm, and if hovering at sea level when the ambient temperature is 100 degrees F that engine only produces 47.3 hp @ 5500 rpm.

2.2 Dynamics of Air & the Airfoil

This section is separated into three parts, subsection 2.2.1 briefly introduces you to those aspects of air dynamics related to laminar air flow over both a rotating and non-rotating object moving thru it and the individuals who helped define the characteristics of that flow. Bear in mind the complexity of this subject is such that 6 to 8 year doctoral programs are offered in Universities to enable one to obtain a PhD in Aeronautical Engineering. For simplicity I will only present air dynamic detail which will be relevant to subsonic rotor blade airflow and not topics such as turbulent skin friction, first-order compressibility effects, hypersonic boundary layer flow, or aerodynamic heating etc.

My comments on the subject are included in subsection 2.2.2.

In subsection 2.2.3 the Magnus Effect on a cylinder is mathematically defined by Kutta and Joukowski, the concept of circulation presented and the equation for the lift it generates stated. Subsection 2.2.4 titled Airfoil Theory will definitely challenge your mind; don't be alarmed if you can't grasp the involved math since the subject is presented in a simpler form in Chapter 8 "The Calculations". This subsection is devoted to the purist where the object is to mathematically formulate an equation to give the exact flow pressure at each point from the forward to trailing edge of a rotor blade. The information is used not only to determine how much weight the rotating rotor blade can support but also where those forces act so the internal structure of the blade can be designed to safely handle the resultant stresses.

My approach in this subsection for your understanding of the development of this formula is presented in the exact way I learned it, in small mathematical building blocks. Ultimately what we are attempting to do is to take the real world pitched spinning baseball and model it mathematically so we can find out the radial pressure distribution, which causes it to curve the way it does. When that is accomplished we physically alter a slightly larger circumference of the original ball into the shape of an arbitrary wing section and generate tie in equations to obtain the pressure distribution over this particular wing section profile. A really major point here from my perspective is we are modeling a spinning baseball and that's it!

2.2.1 Prandtl, Bernoulli, the Magnus Effect and Newton

In the previous subsections atmospheric air was discussed in great detail from its composition, thermodynamic properties, the psychrometric chart, density change as a function of altitude, and its impact on engine performance. Now lets describe what happens as an object moves through it.

Prandtl was a significant player in the science of fluid dynamics and his work covered such topics as boundary layer equations, supersonic gas expansion as a function of flow radius, particle travel distance to attain the turbulent shear stress for the given velocity gradient. His work on the relationship between the "diffusion of momentum" with respect to the "diffusion of heat" was named the Prandtl number or "Pr". His most significant contribution to the field of fluid mechanics occurred in 1904 when he suggested that flow around an object could be differentiated into two separate regions. The region traveling so close to the object that friction becomes a factor and a second region above it where frictional effects can be neglected. Actually this concept is not that difficult to absorb since as discussed in the previous section if air has a value for its resistance to flow, its viscosity, (Table 3 for values) then there would be some friction occurring local to the surface of the airfoil (skin friction) and some resistance of the air molecules to move out of the way of each other due to the "Van der Waal" forces keeping the molecules together.

It is logical that at some point above the moving object the frictional forces generated at the surface would no longer impact the flow. Prandtl defined the flow pattern over the object's surface ranging from simple laminar, to turbulent, on up to supersonic flow.

Based upon Prandtl's work and others in the field it was apparent that to maximize rotor design the airfoil tip velocity should operate in the region where air's compressibility would not be a factor. This would mean subsonic flow since pressure variations generated while flowing over the airfoil are small compared to the absolute pressure and at low Mach numbers the effect on viscosity minimal. The angle-of-attack (to be discussed in great detail in the following chapters) should not exceed the maximum limit for laminar flow to exist over its surface.

The next aerodynamic topic is Bernoulli's Principle. The easiest way to visualize how the Bernoulli Principle works is to hold one edge of a small 1 inch by 3-inch strip of notebook paper next to your lower lip such that the free end droops towards the ground. Now, blow gently over it. Surprisingly instead of the force of your breath pushing this strip of paper further down it begins to rise. The harder you blow the closer this piece of paper approaches the horizontal plane. This occurs because when the air velocity increases over the top surface of the paper the dynamic pressure decreases relative to the air pressure on the lower surface. Taking this one step further, since pressure = force divided by area and the strip area is the same for both the upper and lower surfaces then the force on the upper surface (force upper =

pressure upper x area) is less than the force on the lower surface (force lower= pressure lower x area) resulting in a net upward force. The fact that the paper rises indicates this force is large enough to overcome the force of gravity acting to make the paper droop. Bernoulli's Principle, "simply stated is" the higher the relative velocity of a streamline flow the lower the resultant pressure."

As for Newton's addition to understating air's dynamics lets restate Newton's Third Law "for every action there is an equal and opposite reaction". Since we already know atmospheric air has mass and a corresponding change in its density as a function of altitude, it stands to reason that an opposing force will meet objects pushing air molecules out of the way.

Discussion of the Magnus Effect is more complicated. The Magnus Effect is a force generated by the circulation of air at the surface of the object. One example of its effect can best be visualized when a pitcher adds a significant spin to a pitched ball. The surface air rotation of the ball causes it to curve its path on the way to the catcher. The spinning ball as seen in the illustration has two airflow separation points, one in front of the ball where the streamline splits to go over and under the ball, and one in the rear of the ball where the airstream unites. The one in front is called the upwash stagnation point and the one in the rear the downwash stagnation point. In this forward moving clockwise rotating ball the airstream proceeding over the top of the ball experiences an increased local velocity (decreased pressure), and at the lower surface a decreased local velocity (increased pressure). The net effect being a sideward force on the ball, called the Magnus Effect, which diverts the ball's straight-line trajectory as function of the amount of spin initially induced and for spin's duration. The mathematical formulas that calculate the force developed based upon airflow "Circulation" and the subsequent Magus Effect were established by Kutta and Joukowski and are explained fully in Section 2.2.3.

Diagram 2.2.1-1   The Curve Ball and the Magnus Effect

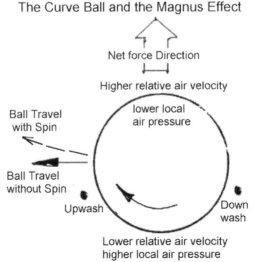

Net force Direction

Higher relative air velocity

Ball Travel with Spin

lower local air pressure

Ball Travel without Spin

Upwash

Down wash

Lower relative air velocity higher local air pressure

2.2.2 Author's Notes & Comments:

I have added this subsection because when I researched the subject on lift and the subsequent topic of Airfoil theory nothing on the subject was straightforward. In fact in the FAA "Helicopter Flying Handbook (2012)" here's what they had to say about lift: The basic concept of lift is simple. The details of how the relative movement of air and airfoil interact to produce the turning action that create lift is complex and from FAA's "Pilot's Handbook of Aeronautical Knowledge" (2008) we read: Although specific examples can be cited in which each of the principles predict and contribute to the formation of lift, lift is a complex subject. How could I design my helicopter if I didn't fully understand the principles behind the actual calculation of lift. I added this subchapter to fill you in on all my research to that end. The information regarding the theories on airfoils is a very complicated one. It is not really necessary for the reader to grasp all these conflicting ideas in order to proceed so if your brain becomes "full" feel free to move on. In later sections the aspects related directly to our project will be simply stated.

2.2.2.1. The Patent:

I'll start my journey into the subject by telling you my personal real world experience dealing with laminar airflow. Of course I studied the subject in my college fluid dynamics class but then again there is a lot to be said about the real world visuals. At the time our company had just completed the assembly of a new 1000 feet per minute plastic film winding machine and I was assigned the task of "starting it up" in our shop. What that really meant was switching on the power button, taking a step back, then seeing what happened next! This film being very thin and the driving force web tension light, at around 400 feet-per-minute film speed, all the support rollers dictating the films path began slowing down. At first this didn't appear to be a problem since the film still traveled in the desired direction. Those dynamics changed quickly as these bearing mounted guide rollers, which numbered around ten, began erratically accelerating back up to the line running speed. This jolted the thin film causing the film to be locally stretched beyond its elastic limit and generated an inordinate amount of objectionable marks across the film. This was unacceptable and when I reported the status of the startup observations to the chief engineer he determined the best corrective action would be to positively drive all the rollers. This was done and the machine shipped.

Now, let me detail the physics the problem of the idler roll interface with the film web. I will begin by saying every physical entity in the world has reluctance to move when acted on by a force. Apply a bigger force and that physical entity will move faster in spite of its reluctance. For example bring a 3000-pound car with a 100 horsepower engine to the drag strip. That car may take over twenty seconds to accelerate to 80 mph. Meanwhile, keeping all other factors equal, equip that same car with a 300 horsepower engine and that time drops into the fifteen-second range. This reluctance of an object to move is called its inertia. As with the car example these idler rolls also had a "reluctance" but in this case it was "to rotate". This reluctance or inertia of the idler rolls is a function of its mass and where that mass is positioned with respect to the objects centerline. There is a point in idler roll design where you can only get the mass down so much, factors such as weight (diameter versus wall thickness), how much deflection can the idler roll exhibit before the web traveling over it starts to bunch up at the most deflected point, what type of coating to use on the surface, the type spec bearing, etc. Lets just say our idler rolls were optimized and as the winder started up the frictional force generated by the web contacting the idler roll surface was sufficient to enable the idler roll inertia to be accommodated without stretching the web as winder speed increased. As the idler roll angular velocity increased so did the resultant viscous force of the air traveling adjacent to the surface of the idler roll. As a result, at around 400 ft/min, the increased strength of this viscous force enabled a layer of air (referred to as a laminar layer) to completely surround the circumference of the idler roll and push the film off its surface. Rather quickly these idler rolls, without the driving force of the web, started to decelerate due to bearing friction. At some point in this idler roll slow down the strength of viscous force which created the laminar layer decreased to the point that the original web tension driving force now dominated and once again started to drive the roller. Unfortunately this instantaneous application of web driving force to the now slower rotating idler roll was more inertia than the film's elastic strength could accommodate and it stretched local to the disturbing force.

During the brief period of time I spent with the startup of this machine I did do one experiment that was interesting. I noticed that if I placed my small 6 inch metal ruler, which I typically carried in my shirt pocket while in the shop, at various positions and angles along the circumference of the rotating idler roller / film interface I could prevent the laminar layer from forming. This action kept a small contact area of the film intact with the idler roll. I wasn't sure at the time how to apply such knowledge since the chief engineer's solution enabled this already behind schedule machine to ship. Little did I know that years later what I observed that day would bring me one of the most satisfying engineering accomplishments of my career.

It was close to ten years later that I applied the knowledge learned from that ruler experiment. At this point in time our company received a multimillion-dollar order for several cast film lines from a Canadian manufacturer to produce film to be used for restroom toilet seats. Yes, I know what your thinking, but what can I say it's still engineering. Anyway, our expertise was "Blown film", a process where polymer film extruded from a circular die is inflated into a tubular bubble and pulled vertically upward by a tower mounted set of nip rollers in order to cool down before being converted into large diameter film rolls on floor mounted winders. "Cast film", in contrast, is extruded from a slot die of some predetermined width onto chill rolls for cooling before entering the winder. At the time the current technology demanded that a device be used to enable a vacuum be maintained between the film just as it was extruded from the slot die and where it first contacted the chill roll surface. This vacuum chamber was designed to do just one thing – keep laminar airflow from getting between the soft molten extruded film and the chill roll surface. If ambient air did get trapped there, localized hot spots occurred across the polymer's longitudinal surface and the resulting film basically reduced to scrap.

I was in charge of this project and confident my team could design a decent vacuum chamber. I was also cognizant the engineering time required would be extensive, the machining cost exorbitant, the thermal expansion differentials difficult to calculate and the controls required to keep a vacuum uniform across a 100 inch plus width of molten film being drawn off at a rate of 1200 ft/min would be a daunting task. In fact the main cause of down time for cast film lines at the time occurred during startup when the vacuum box would in fact suck-in the molten extrusion. Once this happened the vacuum box had to be removed from the machine and the mess carefully removed without damaging any of the critical airflow surfaces. To minimize my exposure to failure in the design of this vacuum box I authorized the expenditure of $1000 to bring in a consultant for the day (at the time this was a lot of money for a consultant for one day). I decided for that amount of money I would invite all the significant players to the meeting which covered the whole spectrum of our companies organizational functions, starting with sales and ending with the assembly foreman. When that meeting was over I soon realized that what we had to do was extremely complex and the money we charged the customer for that option would not be enough to cover my estimated engineering costs never mind the cost / profit mix for the completed assembly.

It was during that consultant's talk that I remembered my simple experiment with the ruler. The next day I assigned Eugene, who by the way was one of the most brilliant men I have ever had the pleasure of working with, the task of experimenting with laminar layer deflection based upon my earlier observations, and the operating parameters of the cast film line sold. Within one day he and Carlos, the machine shop foreman (also a great craftsman) had a working test bed from which to conduct the necessary experimentation and subsequent data collection. Let me just interject something about engineering in general, sometimes no matter how much time, effort and work goes into a specific design there is no real guarantee it will be the best solution. Periodically there are those times when that solution turns out perfect and defies critical review. When that happens the solution itself just generates its own euphoria not only to you personally but to everyone involved. That's exactly what happened with the air deflector solution for the elimination of laminar air from the chill roll/ cast film interface. It turned out perfect right out of the box. It resulted in a very generous profit with the company patenting it in Eugene's and my name, and the concept later brought about a paradigm shift in the industry. I have included a copy of the patent and key diagrams in the Appendix because maybe my simple ruler experiment and the work we did may inspire you somewhere in your future.

2.2.2.2. The FAA and NASA:

When I first purchased the FAA (Federal Aviation Administration) Rotorcraft Flying Handbook (FAA-H-8083-21) originally issued in 2000 they referred to Lift being a sum of forces generated by Magnus Effect, Bernoulli's Principle (Venturi Effect), and Newton's Third Law of Motion. Before I start this discussion bear in mind Bernoulli and Newton never actually tried to explain aerodynamic lift on an object, others did. I will also quote NASA (National Aeronautics and Space Administration) for any analytical supportive substance.

a1. FAA statement: In summary, the production of Lift is based upon the airfoil creating circulation in the airstream (Magnus Effect) and creating differential pressure on the airfoil (Bernoulli's Principle)].
a2. My counter point: Shown on the below is circulation about an airfoil:

Diagram 2.2.2.2-1

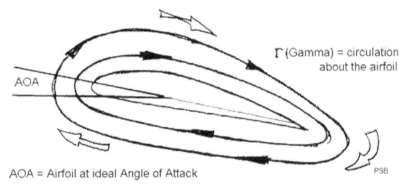

Γ (Gamma) = circulation about the airfoil

AOA

AOA = Airfoil at ideal Angle of Attack   PSB

In the previous section using the Magnus Effect to explain how the curve ball curved was simple to understand since the force generated to move it was based upon differential pressures caused by different air flow rates generated by the physical spin to the ball. Question in my mind is, how do you put a spin on an airfoil to generate the airflow shown in the above diagram??? This requires a lot more space to discuss so I'll just jump ahead to FAA's statements on "Bernoulli's Principle" and "Newton's Third Law" and pick up the "Magnus Effect" discussion later in section 2.2.2.6 of this chapter.

b1. FAA Statement: Bernoulli's Principle or the "Venturi Effect / Theory" as quoted from the both the old and new FAA helicopter manuals [as air velocity increases through the constricted portion of a venturi tube, the pressure decreases.... The combination of decreased pressure on the upper surface and increased pressure on the lower surface results in an upward force.]

Diagram 2.2.2.2-2

"Venturi Theory"

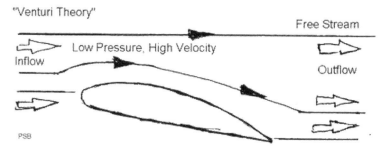

b2. NASA: Before I start explaining what NASA defines as the actual "Venturi Theory" be aware the FAA's interpretation of the "Venturi Theory" differs in that they also include the Magnus effect and Newton's law as "Lift" contributors.

NASA refers to the "Venturi Theory" as "Incorrect Theory #3". NASA's position on the subject is that the airfoil is not a venturi nozzle, which is what the "Venturi Theory" is based upon. Half of the "nozzle" is missing to create the desired results. Experimentation shows velocity decreases as you move further from the airfoil surface ultimately approaching the free stream velocity. The velocity does not increase as it would along the centerline of an actual venturi nozzle.

Another flaw in the Venturi analysis is the lift generated by flat plate in an airstream. The nozzle part does not exist yet lift is generated. Another suggestion posed is to create a nozzle- like effect by tilting the flat plate downward, but that experiment resulted in negative lift being created. When at this negative angle of attack the velocity actually slows down contrary to the theory's prediction. Keep this thought in mind when I discuss this topic further in my boundary layer dialogue.

Additionally in stating this Venturi Theory nowhere is the effect of airflow over the lower airfoil surface mentioned. Obviously there is more to it than just pressure and velocity changes over the upper surface. The implication of this theory is that you could have any lower airfoil shape and it would not matter. Obviously this is incorrect as the lower surface does contribute to lift.

As a result of this invalid assumption if you applied this theory to predict the velocity profile over the airfoil top surface and then utilized Bernoulli's equation to calculate the pressure so as to generate a resultant force distribution and obtain a lift number, that number for a given airfoil disagrees with actual measured results.

c1. FAA's Statement: Their reference to Newton (third law of motion) was as follows: ["for every action there is an equal and opposite reaction", the air that is deflected downward also produces an upward (lifting) reaction], exact quote. Later in the section they go on to say this deflected air represents only a small contribution to total Lift.

Diagram 2.2.2.2-3

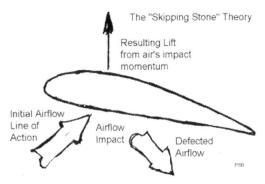

c2. <u>NASA</u>: The fact FAA's mentions that only a small percentage of lift is generated as a result of air molecules deflected downward off the lower surface of the airfoil gives some possible credence to their inclusion of it to the total lift summation. On the other hand, NASA points out that in some reference books it is stated "all" the airfoil lift is generated this way and they refer to it as the "Newtonian Theory of Lift". NASA states this is in fact in error and to not confuse readers, the above molecular action-reaction explanation is referred to as the "Skipping Stone Theory." Predictions of lift using this flow model are typically wrong because top surface airfoil flow is ignored. It should be noted that the correct Newtonian Theory deals with airflow turning.

You just can't eliminate the effects of airflow "over or under" an airfoil surface from any lift analysis. This "Skipping Stone" theory doesn't include any reactions taking place on the upper surface and assumes all of the airflow turning and subsequent lift is generated by the airflow interaction on the lower surface alone. Based upon NASA's experiments the upper surface in a lifting airfoil contributes more airflow turning than the lower surface when considering the nature of the resultant downwash. This theory can't explain or predict this result.

Using the same analogy as before in the Venturi theory but now only considering the top of the airfoil, we could have any shape upper surface and the lift generated would be the same. Of course this is not the case. In fact there are spoilers used on the upper surface of commercial aircraft wings to maneuver an aircraft by altering its lift. Additionally, the theory does not predict the negative lift generated when the angle of attack is negative because all momentum to the airfoil generated by air impacting it is ignored.

Bottom line is lift predictions based on these theories are totally inaccurate. The main problem is they do not reflect what happens when an object passes thru a real fluid, one with physical properties such as viscosity. In reality, at normal flight conditions all surfaces of an airfoil contribute to the lift producing effect by turning a moving viscous fluid.

What happens when you're not in normal flight and the air density is very low, such as at the threshold of outer space (altitudes above 50 miles)? Based upon what happens when the space shuttle begins the early phases of its re-entry into the earth's atmosphere at hypersonic speeds (above 10,000 mph) the theory is very accurate. The key point here is air density (number of molecules per unit volume) and its viscosity (molecular attraction) determine whether or not to use this theory.

2.2.2.3 More Conflicts:

I have to say having two respected government agencies like the FAA and NASA offering conflicting views on how lift was produced definitely disturbed me. I was determined to learn more, after all NASA did put a man on the moon in 1969. There was one more theory NASA debunked which was not mentioned in the FAA documents it was, Incorrect Theory #1 or commonly referred to as the "Equal Transit Theory". Here's the way this one goes:

a. When generating Bernoulli's calculation of Lift, both upper and lower airfoil flows must meet and be equal in order to satisfy the condition that the total energy of the air be constant. Lets think about that for a second.

Diagram 2.2.2.3-1

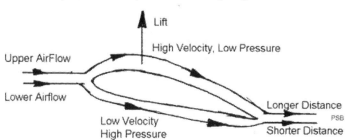

"Longer Path" or "Equal Transit" Theory Diagram

If molecules must accelerate when traveling over the top surface in order to cover the greater path length to reach the trailing edge mating point, how do these molecules decelerate to match the speed of the slower moving molecules traveling the shorter lower path?

Slowing these faster moving molecules down to match those of the lower surface would occur but it would definitely take place aft of the trailing edge when the kinetic energy of the molecules was dissipated by viscous forces interacting with the lower flow stream and the surrounding air. The resultant path of that mating flow was not necessarily parallel to the original airflow direction due to the resolution of the potential differences in static pressures. What is described above is referred to as the "Equal Transit (time)" Theory, where the time for air to travel over the top of the wing is equal to the time it takes for air to travel below the wing in order for both air streams to match at the trailing edge. This implies that the air must travel faster over the wing's top surface because the path length is longer. This theory is now referred to as the "Equal Transit-Time Fallacy" (NASA's incorrect theory #1) because neither the hypothesis has any theoretical basis nor can it be substantiated experimentally. Yes, it is true the air moves much faster over the top surface but both flows do not catch up at the same time.

Additional NASA's comments: The real show stopper with this theory is that a flat plate can generate lift. More realistically a number of symmetrical airfoils are in use, especially in early helicopters, which provide lift. Both these examples have airfoil shapes where the upper and lower surfaces are equal. You also have modern, low drag airfoils in which the lower surface is longer than the top. Then you have those air show stunts where aircraft actually fly inverted. None of the above is explainable by the theory. The main point here is that it's not the distance traveled but the way that airflow is turned in the process that produces the lift.

Lets analyze this further. This theory tries to generate a value for the air velocity over the airfoil based upon the assumption that two molecules splitting up at the leading edge will meet at the trailing edge. Experimentation shows this does not happen. In fact actual velocity over the top surface is higher than predicted with this theory. Like the other incorrect theories the lift predicted does not agree with actual experimental measurements. In this case because the velocities are understated the lift predicted is actually less.

b. Another theory I found upon my Internet research was labeled the "Coanda Effect". This refers to the tendency of a fluid jet to stay attached to an adjacent surface that curves away from the flow, and in the process entrains ambient air into the flow. An example of this is using a directed air jet over the upper surface of a wing. The result is the wing deflects up, and as a result one would say the Coanda Effect generates lift. The problem with this example is it does not represent real world flow over the top surface since the air jet creates a complicated, vortex-laden turbulent mixing layer. Pushing this Coanda effect one step further, in some explanations this effect is extrapolated to say this it is the reason why a boundary layer adheres to a curved surface. In this interpretation the Coanda effect is used to explain why airflow remains attached to the top surface of an airfoil. In reality in the field of aerodynamics the boundary layer airflow's attachment to the surface of an airfoil, or any traveling surface is due to viscosity and not the Coanda effect.

c. Then there was the "Potential flow theory". The only way to make the equations work is if the flow at some singular point had an infinite value associated with it.

Diagram 2.2.2.3-2

Potential Flow Theory Diagram of Impossible Infinite Airflow at Trailing Edge

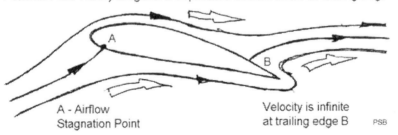

A - Airflow   Velocity is infinite
Stagnation Point   at trailing edge B   PSB

Really now, infinite velocity? It was then that I realized there had to be a better explanation out there.

2.2.2.4. Thing We Do Know:

Before we go any further on this exploratory trip into the nature of lift lets point out some of the things we do know:

In this book Lift is defined to oppose Weight. In its technical definition it can be in any direction since it is calculated with respect to the direction of flow rather than to the direction of gravity. For example when an aircraft is banking in a turn the lift is tilted with respect to vertical. In the case of a sailboat the lift generated may be largely in the horizontal direction.

When air passes over an airfoil it is constantly changing direction and follows a curved path. Associated with an airflow's changed direction, a reaction force is generated opposite to the directional change (Newton's second Law). When airflow follows a curved path, it generates a pressure gradient perpendicular to the flow direction.

The power used to move an airfoil forward is based upon the rate of force required to push air out of the way of its path. The force component perpendicular to the free stream is called Lift and the parallel component called Drag. The equal and opposite force on the surrounding air creates air's momentum transfer called "downwash". In fact, measuring the momentum of this air is another way to determine the Lift on an airfoil.

Bernoulli's principle does not explain why the air flows faster over the top of the wing. If those airflows around the wing are established experimentally then Lift can be determined based upon pressures using Bernoulli's principle and Conservation of Mass.

On occasion one may notice an assumption being made when using Bernoulli's principle that things take place at "constant energy" or in conditions of "incompressible flow". Remember these two simplifications do not necessarily support real-world airfoil conditions. For example an aircraft in level flight is adding energy to the flow while a sailboat that is accelerating is removing energy from the flow, so energy is not necessarily constant. Additionally, compressibility effects must be factored in when dealing with transonic speeds.

2.2.2.5 The Laminar Flow and the Boundary Layer

I believe I have some credibility on the subject based upon my patent on "laminar layer air deflection" that was developed to solve a real world problem associated with eliminating it. In that patent the strength of a viscosity based skin friction laminar airflow boundary layer was established. The subject of the patent was of a stationary spinning cylinder, once the cylinder surface tangential speed reached about 400 fpm a noticeable boundary layer pressure was formed and uniform about the entire circumference (fully detailed at the end of this book under "Patent"). No lift is generated since there is no net differential pressure anywhere on the cylinder surface and for the most part this boundary layer thickness is constant. What this means is the physical surface whether it be a spinning ball, cylinder or a moving airfoil has an extended boundary layer pressure surface whose defining characteristics are a function of airspeed (for subsonic flow). As sited above with the rotating cylinder example, this boundary layer exists whether lift is being created or not.

The diagram below shows how the boundary starts at the leading edge of the flat plate, where due to air's viscosity and resultant skin friction the air velocity "v" closely matches the relative velocity of the plate. Then as you progress further along the plate the free stream velocity "V" creates sufficient stress on the adjacent air flow "v" to have it increase its speed until "v" equals "V" at some distance above the plate surface. This boundary layer can be described technically as that airflow layer (as determined by the Reynolds number) where viscous forces no longer distort the surrounding non-viscous flow. Based upon the fact that the boundary layer exists somewhere between "airflow at the surface" and the "free stream velocity" one can calculate the pressure differential at the layer based upon the velocity difference along the layer.

Diagram 2.2.2.5-1

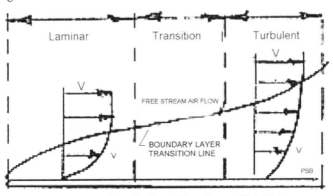

Boundary Layer Creation from Flow Over a Flat Plate

One more point: friction is calculated by determining the force perpendicular to the plate called the normal force (pressure times area in this case) times the coefficient of friction (for air moving over a flat plate). There are two types of friction "static and dynamic". Static friction dominates until things start moving. Once in motion the "coefficient of friction" typically drops in value and the normal force required to keep things in motion decreases.

At some airflow rate over the leading edge of a flat plate the conditions that promoted the start of the boundary layer continue to maintain that layer, even though the initial surface air that started the boundary layer moves rearward.

In actual flight where the wing is traveling into ambient air experimentation has determined that as the pressures increase in the direction of flow along the airfoil, a general flow rate deceleration occurs. The rate of deceleration that occurs near the "outer edge of the boundary layer" follows Bernoulli's law.

"Within the boundary layer" and specifically near the contact surface simple formulas such as Bernoulli's do not apply due to the complexity of the viscous forces present. The effect of these viscous forces "within the boundary layer" causes loss of particle speed relative to those particles at the outer limits of the layer (as shown in diagram 2.2.2.5-1). This reduction of particle velocity reduces its kinetic energy (a velocity based energy form where kinetic energy = ½ x mass x velocity^2) such that there is an overall loss of the boundary layer kinetic energy. This in effect reduces the ability of those particles to flow in opposition to the adverse pressure gradient.

Now lets look at airflow around a symmetrical airfoil pointed directly into the flow path such that no lift is created. The graph shown has "surface velocity / free stream velocity" as a squared term but in spite of that you can readily see the trend of the air flow velocity around the upper and lower surfaces.

This graph is also important in that in experimental work the boundary-layer thickness is taken as the distance from the surface to the point where $(v/V)^2 = 0.5$.

Graph 2.2.2.5-1

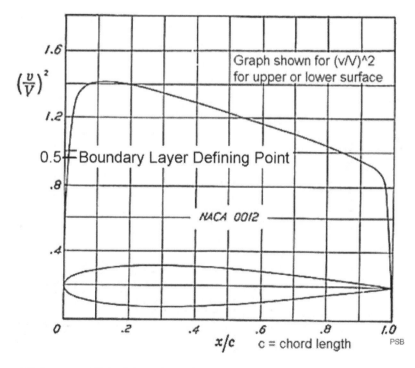

We have established that a boundary layer exists. For aerodynamically smooth wings under low or moderate lift conditions the laminar boundary layers start at the leading edge extending to the relative position of the first minimum-pressure point on the upper and lower surfaces of the airfoil.

In addition to knowing that the boundary layer exists, if you can determine the surface skin friction coefficient and the velocity distribution through the boundary layer (definitely need some sophisticated instruments for this) then by using equations and relationships developed by Karman, you can obtain boundary-layer thickness versus pressure distribution graphs.

My point is: The significance of this boundary layer discussion is that due to its presence and the associated reduced velocities, the airflow (streamlines) around the airfoil are displaced from where you thought they would be.

2.2.2.6. The Magnus Effect and the Circulation revisited.

Here's my take on the aerodynamics of ambient airflow around an airfoil or the "Magnus Effect and Airfoil Circulation". Lets go back to the Equal Transit Time theory for a moment and try to resolve the issue of what actually happens to the upper and lower airfoil flows when they intersect at the trailing edge. We will refer to this as the trailing edge vortex flow and to grasp the aerodynamics of this we will check out the Kutta-Joukowski hypothesis for a mathematical explanation. Before starting this part of the discussion remember earlier in this book when I mentioned the complexity of something as simple as pouring milk into a cup of hot coffee and analyzing the dispersion patterns. Take that same cup of coffee and run a spoon thru it such that the edge of the spoon simulates the leading edge of an airfoil. The coffee wake that follows immediately aft of the traveling spoon creates a counterclockwise circulating vortex stream which attempts to get back to the spoon's trailing edge before trailing off (note you may have to have to do this several times at various spoon attack angles to note the flow circulation variations). The Kutta-Joukowski hypothesis is analogous to this coffee example in that the all the kinetic and potential differences that exist at the airfoil trailing edge leave some sort of void whereby at the stagnation point of the downstream air flow a condition exists that some of this vortex flow travels back to the trailing edge in a counterclockwise circular pattern.

Now this point is easy to understand if you actually did the spoon experiment and saw the resulting swirls. Even the point that this post circulation is counterclockwise is not hard to fathom since the lower surface airfoil flow has the higher static pressure, which would add an upward force to push the higher velocity flow (lower static pressure) exiting the top surface in a counter clockwise path.

What isn't as easy to grasp is the part where Kutta and Joukowski state that if there is a counterclockwise air flow at the trailing edge then for all the system dynamics to be equal there must be a clockwise air flow about the surface of the airfoil (note that Kelvin's theorem supports this concept in that the total circulation in the fluid, clockwise and counterclockwise flow, must remain zero). Now, really? They want me to believe that for a clockwise airflow to exist around the airfoil the flow on the upper surface travels with the general air flow direction, this I can believe, but that would mean the counter rotating air traveling on the lower airfoil surface would have to travel against the general air flow direction. That I can't believe. According to my Engineer Training Manual written by M. R. Lindeburg and endorsed by the National Society of Professional Engineers, my thought about that lower airfoil flow "not occurring" is supported when they write "in the real world this clockwise rotating airflow about the airfoil does not exist". This is where it gets interesting. This concept of the "Magnus Effect and the Airfoil Circulation" as suggested by Kutta and Joukowski is the foundation of "Lift" in subsonic aerodynamics. Wow, what was I missing here?

Right around now it would be a good idea to look at the actual pressure distribution of a real airfoil in motion (noting that the higher the local pressure the faster the airflow velocity). Looking at the diagram nowhere is there a circulatory airflow about the airfoil!

Diagram 2.2.2.6-1

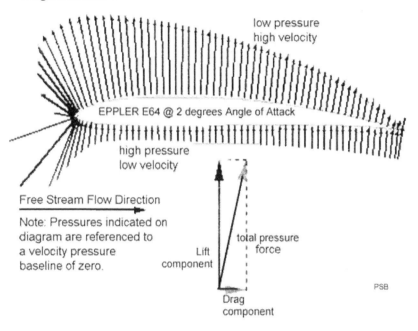

Comparing airflow travel around an airfoil versus that corresponding to a rotating cylinder is definitely unique when you look at it from the perspective of "Airfoil Theory" and must make the jump to accept the concept of air "Circulation". Circulation is the contour integral of the tangential velocity of the air on a closed loop (or referred to as the circuit) around the boundary of an airfoil. You could also say it is the total amount of air spinning (vorticity) around the airfoil while moving forward. The theory goes that this clockwise air circulation about the airfoil boundary layer sets up the counterclockwise circulating shed vortex aft of the airfoils trailing edge. Much like the spoon in the coffee example discussed earlier. The fact that you can see the trailing edge counterclockwise circulating air leaving an aircraft wing on a humid summer day seems to support the theory.

Here is my detail comment on the subject! In the spoon and coffee example you do see shed spoon vortices that trail the moving spoon and they have a counterclockwise circulation, same goes for the aircraft wing on a humid day observation. Point is you do not see any coffee circulating the spoon or air circulating the wing.

In mechanical engineering we typically refer to Castigliano's formulas to calculate stress on a part based upon the distortion energy it takes to deform it. Appling that thought process to the airflow about an airfoil, say with an 8-degree angle of attack, we have air being distorted out of the way of the physical airfoil both above the wing and also below it as the airfoil pushes through it. Citing the Law's of Conservation of Energy, Mass, Momentum and Angular Momentum and any others that may be applicable, all this distortion energy created by the airfoil pushing the air out of the way has to be accounted for. Obviously, you have a good part of that energy consumed with the airfoil lift / drag dynamics, but obviously there is sufficient distortion energy left over both above and below the airfoil to create the counter rotating swirls of air filling the void as the fast moving airfoil moves forward.

Looking back at the tailing edge of the Eppler E64 diagram shown previously, the pressure distribution suggests the airflow leaving both upper and lower surfaces have a certain mismatched positive velocity associated with it. Therefore, one can conclude there is no circulation, as I understand it to be, associated with the traveling airfoil and if the pressure differential at the trailing edge is insufficient to create the resulting counterclockwise swirls, the residual distortion energy in the control volume we are considering does.

At this point you're probably wondering if my take on the distortion energy concept has any aerodynamic support from that community. In fact it does in the form of the Navier-Stokes equations where my concept of "distortion energy" is presented as the sum of the "viscous shear stresses" in the air, which creates the momentum shifts. Unfortunately to model an airfoil using the Navier-Stokes equations with any sort of angle of attack associated with it requires numerical calculations that only a computer program can solve. The simple Bernoulli or Newton equations for flow impacting an airfoil do not exist. Therefore the FAA statement "The basic concept of lift is simple but the details of how the relative movement of air and airfoil interact to produce the turning action that create lift is complex" makes sense.

Now that I have stated my position lets see if there is any common ground between my opinion and that of airfoil "Circulation", which like I said before is the basis of Airfoil Theory. Lets look at the following graph of an NACA 4412 airfoil with an 8-degree Angle of Attack, where the ambient air pressure and reference velocity is set at zero.

Graph 2.2.2.6-1

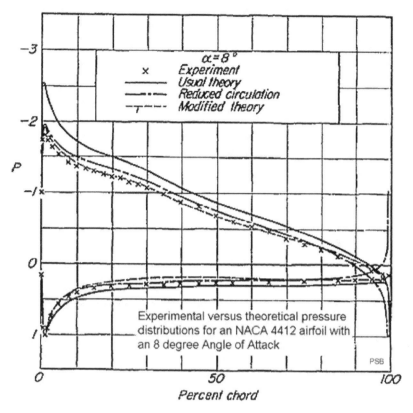

Experimental versus theoretical pressure distributions for an NACA 4412 airfoil with an 8 degree Angle of Attack

In the above graph an interpretation of how that pressure distribution could develop is shown in the following sketch.

Diagram 2.2.2.6-2

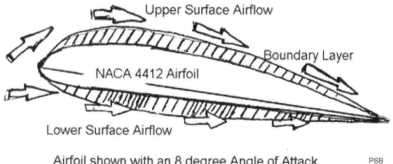

Airfoil shown with an 8 degree Angle of Attack

In this diagram of the asymmetrical NACA 4412 airfoil I have sketched in an exaggerated boundary layer in order to better visualize the airflow streamlines.

- The airflow forced to travel above the leading edge as a result of the impact has an instantaneous burst of velocity creating the spike of low pressure as plotted on the graph. As this high velocity airflow crests the leading edge, the boundary layer formed is relatively large since its thickness is proportional to that velocity. As this deflected airflow progresses further downstream its velocity gradually decreases and the negative pressure drops. Initial interaction with the upper surface "boundary layer" takes place but at a greater distance from the surface where its velocity is greatest and the interaction minimized. All this scales down the further distance traveled along the airfoil.

- The high positive pressure at the lower surface of the leading edge can be attributed to the corresponding location of the impacting ambient air. On the lower surface, the airflow actually impacts or penetrates the "boundary layer" and I suspect, based upon the positive pressure numbers shown on the graph not only does that airflow slow down relative to the free stream velocity but a certain amount of momentum is transferred. Remember, inside the boundary layer strong viscous forces are present and those will interact with any air entering to reduce air particle velocity.

Now getting back to the graph. The first thing you notice is the extremely high pressure at the leading edge indicating virtually zero air velocity rearward. Upon closer inspection the forward stagnation point appears to be slightly aft of the leading edge as shown on the lower surface pressure line. The next apparent pressure spike occurs at the trailing edge. Between those two extremes there is transitional continuity in both the upper and lower airfoil surfaces. Explaining this further, from the forward stagnation point and progressing over the top airfoil surface the flow velocity progressively decreases until the trailing edge is reached. Then starting from the trailing edge and moving along the lower surface forward, the velocity steadily increases to a point, but is still less than the free stream velocity. It then drops off at the forward stagnation point.

You could infer that starting from the forward stagnation point and progressing clockwise there is some velocity continuity in that each adjacent airfoil surface point has decreasing velocity.

Recalling that "Circulation" is defined as the contour integral of the tangential velocity of the air on a closed loop around the boundary of an object, I could accept the fact that even though nowhere does any airflow actually circulate clockwise about the airfoil in the real world boundary layer, the math could be developed "within limits" such that adding a clockwise circular airflow velocity about an airfoil corresponding to a specific 8 degree Angle of Attack, "THEN;" placing that airfoil into an airstream with a uniform velocity would result in the pressure graph shown above (no downstream airflow dynamics are included in this superposition of airflows).

With this understanding of the possibility of "SUPERIMPOSING" airflows to be able to generate some meaningful airfoil pressure distributions, I will submit to discuss the development of Airfoil Theory starting with Kutta and Joukowski & the cylinder.

One last comment, I believe the FAA should perhaps revise their explanation of how lift is produced by eliminating any reference to the "Venturi Principle" because the correlation is inaccurate. They should also not include the "Magnus Effect" because the concept of Circulation taken out of context implies lower airfoil surface airflow "counter" to the airstream flow and that just doesn't happen. In addition it can only confuse the issue for any reader.

Any reference to the "Magnus Effect" in conjunction with an airfoil should be labeled strictly as a theoretical explanation of experimental results. In other words, it could be described as a theoretical "mathematically generated" counterclockwise tangential flow about an airfoil boundary such that when "added" or "superimposed" onto a uniform airflow it results in a pressure distribution similar to that found experimentally. That resultant pressure distribution can then be converted into a force vector via Bernoulli's principle. The vertical component of that force vector is called Lift and its horizontal component Drag. To put it simply, reality was redefined to allow for a mathematical result that matches data collected.

2.2.3 Kutta and Joukowski –the cylinder

Before we tackle this next section I use the Greek font in the actual equations and in the text the written name given to the font. This is because the software used in preparing this book does not easily lend itself to the use of the symbols and yet I want to be true to the accepted mathematical nomenclature for this work.

At this point I realized presenting the analytical side of this topic, which by the way is officially called Airfoil Theory, was not going to be easy. I decided to start from the beginning like the researchers did with the discussion of a simple "stationary" cylinder of finite length placed into airflow where its longitudinal axis is perpendicular to the direction of flow to duplicate the spinning ball. In this case the net result of the flow produces no lift because the flow over the top and bottom surface is equal.

Diagram 2.2.3-1

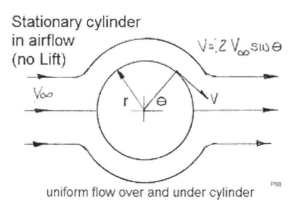

Stationary cylinder in airflow (no Lift)

$V = 2 V_{\infty} \sin \Theta$

V_{∞}

uniform flow over and under cylinder

If that cylinder rotates clockwise to the direction of the flow, then the velocity of the flow over the top surface is greater than that at the lower surface and Lift is generated. This, of course is fact. The question is what is the lift contribution at each point along the circumference? To get that answer the change in velocity at each point around the surface must be taken into account and the setup for this calculation can be shown below as a closed curve around the body. Here for every differential increment dl (a small distance looked at on the curve) the corresponding tangential component of velocity is calculated. Please be aware of the importance of this mathematical concept of calculating velocity at a point on a surface. Eventually, through the mathematical journey in this chapter we will get to the point where we can calculate the actual pressure at each point on an airfoil surface. Totalizing this pressure distribution gives us the Lift capability of the airfoil and also the load distribution.

Diagram 2.2.3-2

The concept of circulation about a point

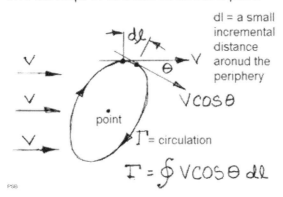

dl = a small incremental distance aronud the periphery

$V\cos\theta$

point

Γ = circulation

$$\Gamma = \oint V\cos\theta \, dl$$

This equation sums up all the dl values for the circulation around the arbitrary closed curve drawn around the point.

The term given to this relationship is referred to as Circulation (the upside down L actually the Greek symbol Gamma) and is defined as the integration (summation) of all the differential increments (dl terms) with respect to velocity (v) around the closed curve. The mathematical representation of that relationship is given below, and since the velocity "v" is given as a distance per unit time (ft/sec) and length "l" a distance (ft) then it follows that the units for Circulation are: velocity x distance = (ft/sec) x (ft) or ft^2 /sec.

Looking at the next diagram of the rotating cylinder, lets get a more useful form of equation for Circulation so we can calculate the Lift per unit length. If the radius of the cylinder is "r", its rotational speed is omega (radians/sec) and is traveling thru the air (density equal to roe) at a velocity of "v infinity" then Circulation is shown in the diagram as:

Diagram 2.2.3-3

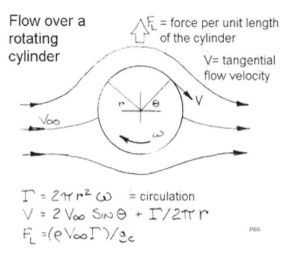

Flow over a rotating cylinder

F_L = force per unit length of the cylinder

V= tangential flow velocity

$$\Gamma = 2\pi r^2 \omega \quad = \text{circulation}$$
$$V = 2 V_\infty \sin\theta + \Gamma/2\pi r$$
$$F_L = (\rho V_\infty \Gamma)/g_c$$

This relationship of the Kutta-Joukowski theorem is called the Magnus effect and is qualified by the following conditions: in the ideal case where the air flow adjacent to the rotating cylinder travels at the same rotational speed as the cylinder, this equation has a maximum lift coefficient of 4 pie or 12.6. In reality rotating devices rarely achieve a lift coefficient above 9.

2.2.4 Airfoil Theory

Airfoil Theory Mathematical Foundation (remember you don't have to understand this to read the rest of this book but you will need this surface pressure distribution detail to design the internal structure of a rotor blade).

A) There is no easy way to present this topic without some sort of previous exposure to advanced mathematics, but I will attempt to delve into this subject nonetheless. Now, if you have never heard of advanced mathematics it's that part of math that explains everyday things and the totally abstract things in life, for example what is the rate of heat transfer from a three dimensional radiator surface, or the charge and current of a series circuit containing an inductor, capacitor, a resistor with an emf (electro motive force) as a function of time using Laplace transforms, or determine what happens when a membrane vibrates. The list goes on and on but these few topics may give you an idea of what is analyzed in the field of advanced mathematics. Now, just to stimulate your mind on the topic, how much is "2 x 2"? That's simple its "4". What is the square root of "4"? That too is simple it's "2". What is the square root of "–4"? You guessed it, "2 times the square root of –1". What name do they give to the "square root of –1"? They symbolize it as "i" and refer to it as an imaginary number. If you think I'm wrong try to find it on any x-y-z numerical coordinate system. The only way you can find an imaginary number is of course on a coordinate system with an imaginary axis referred to as the "complex plane". Numbers plotted here are called complex variables "z" represented in the form (a+ bi), where "a" is the real part and "bi" the imaginary part. Therefore when we write z = x + iy, the real part "x" is plotted as the abscissa of the point and the magnitude of the imaginary part "y" as the ordinate.

If you are still interested in more on the subject lets define some terminology that will be used later on in the actual discussion of Airfoil Theory:

1) Definitions:

•   The Perfect Fluid: lets just say that the KISS (keep it simple stupid) principle is used here. There is no one who has ever completed just one course in engineering who doesn't know what that means. For the record it implies not to over complicate things or solutions. With that in mind, looking at the basic mechanics of flow around an object certain assumptions were made to get justifiable results by stripping away some of the complex details. In that vein the first complexity to be thrown out was air's compressibility. The point here was the flow to be analyzed would be in a range where the flow density change was small such that the pressure variation minute with respect to the absolute pressure.

Next up was air's viscosity. The perfect fluid had none, zero shearing stress, and zero viscosity. Tossing out viscosity was a major simplification because without taking into account air's viscosity the drag produced, as the airfoil moved thru the air could not be calculated, neither could any information concerning separation of flow from the surface. Remember when I stated that Circulation about an airfoil counter to the air flow direction couldn't exist in the real world, well with the elimination of shearing forces Circulation about the airfoil to offset trailing vortices could in theory exist. How was that simplification justified? Basically, because for many actual flows, with exception of the fluid layer adjacent to the surface, the viscous forces are small with respect to the inertial forces and the simplified equations generated reasonable approximations of velocity distributions in the flow ranges under consideration.

The perfect fluid is a homogeneous continuous medium and in the field of flow it cannot be created or destroyed. This means that any amount of fluid entering a discrete element of volume must exit that volume in an equal amount. The only forces acting on this steady flow thru this discrete element of volume are normal pressure forces. Under these conditions all the partial differential equations developed to define this flow when resolved thru the mathematical process of integrating along the path of motion (normally referred to as a streamline) boils down to Bernoulli's equation. Stated next are the partial differential equations developed in accordance with Newton's laws to show the detail math of how this discrete surface pressure is presented for the referenced directions, where u, v, and w respectively are components of velocity in the x, y and z direction. The systematic development of these equations can be found in the "Theory of Wing Sections" book. They are included here just to give example of the look of the equations we are dealing with. Fortunately, all this resolves into Bernoulli's equation, which follows this differential set of equations.

Diagram 2.2.4-1

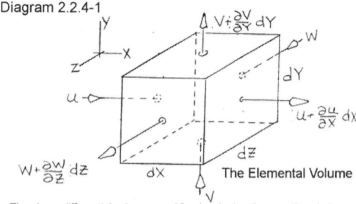

The Elemental Volume

The above differential volume, used for developing the equations below, satisfies the condition of continuity which means the fluid entering any small element of volume equals that exiting that volume.

$$-\frac{\partial P}{\partial x} = \rho \left(\frac{\partial u}{\partial t} + u\frac{\partial u}{\partial x} + v\frac{\partial u}{\partial y} + w\frac{\partial u}{\partial z} \right)$$

$$-\frac{\partial P}{\partial y} = \rho \left(\frac{\partial v}{\partial t} + u\frac{\partial v}{\partial x} + v\frac{\partial v}{\partial y} + w\frac{\partial v}{\partial z} \right)$$

$$-\frac{\partial P}{\partial z} = \rho \left(\frac{\partial w}{\partial t} + u\frac{\partial w}{\partial x} + v\frac{\partial w}{\partial y} + w\frac{\partial w}{\partial z} \right)$$

The above pressure equations after reworked using Newton's laws results in:

$$H = P + \tfrac{1}{2}\rho V^2$$

PSB

This is Bernoulli's equation, where P the first term represents the pressure, the second term the free stream pressure, and H the total pressure.

- Flow Patterns: In developing Bernoulli's equation the differential equations used were established for flow thru a three-dimensional volume defined by the x-y-z planes (see appendix pipe flow example) with the corresponding velocities being u, v, and w. What if we were to look outside the box, so to speak and determine if there is any other way to represent three-dimensional flow without using u, v, and w directional components of velocity? To start, the powers-to-be simply made this a two-dimensional problem and effectively eliminated the z-axis component of velocity by making w = 0. Next they realized that since the Law of Continuity states that any flow elements entering a defined area in the x-y plane must leave in the same amount a single function could be used to define both x and y components of the element velocity at all points in the given flow pattern. This is called the "Stream Function" represented by the Greek symbol "capital Psi" and when this function is equal to a constant, the lines of element flow are called "streamlines." Moving right along in the simplification process, the flow would be restricted to irrotational motion. In this special case once again the flow pattern can be described completely by only considering a point in the given flow. This function is called the "Velocity Potential" represented by the Greek symbol "capital Phi". Based upon the above let's look at a simple two-dimensional flow using the "Stream Function" and the "Velocity Potential" to define it.

a)   The Uniform Stream

Stated below are the "Stream Function (capital Psi)" and "Velocity Potential (capital Phi)." of a uniform flow having a velocity equal to $[a^2 + b^2]^{0.5}$ whose tangent is a/b and is inclined to the x-axis.

Formula 2.2.4-1

$$\Psi := b \cdot y - a \cdot x \qquad \text{stream function (Psi)}$$

$$\Phi := b \cdot x + ay \qquad \text{velocity potential (Phi)}$$

where the x axis velocity component is v

$$v := -\frac{\partial}{\partial x} \Psi = \frac{\partial}{\partial y} \Phi = a$$

and the y axis velocity component is u

$$u := \frac{\partial}{\partial y} \Psi = \frac{\partial}{\partial x} \Phi = b$$

b)   Sources and Sinks:

This concept of flow being created at a defined rate at a "source" (point) and being destroyed at that same defined rate at the "sink" (point) is one of the basic building block concepts used in the theory of wing sections (see appendix pipe flow example where we used a water cannon as a source and whatever water entered the pipe also left it). Here the flow at either point is taken as being uniform in all directions, is irrotational, and conforms to the Continuity Laws at all locations in the flow except at the source points themselves. At those two specific point locations the "Stream Function (capital Psi & shown below)" is a function of both x and y and the tangential component of the velocity is zero, "m" is the rate with which fluid is being created by the source per unit length perpendicular to the plane of flow.

Formula 2.2.4-2

$$\Psi := \frac{m}{2\pi} \cdot \left(\tan^{-1} \frac{y}{x} \right)$$

and the point source flow velocity potential is:

$$\Phi := \frac{m}{2\pi} \cdot \ln\left(\sqrt{x^2 + y^2} \right)$$

c) Doublets:

This is a special case of a source / sink relationship. The doublet consists of an equal strength source and sink such that at some distance between them the product of distance "s" and the source strength per unit length "m" remains constant. The Stream Function (capital Psi) for the flow about a doublet (when "s" small with respect to x and y, and the product of "m" times "s" is constant as "s" approaches zero) is stated just below the diagram. The "s" represents the position of the source on the x-axis

Diagram 2.2.4-2

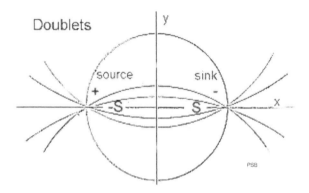

Formula 2.2.4-3

$$\Psi := \frac{m}{2\pi} \cdot \left(\frac{-2 \cdot y \cdot s}{x^2 + y^2} \right)$$

d) Circular Cylinder in a Uniform Stream:

Finally, in this section we start to see how all the above begins to come together. We need to get into this next degree of mathematical depth in order to have a better understanding of how the airfoil section was developed enabling the pressure distribution on its surface to be ascertained. In the last subsection we ended up with the formula for what is called the Magnus Effect, that being lift generated from a rotating cylinder. Now, in order to get to the next phase in our development of Airfoil Theory we will take a closer look at that cylinder with respect to the Stream Function (capital Psi) and Velocity Potential (capital Phi).

Shown below looks like a flow about a cylinder but it is actually the mathematical representation of a superimposed "Uniform Stream (a)" on the flow about a "Doublet (c)" defined above.

Diagram 2.2.4-3   Uniform flow streamlines about a circular cylinder

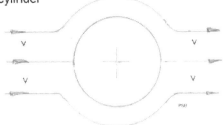

The Stream Function (capital Psi) for this case is:

Formula 2.2.4-4

$$\Psi := y \cdot \left[V - \frac{\mu}{\left[2\pi \cdot \left(x^2 + y^2 \right) \right]} \right]$$

where
$\mu = 2 \text{ m s}$

Where "V" is the free stream velocity and "mu" is the "Doublet" strength.

The math is now headed into the concept of polar coordinates. Basically it's just another way to represent a point in space with reference to the origin. If a point is defined from the origin on an x-y axis in a rectilinear coordinate system as (x, y) then that same point in polar coordinate system can be stated as (r angle θ) where r is the distance to the point from the origin when inclined from the x-axis θ degrees (remember that this terminology is similar to the way we defined an origin based vector only then it was magnitude and direction). Actually, since all our future work will be in reference to points on the surface of the cylinder, using polar coordinates is the preferred way to further develop the representative equations.

With that said the above Stream Function (capital Phi) is restated below in polar coordinate form.

Formula 2.2.4-5

$$\Psi := V \cdot r \cdot \left(1 - \frac{a^2}{r^2} \right) \cdot \sin(\theta)$$

$$a^2 := \frac{\mu}{2\pi \cdot V} \qquad r^2 := x^2 + y^2$$

$$v := -\frac{\partial}{\partial r} \Psi = -V \cdot \left(1 + \frac{a^2}{r^2} \right) \cdot \sin(\theta) = -2V \cdot \sin(\theta)$$

This stream function has an "a" term and an "r" term, when those two terms are equal, the streamline takes the shape of a circle having a radius of "r" and at that circumferential surface $\Psi = 0$. Furthermore, since the definition of a two-dimensional streamline implies infinite depth in the z plane what you see is actually just the circular side of an infinitely long cylinder (mathematically represented of course!).

If we want to determine what the velocity distribution "v" around the cylinder as a function of angle we simply set "r" = "a" to get the tangential component of velocity at each angle.

Obtaining the pressure distribution about the streamline defining the cylinder surface is straightforward since Bernoulli's equation is valid. Below is that equation where H is defined as the total pressure. S is the pressure coefficient which when distributed over the cylinder surface becomes 4 $\sin^2(\theta)$, "p" is the pressure, and "rho" is the mass density of air.

Formula 2.2.4-6

$$H := p + \frac{1}{2} \cdot \rho \cdot \left(4 v^2 \cdot \sin^2 \theta\right)$$

$$S := \left(\frac{H - p}{\frac{1}{2} \cdot \rho \cdot v^2} \right) \qquad S := 4 \cdot \sin^2 \theta$$

e)   The Vortex:

This is another special case, which as hard as it is for me to believe, can be represented not as a simple cylinder in a flow path but as a physical tornado. Everyone has heard of the term "the eye of the storm". In case you haven't it's at the center of the action where everything is rotating about some central imaginary point on some imaginary filament extending into the atmosphere. More technically stated as a flow pattern of fluid elements, which move in a circular path about a point. This flow is irrotational except at the very center. The rotational aspect of this flow introduces the Circulation term (Gamma) into the Stream Function (capital Phi) and is written as follows:

Formula 2.2.4-7

$$\Psi := \frac{\Gamma}{2\pi} \cdot \ln(r) + c''$$

where "c" is a constant.

To see how the pressure calculation is developed on an infinitesimal element look at the small increment of area shown below in polar coordinates:

Diagram 2.2.4-4

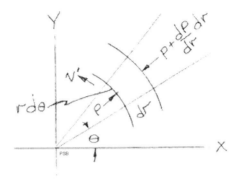

Formula 2.2.4-8

with the differential being:

$$\left[p - \left(p + \frac{d}{dr} p \cdot dr \right) \right] \cdot r \cdot d\theta$$

which when differentiated is a form of Bernoulli's equation

$$H := p + \frac{1}{2} \rho v'^2$$

which implies Bernoulli's equation is applicable throughout the flow field.

f)   Circular Cylinder with Circulation:

In case d "Circular Cylinder in a Uniform Stream" the mathematical representation of a superimposed "(a) uniform flow stream" on the flow about a "(c) doublet with a uniform flow from source to sink" was given. In similar fashion the flow representation of a "Circular Cylinder with Circulation" is mathematically generated by the superposition of "(e) vortex flow about a point" onto the "(d) uniform flow about a circular cylinder." This theoretical development of flow around the cylinder will be used in the next section to show how this flow can be mathematically manipulated to represent the flow about airfoils of arbitrary shape at various angles of attack.

The stream function for this type flow is:

Formula 2.2.4-9

$$\psi := V \cdot r \cdot \left(1 - \frac{a^2}{r^2} \right) \cdot \sin(\theta) + \frac{\Gamma}{2\pi} \cdot \ln\left(\frac{r}{a} \right)$$

The diagram below shows how the flow about the cylinder changes, with respect to case "d", when Circulation is included. It was referred to as the Magnus Effect section 2.2.3 and, as there, lift on the cylinder is produced. Using Bernoulli's equation at the cylindrical surface (where $\Psi = 0$) the equation for that pressure distribution is found to be:

Diagram 2.2.4-5

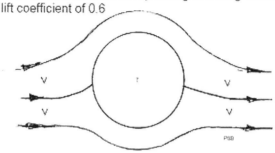

Streamlines around a circular cylinder in uniform flow with circulation corresponding to a wing-section lift coefficient of 0.6

Formula 2.2.4-10 Section Lift "l" Equation

$$H := p + \frac{1}{2} \cdot \rho \cdot \left(4 \cdot V^2 \cdot \sin^2 \theta - \frac{2 \cdot V \cdot \Gamma \cdot \sin(\theta)}{\pi \cdot a} + \frac{\Gamma^2}{4\pi^2 \cdot a^2} \right)$$

reworking the above equation we obtain the pressure coefficient S

$$S := 4 \cdot \sin^2 \theta - 4 \cdot K \sin \theta + K^2 \qquad\qquad K = \Gamma / 2\pi \, a \, V$$

setting θ equal to $\pi/2$ sets S to the cylinders line of symmetry and the lift "l"generated there by the components of pressure normal to the stream flow is

$$l := \frac{1}{2} \cdot \rho V^2 \cdot \int_0^{2\pi} S \cdot a \cdot \sin \theta \; d\theta$$

$$l := \frac{1}{2} \cdot \rho V^2 \cdot \int_0^{2 \cdot \pi} \left(4 \cdot a \cdot \sin^3 \theta - 4 \cdot a \cdot K \sin^2 \theta + a \cdot K^2 \cdot \sin \theta \right) d\theta$$

$$l := \rho \cdot V \cdot \Gamma$$

This implies lift is determined by the Circulation and not on the shape of the body (this is the Magnus Effect as stated in section 2.2.3, only in this presentation the actual evolution of the detail math is included).

• Complex variables: Actually, for a better understanding of "complex variables" and the "complex plane" I like to think of it as talking to anybody from either Brooklyn or the deep South. In this situation part of what you hear is the basic words or the "real part" of the conversation and then there is the strange dialect "the imaginary part" which taxes your brain to keep up with the words being said. The extreme case would be a conversation with someone speaking a foreign language. In this situation translation is required. The same is true with complex numbers. Under normal circumstances our conversations take place in rather a routine manner similarly, in mathematics typical calculations take place in the x-y-z plane, the plane where physical motion takes place. In complex conversations where a certain degree of mental translation must take place there is the mathematical equivalent where one can make the transformation from the x-y-z plane to various other planes, including complex planes and back again. Before we get too far ahead of ourselves let's explain some more basics about complex numbers and how it applies to airfoil sections. Our main objective for all this work thus far with the cylinder flow is to get an airfoil shape we can work with. If it was easy those brilliant mathematicians would have presented the equations already without having to resort to doublets, cylinders and vortexes. Unfortunately, now our calculations must enter the world of the complex plane.

In the introductory paragraph of this subchapter the complex variable z was defined as $z = x + iy$. The real part "x" is plotted as the abscissa of the point and the magnitude of the imaginary part "y" as the ordinate. Represented in polar coordinates z takes the form:

Formula 2.2.4-11

$$z = x + i\,y,$$

this expression in polar coordinates becomes:

$$z := e^{i \cdot \theta}$$

Now lets looks at a complex varible defined as w

$$w := \phi + i \cdot \Psi$$

and is a function of z. That relationship is written as $w = f(z)$. All the various flow relationships given previously in the two-dimensional x-y plane can be mathematically represented in the complex plane and for the specific example of flow about a "Circular Cylinder with Circulation" (case f) the complex variable "w" now defined to represent that flow in the complex plane is:

Formula 2.2.4-12

$$w := V \cdot \left(z + \frac{a^2}{z} \right) + \frac{i \cdot \Gamma}{2 \cdot \pi} \cdot \ln\left(\frac{z}{a} \right)$$

(Noting that the first term is the real part and the second term the imaginary part)

•   Conformal Transformations: Lets start this topic with an example of a conformal transformation, such as the development of a road map or atlas. In this case the spherical earth is presented as a flat surface referenced as the Mercator projection and is used in every GPS device. The main point in a conformal transformation is that the detailed shape of infinitesimal area elements "remains the same" but the "shape of finite area can be altered." This is referred to as the "F" plane. There the equipotential lines and the streamlines form an orthogonal grid similar to the typical x-y coordinate plane. For example if "w" represents a possible flow pattern in the z plane, its equation could be written as w = f (z) where complex variable z = x + iy and the coordinates are x and y. In the (Zeta) plane this same flow pattern would be written as w = g (Zeta) where the complex variable Zeta = "Xi"+i"Eta" in which case the coordinates are stated as Greek letters Xi and Eta. The key here is that both functions must be set equal to each other in order to find points that correspond to one another in each plane.

Formula 2.2.4-13

$$f \cdot (z) := g \cdot (\zeta)$$

Typically, to make the conformal transformation from plane to plane, the flow function in the "z" plane is already known and the corresponding flow function in the "Zeta" plane must be determined. To do this the "relationship" between the two functions must be solved. For example for flow about a "Circular Cylinder" in the "z" plane the tie-in relationship for Zeta is:

Formula 2.2.4-14

$$\zeta := z + \frac{a^2}{z}$$

For the visual see the next diagrams, where for example points "e" and "e1" in the "z" plane lie on the circumference of the cylinder while in the "Zeta" plane both those same points coincide on the horizontal axis.

Diagram 2.2.4-6

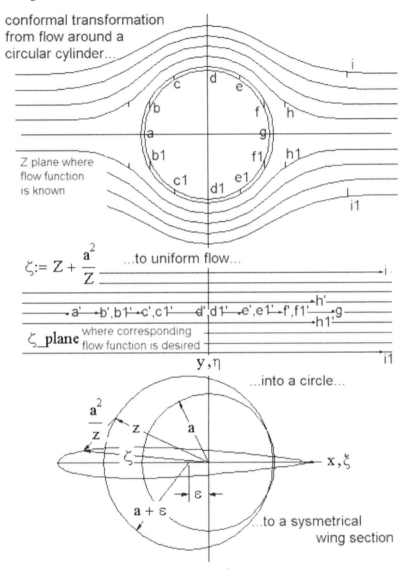

conformal transformation from flow around a circular cylinder...

Z plane where flow function is known

$$\zeta := Z + \frac{a^2}{Z}$$

...to uniform flow...

where corresponding flow function is desired

ζ_plane

...into a circle...

$\frac{a^2}{Z}$

...to a sysmetrical wing section

In diagram 2.2.4-6 with the airfoil shape, the "a radius" is the same "a radius" found in the top diagram where the Z plane flow function is shown. A Vortex such as a tornado is held together by the strength of the centrifugal force generated by the flow "Circulation" about its circumference balanced by the negative pressure at the center. The rotational center of the Circulation lies on the vortex filament, which defines the shape of the tornado. Applying this knowledge of Vortex Strength and its companion

Circulation we now find out if we generate enough Circulation about an airfoil to get the pressure stagnation point located near the rear of the airfoil to line up with the trailing edge. This is called the Kutta condition of the Joukowski analysis. Basically, this Joukowski transformation from cylinder plane to airfoil plane has two problem areas at the leading and trailing edges, which the Kutta condition addresses.

Graph 2.2.4-1

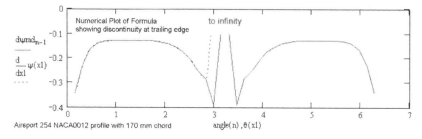

Airsport 254 NACA0012 profile with 170 mm chord

Note in the graph above the calculation shows the infinity spike if the Joukowski analysis doesn't include the Kutta condition adder.

These two undefined problem areas Joukowski called poles which in the cylinder plane occur at $x = 1$ and $x = -1$. Here's the twist. If during the transformation, the cylinder is setup to touch a pole, a sharp edge is formed on the airfoil. If both poles stay inside the cylinder a closed body is formed in the airfoil plane. If this cylinder moves up or down from the x-axis camber is introduced in the airfoil and moving the cylinder left or right changes the thickness distribution.

Finally, we are getting to the mathematical transformation of a "Circular Cylinder with Circulation" into our long sought after wing section. Here are the boundary conditions needed in order to develop the above wing section diagram.

• The center of the circle must be on the "x" axis so the resulting wing section is symmetrical.

• The variable "Zeta" is substituted into the flow expression "Circular Cylinder with Circulation" whereby this new cylinder has a radius slightly larger than that of the original and is positioned such that its circumference coincides with the original at a point where $x = a$.

• If the new radius is larger by a real distance "Epsilon" then the larger cylinder has its center at $x = $ -Epsilon and its radius = a + Epsilon.

• Z* is the complex variable for the flow about a cylinder whose center is shifted from the center of coordinates.

Taking all this into account, where the previous representation of the complex variable "w" defined the flow of a "Circular Cylinder with Circulation" (case f) this complex variable "w" now represents that flow of a "Larger Diameter Cylinder with Circulation" and becomes:

Formula 2.2.4-15

$$w := V \cdot \left[z^* + \varepsilon + \frac{(a+\varepsilon)^2}{z^* + \varepsilon} \right] + \frac{i \cdot \Gamma}{2\pi} \cdot \ln\left(\frac{z^* + \varepsilon}{a + \varepsilon} \right)$$

ε = epsilon

B) Airfoil Theory

Airfoil theory is based upon Circulation. In this section the concept of Circulation is elaborated on to define the Joukowski airfoil and ultimately obtain equations to generate the resultant pressure distribution over any airfoil surface to determine the lift coefficient.

[At this point I'll jump ahead a little and show you three flow conditions impacting a Cylinder with Circulation in a flow stream and its equivalent Airfoil Conformal Transformation. All are shown in an actual airflow.]

Diagram 2.2.4-7 Cylinder Flow Conformal Transformation to Wing Section as a Function of Angle of Attack

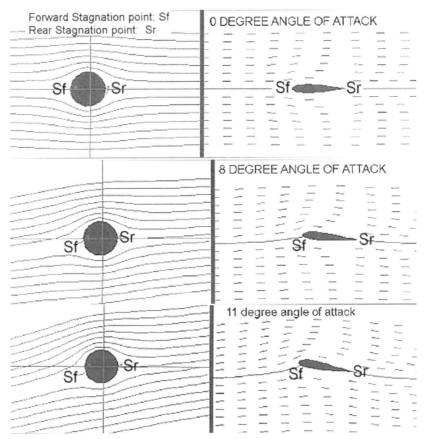

Now, lets bring the math up a notch. Bernoulli's equation gives us total lift on an airfoil and the Kutta and Joukowski equations gives us the discrete static pressures at each point on the surface of any defined airfoil shape as a function of air velocity. This type of specific airfoil static pressure detail is used in the structural analysis of the rotor blade. The equations are developed off of Joukowski's hypothesis, which boils down to stating that in the real world "air flows do not have infinite velocities." That seems to make sense to me, after all who ever heard of anything in life having an infinite quality associated with it. In the previous section "Circulation about a Cylinder" was defined and how one could make conformal transformations from x-y planes to various complex planes. Although not stated then the fundamental theorem of Circulation states that its value obtained for any closed curve is always the same. With that in mind our mathematical journey takes us further into the realm of advanced mathematics and the airfoil.

Based upon Kutta and Joukowski the simplest way to go from the flow function (Circular Cylinder with Circulation) to obtain the velocity at any point on the wing section was to take values directly from the "z" plane then find them on the "Zeta" plane. The equation to do just that is stated below:

Formula 2.2.4-16

$$\frac{d}{d\zeta}w := \frac{d}{dz}w \cdot \left(\frac{d}{d\zeta}z\right) = \left[V \cdot \left[e^{-i \cdot ao} - \frac{(a+\varepsilon)^2 \cdot e^{i \cdot ao}}{(z+\varepsilon)^2}\right] + \frac{i \cdot \Gamma}{2\pi \cdot (z+\varepsilon)}\right]\left(\frac{z^2}{z^2 - a^2}\right)$$

Unfortunately, in the expression $(z^2/(z^2 - a^2))$ when "z" approaches "a" (at the trailing edge of the wing section where z = a) this expression becomes $(a^2/(1/\text{infinity})$ which can also be written as $(a^2$ times infinity) which simplifies to just (infinity). The incremental flow at that point becomes infinite which according to Kutta-Joukowski cannot happen in the real world.

To alleviate this dilemma the Kutta-Joukowski Condition was devised whereby the value given to the Circulation is such that at z = a, the first bracketed expression equals zero (zero times infinity is still zero). The physical ramification of this is: "smooth flow at the trailing edge". Let's look at that first bracketed expression when set to zero and find the equation for the Circulation, which satisfies the condition.

Formula 2.2.4-17

$$0 := V \cdot \left[e^{-i \cdot ao} - \frac{(a+\varepsilon)^2 \cdot e^{i \cdot ao}}{(a+\varepsilon)^2}\right] + \frac{i \cdot \Gamma}{2 \cdot \pi \cdot (a+\varepsilon)}$$

The equation for Circulation is therefore:

$$\Gamma := 4\pi \cdot (a+\varepsilon) V \cdot \sin \cdot ao$$

At the leading edge Zeta = -2a and at the trailing edge Zeta = 2a which means the total chord length is 4a, and the Lift Coefficient "Cl" equals:

Cl = 2 x pie (1 + epsilon / a) sin (alpha $_0$)

Diagram 2.2.4-8

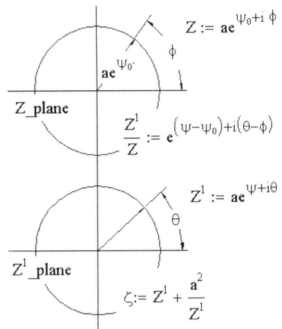

Diagram of the mathematical transformation used to derive airfoil shapes and pressure distributions

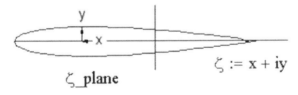

The relationship dw / dZeta stated above in rectangular coordinates can be given in polar coordinates as:

Formula 2.2.4-18

$$\left|\frac{d}{d\zeta}w\right| := v = V \cdot \frac{\left(\sin\left(ao + \phi\right) + \sin\left(ao + \varepsilon_T\right)\right)\cdot\left(1 + \frac{d}{d\theta}\varepsilon\right)\cdot e^{\Psi_0}}{\sqrt{\left(\sinh^2\cdot\Psi + \sin^2\cdot\theta\right)\cdot\left[1 + \left(\frac{d}{d\theta}\Psi\right)^2\right]}}$$

To summarize, the above equation was based upon the Kutta-Joukowski theory for calculating the pressure distribution on the surface of an arbitrary wing section when subjected to the flow of a perfect fluid. This equation produced numbers that were close to experimental data but the discrepancies between the two increased with increasing angle-of-attack. This of course was due to the viscous effects produced by real fluid flow, such as the introduction of drag and the boundary layer distortion, which effectively changes the shape of the section. Since the basic equations were sound some edits were made to the equation to include the effects of real fluid flow. First up was Pinkerton, who realized that the value for the Circulation had to be altered if the flow no longer followed the airfoil surface but instead some new flow path introduced by the boundary layer distortion (remember I've been talking about that laminar layer influence since the beginning of this topic). If the new path was followed the infinity term was reintroduced which had to be eliminated. The next development was by Theodorsen and Garrick. They rewrote the ideal v / V formula such that the velocity distribution over the wing section was based on the shape parameters the "Stream Function (capital Psi)" and the "Velocity Potential (capital Phi)." Their equation is stated below:

Formula 2.2.4-19

$$\frac{v}{V} := \frac{\sin(ao + \phi) + \sin(ao + \varepsilon_T)\cdot e^{\Psi_0}}{\sqrt{\left(\sinh^2\Psi + \sin^2\theta\right)\cdot\left[\left[1 - \left(\frac{d}{d\phi}\varepsilon\right)\right]^2 + \left(\frac{d}{d\Phi}\Psi\right)^2\right]}}$$

In this equation we obtain the relation between the Velocity Distribution over an airfoil section as a function of its shape parameters the Stream Function (capital Psi)" and the "Velocity Potential (capital Phi)." To obtain the Velocity Distribution for a specific airfoil such as an NACA0012, values for the Stream Function and Velocity Potential are chosen based upon those used to define this shape by NACA. If not you must come up with your own shapes by assuming suitable values of d(epsilon)/ d(phi) as a function of phi.

Varying the section angle of attack (ao) in the above equation will give you the Velocity Distribution for that angle.

Although we have not yet discussed how to get from the "Airfoil Velocity Distribution" as a "function of Angle of Attack" to a corresponding Lift Coefficient I do believe it appropriate to show you how the above-modified equation compares with experimental data. Shown next is a generic airfoil where the Lift Coefficient, Cl (calculated from the above modified equation) versus Angle of Attack is plotted along with experimental data. As you can see the curve generated from the modified equation and the experimental data closely match.

Graph 2.2.4-2

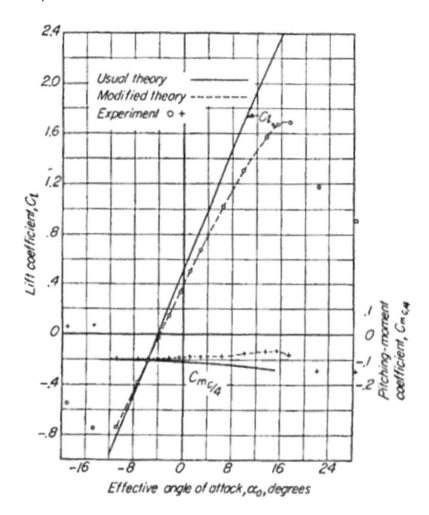

Now that we have an equation for a Velocity Distribution how do we get from the calculated Velocity Distribution to the Lift Coefficient for a NACA0012 profile? Our starting point will be where the airflow separates near the nose section to travel over the upper and lower surfaces of the airfoil. At that flow-separating position, called the "Stagnation Point" the flow velocity is reduced to zero (note for the record the stagnation point does shift with changes in angle of attack). After that point the flow velocity increases then trails off.

Bernoulli related velocity and pressure in his equations on compressible and incompressible flow. Basically, those equations state when velocity increases, local pressure decreases and vice versa. The connection we want is to convert the Velocity Distribution to a Pressure Distribution and to use that to get to a Lift Coefficient. To achieve this Bernoulli's equation is rewritten to use the local point velocity on the airfoil to obtain its pressure equivalent. That equation is:

P(local)=
P(stagnation)+[(density/2)(V(ambient)^2)(1−(v(local)/V(ambient))^2)]
In this calculation of P(local)
P(stagnation) = the ambient air pressure
     Density = ambient air density
     V(ambient) = the free stream velocity
     v(local) = air velocity at specific point on airfoil surface

The key to solving this equation for P(local) first requires solving Formula 2.2.4-19 to obtain the "[v(local)/V(ambient)]^2 term" (for values starting at the leading edge to the airfoils trailing edge). The tabulated results for this term resulting from that calculation is shown below for an NACA0012 airfoil with a zero Angle of Attack. The graph that follows plots that term versus chord length.

Figure 2.2.4-1

x (per cent c)	y (per cent c)	$(v/V)^2$	v/V	$\Delta v_a/V$
0	0	0	0	1.988
0.5	.....	0.640	0.800	1.475
1.25	1.894	1.010	1.005	1.199
2.5	2.615	1.241	1.114	0.934
5.0	3.555	1.378	1.174	0.685
7.5	4.200	1.402	1.184	0.558
10	4.683	1.411	1.188	0.479
15	5.345	1.411	1.188	0.381
20	5.737	1.399	1.183	0.319
25	5.941	1.378	1.174	0.273
30	6.002	1.350	1.162	0.239
40	5.803	1.288	1.135	0.187
50	5.294	1.228	1.108	0.149
60	4.563	1.166	1.080	0.118
70	3.664	1.109	1.053	0.092
80	2.623	1.044	1.022	0.068
90	1.448	0.956	0.978	0.044
95	0.807	0.906	0.952	0.029
100	0.126	0	0	0
L.E. radius: 1.58 per cent c				

NACA 0012 Basic Thickness Form

Graph 2.2.2.5-1 (shown previously in the discussion of the boundary layer)

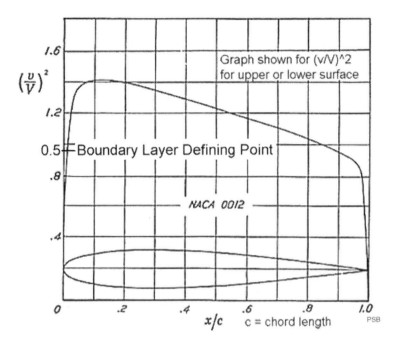

In the above graph positive values of $(v/V)^2$ indicates that v = v(local) velocity is greater than V = V(ambient) velocity. It follows that the higher the local velocity the lower the local pressure.

Lets take another look at the Eppler-64 airfoil, which shows the resultant airfoil pressure distribution versus ambient air pressure :

Diagram 2.2.4-9

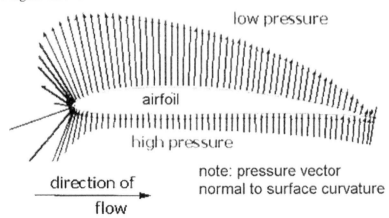

Now for some refresher material. Force is a vector having magnitude and direction. Pressure is a scalar quantity and in this case is simply related to the momentum of the fluid. We also know pressure acts perpendicular to the airfoil surface. To obtain the net mechanical force over the entire airfoil we must add all the incremental force vectors acting on the upper and lower surfaces (Force = pressure x Area). Finally, in the limit of adding infinitely small sections where "delta A" is the incremental area normal to the surface we obtain the resultant Force vector on the airfoil. As you can tell by the Eppler-64 diagram setting up this calculation is not easy since it's a function of the airfoil curvature. On the plus side computers easily crunch numerical calculations of this type especially when the equation of the NACA0012 airfoil is already defined (discussed in Chapter 8.2.3 "Airfoil Selection- the Decision"). Consequently, once this force is obtained, it can be further broken down into two components, the vertical component called LIFT and the horizontal component Drag.

The formula for the Lift Coefficient, "Cl" (based upon two-dimensional flow over a infinite span non-varying cross-section airfoil) is:

$$Cl = Lift / (0.5 \text{ x density x cord x } U^2),$$

where "density" is that of the fluid, "chord" refers to the airfoil chord that the "Lift" was calculated over, and U the resultant section velocity vector of the airfoil. Now that you understand the basic procedure for calculating the "Lift Coefficient, Cl" for a specific Velocity Distribution occurring at a set Angle of Attack lets continue to move forward.

What I have presented here attempts to explain how the theory evolved. I started with flow about a simple rotating cylinder, then generated a formula for that Circulation and followed with some mathematical macerations to convert all that information to a velocity profile about a generic symmetrical airfoil section in an effort to finally predict aerodynamic characteristics of an NACA0012 airfoil.

I can't begin to tell you how much depth there is on this subject on the Internet. Anything I've written that you don't understand can be found there in the form of some University level lecture on the subject. I'd suggest any query start with NASA itself.

There are some major limitations to our two-dimensional airfoil Velocity Distribution formula even with no variation of chord cross-section, or lift across the span. The condition that each airfoil section acts independently of its adjacent section except for induced downwash just doesn't happen in 3-dimensional flows. It gets even more complex when that airfoil rotates about a main rotor centerline. In this case even small spanwise pressure variations tend to generate large crossflows in the boundary layer, which ultimately tends to flow to the region of lowest pressure instead of following in the stream direction. There are also complications when dealing with conditions generating a strong trailing vortex. Both these situations cause the rotor blade characteristics to become unpredictable especially when nearing the maximum Angle of Attack. Fortunately some of these velocity perturbations are minimized when dealing with high aspect (span/chord) wings, of which our rotor blade can be considered. They are there nonetheless.

Here comes the major let down. I will not be using any formulas presented in this section for the Lift calculations because of the major problems stated with the 2-D conversion to a three dimensional rotating plane. What I will be using is the Energy Balance Equations dedicated specifically to "HELICOPTERS." The Airfoil theory equations developed here will be used to calculate the pressure distribution as a function of angle of attack for the airfoil shape selected. This data will form the basis for designing the internal structure of our rotor blade.

This completes the math portion of Airfoil Theory and how it evolved. This is not the end of the subject. In Chapter 8.2.3 "Airfoil Selection – the Decision", we will visit NACA's experimental data on the subject in order to finalize the helicopter rotor blade shape.

Chapter 3. The Helicopter

3.1 General Information:

Helicopters and V/STOL (Vertical or Short Take-Off and Landing) aircraft: Although both fixed and rotary-winged aircraft require high velocity air flowing over airfoils to produce lift, the way that airflow is generated by these two aircraft types is completely different. In fixed-wing craft the wings cannot move faster than the fuselage therefore to produce lift the aircraft must maintain considerable forward speed at all times to obtain the necessary wing Lift producing airflow. In contrast the helicopter rotor blades (wings) rotate at a relatively high rpm about a fixed point on the fuselage to obtain the desired airflow. Obviously there is no relationship between blade speed and a helicopter's fuselage speed. The helicopter, employing one or more horizontal rotors to give both Lift and translation, can rise and descend vertically from the ground, hover over a spot on the ground, and fly backward and sideward as well as forward. With these flight characteristics, the helicopter can land virtually anywhere there is a clearing about the size of a tennis court. V/STOL are aircraft having certain features of both fixed-wing aircraft and helicopters, typically designed around mission specific objectives.

Now it's time to get to the actual helicopter to be built, the AIRSPORT-254.

3.2 Helicopter Subassemblies

The AIRSPORT-254 is designed as an ultralight full function helicopter. It is comprised of the same basic components found in larger commercial and military helicopters being built today. As an ultralight it is subject to a dry weight limitation of only 254 pounds. As a result of this weight restriction some of the basic components such as the cockpit are rudimentary in nature consisting of not much more than a narrow windshield anchored to the aluminum framework with a small single seat. In spite of the weight limitations this particular helicopter is well designed for optimum functionality. The subsystems comprising the typical helicopter are as follows: Main Frame, Main Rotor, Flight Controls, Swashplate, Tail Rotor, and the Powerplant including the Drive Train so it seemed logical to follow that format in the AIRSPORT-254.

3.2.1 Main Frame:

The "Main Frame" is the structure on which all other subsystems mount. It consists of the cockpit, tail section and landing gear all designed to handle a hard landing generating up to 4 g's (4 times its gross weight).
CAD Dwg 3.2.1-1

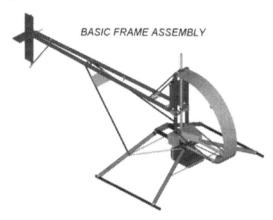

BASIC FRAME ASSEMBLY

3.2.2 Main Rotor:
CAD Dwg 3.2.2-1

This helicopter utilizes a single "Main Semi-Rigid Rotor" system. Detail analysis of the system is discussed in the AIRSPORT-254 systems chapter. Briefly described, this rotor system allows for flapping, feathering and blade lead-lag adjustment. It consists of two rotor blades clamped rigidly into position to the main rotor hub such that their final setup mirrors each other. This completed subassembly is then mounted on centerline to the main rotor drive shaft via a trunion bearing or teetering hinge. In flight this rotor design enables the blades to seesaw or flap together. When resultant air dynamics force one blade to flap up the other goes down. This seesaw motion represents a force imbalance on the system and is dampened, or rather automatically stabilized by feathering which adjusts the pitch angle of the blade. Currently, the feathering rotation is supported by steel thrust and needle bearings to accommodate the design loads but work is underway to incorporate elastomeric bearings. Lead-lag adjustment of the blades is performed as follows. Each blade mounts to the main rotor hub with a single Mil Spec bolt at the end closest to the main rotor drive shaft. At the far end of the main rotor hub a blade positioning clamp is located housing adjustment screws which enable each blade to be angled forward (lead) of theoretical position in the rotor disc plane or aft (Lag). After these adjustments are made the rotor blade clamp bolts are torqued fixing the final blade position. Lead-Lag adjustment enables additional dynamic control over inherent blade distortion during flight operation. The rotor requires an almost perfect set of matching blades to function properly. The weight difference between the two is typically in the range of only a few grams. There will be more on this subject when we get to Chapter 8.

3.2.3 Flight Controls:

CAD Dwg 3.2.3-1

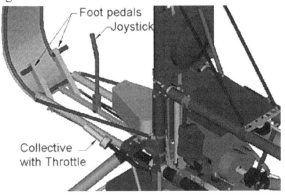

The four basic Flight Controls required to fly a helicopter are the "Collective" pitch control lever, the twist grip "Throttle" (which is mounted on the end of the collective pitch control lever and includes the starter pushbutton), the "Cyclic" pitch control joystick and the floor mounted "Anti-torque Pedals". The Collective changes the pitch of both main rotor blades simultaneously while the Cyclic alters the main rotor blade pitch in selective fashion. The Throttle regulates engine speed and the foot-operated Anti-torque Pedals varies tail rotor blade pitch for fuselage rotational positioning about the helicopter's centerline.

Here's an example of how the controls are interactive. If you want to lift off and just hover over the same spot you would first use the Throttle to increase rotor speed to the design rpm. Next step is to increase the angle of attack on all blades by raising the Collective lever so both rotor blades Lift uniformly. At that point two dynamic events occur which must be addressed simultaneously - immediate loss of engine rpm due to increased torque requirements and fuselage rotation about the helicopter centerline due to the generation of that increased main rotor torque by the newly angled rotor blades. These two conditions are corrected by increasing the Throttle to get back to the design rpm and at the same time using the Foot pedals to maintain the desired fuselage position. Needless to say all this requires a lot of eye-hand-foot coordination. Forward flight gets even more complicated and will be discussed as we proceed.

3.2.4 Swashplate:

CAD Dwg 3.2.4-1

The main function of the "Swashplate" is to transfer fuselage mounted control rod inputs to the rotating main rotor assembly. The lower section of this device is in the same fixed reference plane as the main frame and contains a lower-plate mounted to a spherical bearing which straddles the main rotor shaft centerline. Cyclic and Collective control rod inputs connect to it and manipulate the lower-plate's position. The upper section of the swashplate is tied to the lower-plate via another bearing and rotates with the main rotor shaft. Inputs from the Cyclic and Collective are then translated to the rotating plane where the upper two control rods transfer these inputs to direct the main rotor.

3.2.5 Tail Rotor

The torque to provide Lift produces an opposing torque on the airframe causing it to rotate opposite the blade direction. To prevent this unwanted rotation an opposing torque must be applied. The AIRSPORT 254, as with most single main rotor helicopters, utilizes a variable pitch Tail Rotor to neutralize the undesired counter rotation. The amount of variable pitch required is controlled via the cockpit foot mounted Anti-torque Pedals. The Tail Rotor in this design is driven by a belt and chain drive off the main rotor shaft.

CAD Dwg 3.2.5-1

3.2.6 The Engine:

Before we discuss the engine used for this particular helicopter let's review the types available. Currently helicopters utilize two types of engines - the reciprocating piston engine and the gas turbine. In today's environment the turbine is more or less the standard in the industry for performance orientated helicopters demanding substantial horsepower from a small size. The piston engine while still around is used on smaller helicopters due to lower initial purchase price, reduced maintenance requirements and lower operating costs. Just for the record, as stated in the helicopter history section the first use of a gas turbine was on the Karman K-225 back in the early 1950's.

3.2.6.1a Turbine Engines

Due to the cost involved I didn't consider using one, so I didn't bother to define its function since this topic is so well covered in the FAA manuals I mentioned earlier in this book. The engines I will discuss in detail are the reciprocating type.

3.2.6.1b Reciprocating Piston Engine

The Four Stroke is the most common type of helicopter engine and the only type that the FAA will approve for aircraft use. This engine operates thru four distinct cycles. The following description is for the basic naturally aspirated carburetor engine. Starting with the Intake Stroke, as the piston moves from top dead center (closest position to the cylinder head) downward the intake valve opens allowing a stoichiometric correct combustible air / fuel mixture to be sucked into the void left as the piston moves down the cylinder to the end of it's stroke. When the piston begins its upward motion towards top dead center during the Compression Stroke, the intake valve closes and the trapped air / fuel mixture is compressed. When the piston moves just past top dead center the spark plug is energized causing a high voltage arc to ignite the combustible mass. This begins the Power Stroke. The explosive force of the exothermic reaction drives the piston away from the head as more of the byproducts of the reaction continue to expand. Once the piston reaches the end of its' stroke and begins to initiate upward travel for the Exhaust Stroke, the exhaust valve opens and the burning gas is forced from the cylinder as the piston travels toward top dead center only to repeat the process over again.

3.2.6.1c The Two Stroke engine is described here because at the current time it is one of the most powerful naturally aspirated engines on the market. Its use is legendary in competitive racing from production dirt bikes, jet skis and snowmobiles to other high performance applications. In its basic form there are only three moving parts and that keeps maintenance simple. Another benefit is its' compact size and the fact that it generates twice as many power strokes per process cycle as a Four Stroke engine. The Two Stroke engine combines the Four Stroke Power/ Exhaust Stroke together as well as the Intake/ Compression Stroke to increase the power delivery frequency.

The Two Stroke engine operates as follows: As the piston travels toward "top dead center", the lower section of the piston skirt is high enough on the cylinder wall to expose the cylinder inlet port. When this happens a combustible air/ fuel mixture enters the crankcase. At this point crankcase pressure is lower than atmospheric pressure causing air to be pulled thru the carburetor. Due to the venturi effect of this high velocity air within the carburetor, fuel will flow and mix with the air stream. A transfer passage connects the crankcase space to a cylinder port located just above the piston. When the piston moves to the bottom dead center position this downward motion of the piston into the crankcase space increases crankcase pressure such that when the inlet port is uncovered the high-pressure crankcase air/ fuel mixture forces its way into the combustion chamber.

When the piston starts its upward travel again, the air/fuel mixture is compressed and ignited by the spark plug after reaching top dead center. The combustion of the fuel violently forces the piston away from the cylinder head in what is referred to as the Power stroke. When the piston nears the bottom of this stroke an exhaust port is exposed providing an exit path for the hot gaseous products of combustion to enter the exhaust pipe then leave the engine.

Due to the close proximity of both the inlet and exhaust ports both are going to be open almost at the same time. In order to minimize the possibility of the inlet charge from moving directly across the piston to the exhaust port while trapping some of the exhaust gas near the top of the chamber, a specially shaped piston is utilized. This piston is crowned to deflect the incoming air/fuel charge upward and over the piston toward the exhaust port. This path the inlet charge is forced to travel helps scavenge the cylinder of exhaust gas as the air/fuel mix fills the combustion chamber. Ideally all the exhaust gas will exit the cylinder via a specially designed expansion type exhaust pipe.

The advantage of a two-stroke engine is since no oil is maintained in the crankcase the engine can be mounted in any configuration, even upside down if so desired. Its downside is it is not more efficient than a 4-stroke engine and its spark plugs are prone to fouling (excessive oil on the spark plug tip causing them not to work).

3.2.7 Fuel System:

A helicopter Fuel System is divided into two subsystems, the Fuel Supply and Fuel Control.

3.2.7.1 Fuel Supply: This subsystem contains all the components necessary to store and deliver fuel to the Fuel Control subsystem. It includes the storage tank or tanks, filter, fuel lines, primers and fuel pump. It also includes the power source for the pump whether it is electrical, mechanical or vacuum drive, including shut-off valves and gauges. Starting with the fuel tank this fuel storage device has the typical features found on most commercial trucks such as a vent to eliminate negative pressure being created as fuel is drawn off, an overflow drain or expansion tank to prevent the main tank from rupturing when the fuel expands, a drain valve to allow manual removal of any possible build up of water or sediment in the tank and a fuel gauge sending unit.

The sending unit and cockpit gauge may be calibrated to indicate fuel quantity remaining in pounds and / or gallons or both. Ideally the fuel tank is located as low as possible on the airframe at the helicopter's "Center of Gravity." This ideal position increases flight stability and as fuel is consumed (about 7.5 pounds per gallon) there is minimal impact on the craft's "Center of Gravity."

Almost all helicopters use a pumping system to deliver positive pressure fuel to the Fuel Control system independent of consumption rate with the excess fuel being returned back to the tank. The primary fuel pump is mechanically driven off the engine and operates whenever the engine is running. A backup electrical pump controlled by a switch in the cockpit can be used to either maintain fuel pressure to the mechanical pump or take its' place if the mechanical pump fails. The Airsport 254 fuel pump differs from that described above in that the primary pump is driven from pulsating crankcase pressure developed during the two-stroke engine cycle and there is no backup electrical pump.

Fuel pumped from the tank flows first thru the shutoff valve, either manually operated by the pilot or via remote electric solenoid. The main purpose of this device is to shut off the fuel supply in the event of a fire or potential emergency. Under normal operating conditions this valve is open. From the shut off valve the fuel flows through the filter.

The fuel filter prevents any residual fuel particulate from entering the Fuel Control System. The housing containing the fuel filter is designed to trap water at the bottom of the bowl where it can be manually drained off or emptied. This is possible because water being heavier (higher density) than the fuel will settle to the lowest point of the bowl and out of the way of the fuel flow path provided the amount of water is not excessive. Note that because the tank is vented any negative pressure there brings in atmospheric air which depending on the humidity introduces a certain amount of water to the tank. The problem is exacerbated when the tank is not full during storage and moisture continues to condensate on the tank wall, water being heavier than gas settles to the bottom of the tank. This condensate must be drained off because as soon as the engine starts this water will be atomized back into the fuel supply. This is not a good thing.

A manually operated primer pump is used to get a fuel supply to the Fuel Control Subsystem before startup. This is effective in minimizing engine starter time required to bring fuel to the Fuel Control System. This added manual pump is especially useful if the engine hasn't been started recently on helicopters having a non-gravity fuel tank position (meaning the tank is mounted lower than the fuel pumps and engine).

3.2.7.2 Fuel Control System:

The combustion process whether in a reciprocating or turbine engine requires fuel and ambient air to be supplied in the correct proportion. The Fuel Control System accomplishes this.

• <u>Turbines</u>: Due to the complexity of turbine operation a computer controlled Fuel Injection system is used called the FADEC (full authority digital engine control). This system manages all fuel adjustment functions as input ambient air characteristics change such as, humidity, density, temperature, etc.

• Reciprocating Engines: The fuel control system that delivers the air / fuel mixture to the combustion chambers is either by Carburetion or Fuel Injection.

Carburetors: During engine operation of either a four-stroke or two-stroke naturally aspirated engine a lower than atmospheric pressure is created in the intake manifold. This varying differential pressure existing between the ambient air and the manifold is the driving force for airflow into the engine. A carburetor is designed to utilize this airflow to draw fuel and during the process uniformly disperse it. Helicopter carburetors are calibrated for the correct air to fuel mixture at sea level. This ratio is established with the carburetor mixture setting set at FULL RICH. Newer more sophisticated carburetors automatically reduce the fuel flow as altitude increases since the less dense air decreases the available oxygen. In non-atmospheric compensating carburetors the fixed sea level setting sends too much fuel to the combustion chambers for the amount of available oxygen with increased altitude. This results in incomplete combustion, cooler cylinder temperatures and excess carbon build up. During this operating condition the spark plugs tend to foul reducing power and causing the engine to run rough. There are procedures for manually "leaning the mixture" as altitude increases but since most manufacturers do not recommend it I will not elaborate further other than to say once a "leaning of the mix" is performed at altitude it must be reset back at sea level or the engine cylinder temperature will increase to the point of engine failure.

Carburetor Ice: The low pressure high velocity air flow not only vaporizes the fuel entering the airstream but also lowers the localized temperature during the process so much that ice can form on the internal carburetor surfaces. Carburetor icing can occur at ambient temperatures as high as 100 degrees F and humidity as low as 50%. Typically the condition occurs when the ambient temperature is below 70 degrees F and becomes more probable as the temperature nears 32 degrees F. The same applies if the relative humidity exceeds 80 %. As temperatures drop below freezing the probability for icing decreases.

Fuel Injection: This system is used on all diesel engines due to the high compression ratios on those engines. On gas engines it replaces the carburetor by utilizing a fuel control unit, which pressurized the fuel so it can be direct injected into either the intake port or the cylinder itself. Its main advantage is increased engine performance due to increased dispersion of fuel in the combustion chamber.

3.2.8 Electrical System:

The Electrical System consists of the Power Supply, the Distribution System and the supplied Components. The Electrical System at its most basic level provides the power to start and maintain engine operation whether it is reciprocating or turbine. It supplies current of either 14 or 28 volts to operate console instruments, lights, and communication equipment and in the case of the more sophisticated helicopters highly involved avionics. Although more and more electrical supplied hardware is incorporated in helicopters today with the exception of engine function all helicopters can be flown with these peripheral devices off such as during the event of an electrical malfunction or other operational emergency.

3.2.8.1 Power Supply: Electrical power on helicopters using a reciprocating engine is supplied from either potential energy stored from an on board battery and / or kinetic energy produced by an engine driven alternator or magneto. Turbine helicopters also use a battery for startup but once running the direct gearbox coupled starter continues to spin and functions as a generator to supply electrical current. The battery although primarily used to start the engine can be used to operate some electrical equipment without the engine running such as communication equipment and radios. It is also capable of supplying reserve power should the alternator or generator malfunction. The electrical current produced from either the alternator or generator is controlled via a voltage regulator. This device monitors the helicopters electrical load demand and allows either source to produce current up to its rated output. The Airsport 254 has an onboard 12 volt battery and generates its' current from an engine flywheel mounted magneto. Voltage regulation is via a dedicated piece of hardware as is the engine spark plug coil.

An ammeter utilizing the battery as a voltage reference point indicates the electrical system status. Just after startup an ammeter shows positive current flow as the battery recharges to its normal voltage potential. Once attained the ammeter will stabilize near zero. This means current flow to the battery has virtually ceased and all electrical system demands are being satisfied by the electrical output of the charging source whether alternator or generator. If the ammeter reads negative then the battery is supplying current. This could be due to a malfunction of the alternator/ generator or an excessive electrical load. If a load meter is used this device indicates the actual electrical load being satisfied by the alternator/ generator. The helicopter manual will state the specific RMF (load) to expect during normal operation. The load meter reads zero if there is a loss of electrical current flow from the alternator/ generator.

Distribution: The current generated by the alternator/ generator flows to a bus bar for distribution to the various electrical devices. Current coming off the bus is controlled in two ways. If the current draw is low it may be simply turned "on" or "off" directly from a cockpit instrument panel switch. If the current draw to the device is large then the device is wired such that the heavier gage cables tapped off the bus bar connect to a relay input and from the relay output to the electrical device. To initiate the relay and allow current to flow a second set of relay inputs are connected to the panel mounted switch via small gage signal wires to turn "on" or "off" the device. Use of a relay simplifies the wiring process and saves weight. This relay setup is used for the engine starter on the Airsport 254. In this example heavy gage wires connect from the battery to one relay terminal, and off the other relay terminal a heavy gage wire connects directly to the starter via the shortest path length possible. At the signal wire relay terminals two small 14 gage wires route thru the frame to the inside of the collective lever to the engine starter button. Depressing the starter button energizes the relay magnetic coil that completes the circuit to the starter.

To protect helicopter-wiring integrity from melting down or catching fire, circuit breakers and or fuses are added to each branch circuit connecting to electrical devices. Circuit breakers and fuses are sized to limit the current flow thru the particular wire to its rated value minus some safety value. If an electrical device fails and generates a current overload a fuse type protection would burn out, or a circuit breaker would trip. In the case of a burnt out fuse, it must be replaced to get current to flow thru the circuit again whereas with a circuit breaker it can simply be reset. In either case the cause of the overload condition must be addressed or fuses will continue to blow and circuit breakers trip. Some instrument panels also include caution lights for electrical device failure.

In the Airsport 254 all instruments are fuse protected along with the cooling fan. A relay is used for the starter circuit with a momentary contact pushbutton switch used to energize it. Switches are also used to turn the cooling fan on and select which cylinder ETG (exhaust temperature gauge) is to be monitored. A stand-alone laser is used as a Joystick positioning aid.

3.2.9 Power Transmission:

All engines have some operational point where output power or fuel efficiency is maximized at a particular RPM. Optimizing Main Rotor and Tail Rotor performance dictates that input torque at a specific RPM be maintained. It is the function of the Power Transmission System to match the engine RPM to the design RPM of the Main and Tail Rotor systems including any peripheral accessories. Typically included in the Power Transmission section is the Main Rotor Transmission with associated Drive Train, the Tail Rotor Drive, the Clutch, the Freewheeling unit and any power take-off accessory drives. Normally, a self-contained gear reduction transmission is cooled and lubricated with its own internal oil supply. It has a sight gauge to monitor fill level and a magnetic portion on the drain plug to remove metal debris. More sophisticated units also incorporate chip detector sending units that signal possible damage to the pilot via cockpit warning lights.

All prototype testing has been performed with a two-cylinder, two-stroke, water-cooled, 650 cc, electric start, Kawasaki engine. This engine is mounted horizontally on the Main Frame and is belt driven to a gearbox whose output shaft connects to a one-way clutch. The output from the clutch transmits its power via a chain drive to the main rotor drive shaft and the tail rotor drive. A thermostatically controlled electric fan mounted on the radiator achieves engine cooling.

Chapter 4 Basic Helicopter Aerodynamics

Aerodynamics is the study of the dynamic forces acting on solids resulting from the flow of gaseous fluids in relative motion to one another. In subsection 4.1 The Airfoil, we discuss terminology, in subsection 4.2 The Forces, the dynamic forces germane to helicopter flight are reviewed, in subsection 4.3 Airfoil Specifics, the rotor blade airfoil characteristics are delved into and in section 4.4 The Main Rotor, main rotor operation is examined.

Rotary-Wing Aerodynamics: The aerodynamics of rotary-wing and fixed-wing aircraft is basically the same. Both types of "heavier than air" aircraft employ airfoils to produce lift and both are subjected to identical fundamental forces of lift, drag, thrust, and gravity. The major difference between the two is a helicopter's ability to hover about a fixed position above the ground whereas an aircraft needs a predetermined amount of forward speed to fly. The obvious aerodynamic problem of the helicopter, over its' fixed-wing counterpart, is that when hovering (and in certain flight circumstances) the helicopter rotor's rotation adds unstable dynamic complexities to the resultant air flow impacting steady-state conditions. As a result flying one, especially at low ground speed, requires substantially more pilot control input.

Weight and Lift are closely associated as Weight is the force that pulls any aircraft towards earth and Lift acts to oppose it. Thrust and Drag are related in the sense that Drag is a function of the amount of Thrust produced.

4.1 The Airfoil

Diagram 4.1-1 Aerodynamic Terms of an Airfoil

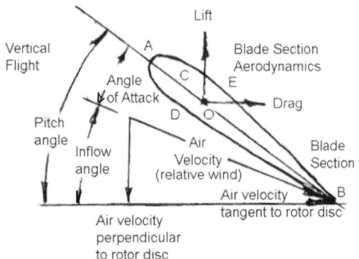

O = aerodynamic center; A = leading edge; B = trailing edge; C = cord line; D = lower camber; E = upper chamber

Helicopter airfoils are typically symmetrical if not completely symmetrical because they provide a wider range Lift generation as the angle of attack is varied. Understanding of helicopter theory of flight must be predicated on a basic knowledge of flight as it pertains to any conventional, heavier than-air aircraft. While the helicopter is capable of many maneuvers impossible to perform with conventional aircraft, and while a number of involved processes take place simultaneously during its flight, it flies for the same basic reason that any heavier-than-air aircraft flies. Moving air passing over its airfoils produces the lift required to keep it aloft. Before discussing that phase of aerodynamics that pertains to helicopters, let's review our basic theory of flight. As stated previously the four major forces acting on any aircraft are lift, drag, thrust, and gravity.

Airfoils used on fixed-wing aircraft are unsymmetrical in shape and are optimized for a specific cruising speed.

The fixed-wing aircraft must accelerate down a runway until a minimum takeoff speed is reached. Typically, for a cargo plane a "Weight and Balance" sheet is prepared to calculate the loaded planes weight. Once the planes weight is established the flight engineer looks in his flight manual to find the minimum takeoff speed required. Once takeoff speed is reached, meaning the airflow over the wings is sufficient to provide the Lift required for flight at that weight, the pilot pulls back on the yoke and is airborne. In contrast a helicopter initiates flight once the ideal rotational speed of the rotor is established. The collective is then raised into position for that weight whereby all the rotor blades angle of attack are increased resulting in a greater Lift producing airflow enabling the helicopter to hover just above the ground.

Definition of airfoil terms:

Absolute angle of attack: an angle of attack measured between a reference line in an airfoil and the position the reference line would have if the airfoil were producing zero lift, i.e., the sum of the geometric angle of attack and the zero lift angle of attack. Also called an "aerodynamic angle of attack.

Angle of attack: this is the acute angle generated between the relative wind and the airfoil chord line.

Aerodynamic center: a point in a cross section of an airfoil or other aerodynamic body or combination of bodies, about which the pitching moment remains practically constant with nearly all changes in angle of attack.

Blade section: a cross section of a blade; the profile of the cross section or the area defined by the profile.

Center of Pressure: the point where the resultant of all the aerodynamic forces acting on an airfoil intersects the chord.

Chord line: is a reference straight line starting at the leading edge and ending at the trailing edge

Camber: refers to the curvature of either the upper or lower surface of an airfoil

Drag: the force which tends to resist an airfoil's passage through the air. Drag is perpendicular to lift and varies as the square of the air velocity.

Leading edge: the forward part of the airfoil to first contact the air mass.
Lift: that component of the total aerodynamic force acting on a body perpendicular to the direction of the undisturbed airflow relative to the body
Mach number: refers to the air speed of the object compared to speed of sound in air.
Mean line: is the locus of point's midway between the upper and lower surfaces. Its shape depends on the upper and lower contours of the airfoil section.
Pitch Angle: this is the included angle generated between the rotor blade and the air velocity tangent to the rotor disc.
Relative wind: the velocity of the air with reference to a body in it.
Reynolds number: relates the chord length, the mass density and the free stream velocity compared to the viscosity of the air.
Span: the length of the airfoil measured from its mounting at the rotor hub to the opposite end.
Trailing edge: the aft part of the airfoil where the upper and lower surface air flows merge
Zero angle of attack: the position of an airfoil, fuselage, or other body when no angle of attack exists between two specified or understood reference lines.
Zero-lift chord: a chord taken through the trailing edge of an airfoil in the direction of the relative wind when the airfoil is at a zero-lift angle of attack.

Both the Angle of Attack and the Blade Pitch Angle have the airfoil Chord line as the upper defining line of the angle. The opposite defining line of the angle, for the Angle of Attack is the direction of the Relative Wind; and for the Blade Pitch Angle it is the horizontal reference plane. They are all related in that using either the collective or cyclic (joystick) to increase or decrease the Pitch Angle there is a corresponding increase or decrease in the Angle of Attack.

4.2 The Forces

4.2.1 Lift

If you read section 2.2 "Dynamics of Air & the Airfoil" you must realize the calculation of Lift on something as simple as a two dimensional airfoil shape set in a ideal uniform air flow is quite complex. Add to that the complexities of the rotational dynamics of a helicopter rotor and you really have a dilemma.

Fortunately, for our calculations for Lift we will use established experimental data for the airfoil we select and Energy Balance Equations developed specifically for helicopter function. The Lift "force" is perpendicular to the Relative Wind. Gravity is the force caused by the pull of the earth. Lift overcomes or balances gravity, depending on the condition of flight. It is the reaction obtained from the flow of air over an airfoil and acts from the Aerodynamic Center perpendicular to the Relative Wind. The

Aerodynamic Center is that point on an airfoil section through which all Aerodynamic Forces may be considered as acting and about which the Aerodynamic Moments are substantially constant.

Diagram 4.2.1-1 Forward Flight (for steady flight A = B)

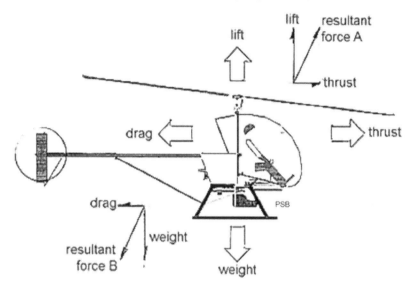

The Relative Wind is the oncoming air flowing parallel and opposite to the flight path of the aircraft. An airfoil is a surfaced body, which responds to relative motion between itself and the air with a useful, dynamic reaction, that generates a resultant aerodynamic force when subject to a moving air stream namely Lift. Your flat hand becomes an airfoil when thrust out the car window traveling at speed. Various hand angles in this moving air stream would cause your hand to rise and fall and force your hand rearward as a function of angle. There are many types of surfaces that can be considered an airfoil. In general airfoils used in helicopters all work the same way under typical flight conditions. In the purest sense a rotor blade is actually an airfoil. Any part of an aircraft designed to produce Lift may be called an airfoil. Generally speaking, all fixed-wing or conventional aircraft utilize airfoils, which conform to the same general shape. They are streamlined in configuration and generally have a greater curvature or "camber" on the top surface than on the bottom. Lift is a variable force. Many things govern the amount of Lift received under stated conditions. For instance, as speed increases Lift increases--but not in direct proportion. Lift increases as the square of the speed. An aircraft traveling at 100 knots has four times the Lift it would have at only 50 knots.

The Angle of Attack, that angle between the Chord line of the aircraft's airfoil and the Relative Wind or flight path of the aircraft, governs Lift. The greater the Angle of Attack, the greater the Lift. This condition prevails until the angle becomes so great that the flow of air over the top of the wing can no longer follow the camber smoothly (laminar flow) and becomes turbulent. The wing is then fully or partially stalled depending on the severity of the angle and the aerodynamic characteristics of the airfoil. The formula for Lift (L) is

   L = 0.5 (density)(velocity^2)(wing area)(lift coefficient)

The density of the air influences the amount of Lift generated. Lift varies directly with density. At 18,000 feet where the density of the air is about half that at sea level, an aircraft will have to weigh only half as much, or else get the "velocity^2" term to be twice what it was. It turns out (1.414 x velocity) ^2 is equal to 2 times (velocity)^2". This means the aircraft would have to travel 1.414 times as fast as it would at sea level to maintain altitude.

   Still another factor influencing Lift is the shape of the airfoil. Generally speaking, the greater the camber or curvature of the wing, the greater the Lift. Lift also varies with the area of the wing. A wing with an area of 200 square feet will lift twice as much at the same angle of attack and speed, as a wing with an area of only 100 square feet, providing proportion of wing and airfoil section are the same.

4.2.2 Weight:

   Your weight (wt) is defined as mass ("m") times the "acceleration due to gravity ("g") or wt = m x g. On earth "g" is approximately equal to 32.2 feet per sec per sec, which means that an object in free fall without any interference will travel 32.2 feet in the first second after being dropped and 64.4 feet after the 2nd second, etc. Astronauts during rocket launch can experience accelerations of up to 5 g's. This means that an 180-pound guy during lift-off has to deal with an instantaneous weight gain responding to 900 pounds (5 x 180). No wonder their faces are so distorted when you see actual footage of these men during rocket launch. By the way if you want to know your mass just divide your weight by 32.2.

   When designing a helicopter there must be sufficient Lift generated to handle the helicopter Weight and Dynamic Loads. For the first case, helicopter Weight is simply everything you are trying to get off the ground. For the AIRSPORT-254 this weight includes the helicopter, 5 gallons of fuel, safety items, checklists / clipboard and the pilot. In this case the amount of Lift required to maintain a helicopter in a Hovering position (maintained vertical distance above takeoff point) or Steady-State flight equals the helicopter Weight. Steady-state flight means a constant straight-line helicopter velocity at a fixed altitude. This is a Special case where at this specific helicopter velocity, typically cruising speed, the efficiency of rotor operation is such that the power required equals that of Hover. This topic will be discussed in detail in the section dealing with the calculation of rotor power.

   As for the second case, Dynamic Loads, in addition to the helicopter Weight certain flight conditions require more Lift generated to accomplish the task. For example for a car to accelerate faster more power is required. The same is true when a pilot tries to reach a higher altitude quicker, or attempts to make a 60 degree bank while sustaining a constant altitude or any of many demanding maneuvers, more Lift is required. This additional Lift, over and above that required to support the basic helicopter Weight defined above, is given the term Load Factor (total rotor load divided by helicopter Weight). In case one, Hovering and Steady-State flight, the Load Factor is one, for the above referenced 60 degree banking maneuver the Load Factor is two which means twice as much Lift is required to be generated by the main rotor system.

As for the Lift to be generated, ambient air conditions such as temperature and altitude are major factors in determining the final Lift available for a given day. If sufficient Lift is available then as the rotor blade angle is increased or varied by the flight controls the helicopter will Hover or be capable of flight maneuvers.

4.2.3 Thrust:

Just as Weight and Lift are paired forces so are Thrust and Drag. Thrust is the name given to the force that enables the helicopter to move in the desired direction (Drag opposes that motion). Thrust and Lift are developed by the rotor blade pitch orientation during each rotation of the main rotor system. If ample Lift is generated then manipulation of the collective and cyclic controls creates Thrust to enable a helicopter to go vertical, sideways, forward, and aft. The ultimate direction and speed of the helicopter is the resultant of the Lift, Thrust, Weight and Drag force components. If the Lift and Weight force components are equal at some altitude and the Thrust is zero then the helicopter will Hover over a stationary point. If a Thrust in the horizontal direction is then added, the helicopter will accelerate in that horizontal direction until the Drag force negates the forward Thrust and the helicopter then assumes constant velocity steady-state flight. If more Lift is added then the helicopter will add an increasing altitude to the flight path until the component forces once again reach equilibrium.

The tail rotor also generates Thrust. It is basically one dimensional in nature and is used primarily to counter-act the main rotor torque and keep the helicopter from spinning about its centerline (Yaw). By varying the thrust of the tail rotor thru the use of the control foot pedals the helicopter can rotate clockwise or counter-clockwise about its centerline, for example from a Hovering position a pilot can orientate the front of the helicopter 360 degrees in either rotational direction via the control foot pedals alone.

4.2.4 Drag:

Simply stated Drag is the force required to move air out of the way of an object moving thru it and since Power equals force (to move the air out of the way) times velocity (how fast the air moves out of the way) then in dynamic terms, Drag consumes power. In terms of helicopter operation it acts not only on the airframe as it moves through the air (Parasitic drag), but also on the rotor blades rotating into it (Profile and Induced Drag). Total Drag includes all three types of Drag that are explained more fully in the

following subsections.
Diagram 4.2.4-1 Drag Pictorial

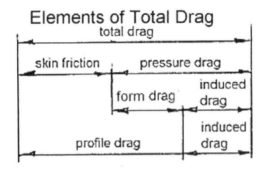

Drag is generated as an object moves through the air. Since both the airframe and the rotor blades move as objects in the air the net effect of Drag is to oppose Lift and Thrust. The diagrams that follow give a quick preview of the topic.

Referencing the next diagram actual total drag is a sum of values typically between pressure drag (case a) and skin friction (case b).
Diagram 4.2.4-2

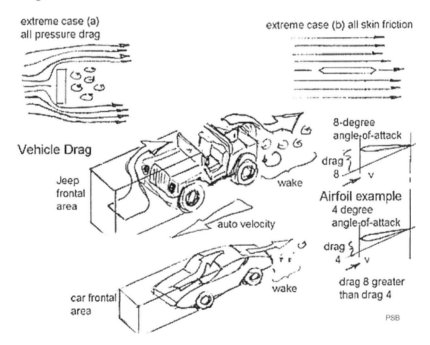

Note: a vehicles drag is calculated by multiplying its frontal area times its specific drag coefficient. In this case of the Jeep, the values for these two parameters are significantly larger than those for the sports car

4.2.4.1 Profile Drag:

When the rotor blade penetrates an air stream it encounters frictional resistance to the motion as a function of rotational velocity. This type of resistance to the motion is called Profile drag. It is relatively unaffected by changes in the Angle of Attack but as the helicopter increases airspeed it also increases. The two components of Profile drag are Skin Friction and Form drag.

a1. Skin Friction: Friction is generated by the airfoil surface roughness where at the microscopic level air molecules create small eddies near these surface irregularities. This produces friction with the streamline flow and induces drag to the airfoil. Stated another way since air is a viscous fluid, has a real value for it's viscosity, and is the reason why laminar flow exists (Prandtl's 1904 observation) it stands to reason that this slower moving boundary layer of air at the surface would impede the airfoil's forward motion. Note that this drag acts parallel to the relative wind.

a2. Form drag:

Diagram 4.2.4.1-1 Form Drag Comparison: all three shapes have the same frontal area but streamlining decreases the magnitude of the form drag by reducing the airflow separation and wake vortices.

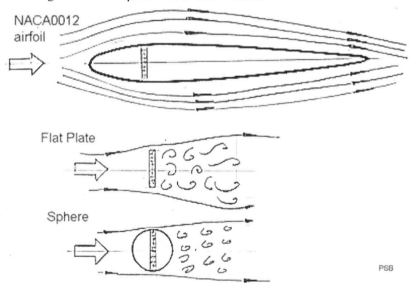

The very shape of an airfoil produces drag. This turbulence-induced drag is created after an object's airflow around its surface separates. The severity of the turbulence is a function of the objects form, that being shape and size, as it interacts with the relative wind.

4.2.4.2 Induced Drag: The resulting airflow rushing into that void generated by the passing airfoil must satisfy any energy imbalance with the surrounding air and the consequence of this is the sight change to the

Relative Wind direction.

When the airfoil Angle of Attack is increased to create Lift a high-pressure area is created (leading edge stagnation point) beneath the leading edge and a low-pressure area created (trailing edge stagnation point) above the trailing edge and at the blade tips. This condition creates a potential energy differential that drives the resulting spiral or vortex type turbulence that fills the void left by the advancing blade, locally deflecting the air stream downward. This condition increases downwash but more significantly it alters the direction of the Relative Wind the airfoil acts on to produce Lift.

This altered Relative Wind is given the term Average Relative Wind and the Lift produced perpendicular to it is called the Total Lift. This Average Relative Wind with respect to the Relative Wind is still near the blade but is angled downward and aft of it. Correspondingly the Total Lift is angled the same amount and it's numerical value is equal to the sum of the vertical Lift force vector and the Induced Drag force horizontal vector.

The dynamics of Induced Drag are such that as the Angle of Attack increases, stagnation point air pressure differentials increase which creates stronger vortices and more drag is generated. Since greater Angles of Attack are required at lower airspeeds and smaller Angles of Attack necessary at higher airspeeds, it becomes readily apparent at low airspeeds Induced Drag is the primary Drag force. The result of this is more power is required at lower airspeeds than at higher airspeeds.

4.2.4.3. Parasite Drag:

Any time an object must displace a mass of air in order to move through it force is required. If the object happens to be a helicopter, the name given to this force is Parasite Drag. The less streamlined the object the more force is required to move the air around it and subsequently the Parasite Drag increases. If the object moves faster, the air displacement offers even more resistance to the motion and under this condition the force required to move the air mass increases with the square of the airspeed. Changing the helicopter velocity by a factor of two generates four times the Parasite Drag. The resistance of parts of the aircraft that do not contribute to lift is called parasite drag, and typical examples are landing gear, fuselage, etc.

4.2.4.4 Total Drag: As the name implies it is the total of all drag forces on the helicopter. Since each drag force varies as a function of airspeed the Total Drag curve when plotted as a function of Drag versus Speed has an optimal point where the Total Drag force on the helicopter is minimized. This point referred to as the L/Dmax (lift-to-drag ratio) indicates that the helicopter has the best Total Lift capability at this speed in comparison to the Total Drag on the helicopter during flight. The opposing resultant force developed parallel to the direction of motion is referred to as the Total Drag.

Graph 4.2.4.4-1 Total Drag versus Airspeed: Shown is a generic plot of the total drag curve as a summation of parasite, profile and induced drag. An actual plot of total drag (in the form of power loss) for the Airsport 254 is found in Chapter 8.

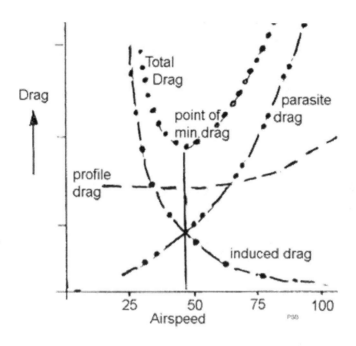

4.3 Airfoil Specifics

NACA back in the forties developed a substantial amount of data for a wide variety of airfoils. They evaluated these airfoils to determine their airflow performance as a function of angle. To get a better perspective lets start with some history. As can be expected wing theory had a beginning, which in this case started with basic observation of early aircraft performance during World War I. During that time extensive testing of wing plan forms (airfoils) was conducted at Gottingen and the development continued after the war in many countries. It was the USA's program operated by NACA (National Advisory Committee for Aeronautics) back in the 1940's that made the most significant contribution to the general understanding of wing theory. It was NACA's systematic experimental approach that enabled families of wing sections to be segregated based upon the effects of camber (the top or bottom curve of an airfoil) and thickness distribution as a function of blade width and to compile experimental subsonic aerodynamic characteristics for these shapes. Adding to NACA's success was the fact that their comparison experimentation was conducted at higher air speeds with the common reference being the Mach number and the Reynolds number. This made the data relevant to all future high-speed craft and very germane to the high rotor blade tip velocities experienced with helicopters. In Section 8.2.3 "Airfoil Selection – The Decision" more is said about the wing section families and those relevant to our project.

Fortunately for us, after tediously laboring over my dissertation on air flow dynamics and the concept of circulation we get a break from the mathematics for a while and simply use the experimental data provided by NACA to learn more about airfoils. In subsection 4.3.1 the differences between a "Symmetrical versus Unsymmetrical" airfoil are evaluated from a Lift versus Drag perspective, and in subsection 4.3.2 we look at airfoil "Angle of Attack versus Stall".

4.3.1 Symmetrical versus Unsymmetrical Airfoils

Basically there are two NACA types of wing sections used on airplanes, those airfoils that are symmetrical about their centerline (same curvature or camber on both upper and lower surface or stated as having a straight "mean line" which is identical to the chord line) and those being unsymmetrical about their centerline (curved mean-line).

Diagram 4.3.1-1 Symmetrical airfoil (ref. NACA 0012): upper and lower camber (surfaces) are symmetrical about chord line.

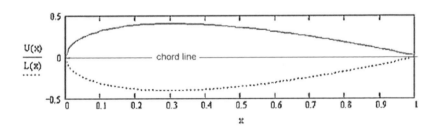

Diagram 4.3.1-2 Asymmetrical airfoil (ref. NACA65,3-818): note that the chord line is straight but the mean line curves, the extent of which mirrors the average thickness.

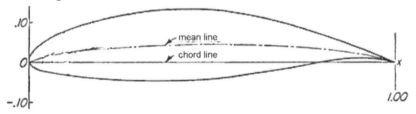

The choice of which to use is determined by the specific operational requirement. In general, airplanes typically use a wing with an unsymmetrical profile because this airfoil type allows for increased lift, greater forward speeds and is more "efficient".

Like the saying goes "a picture is worth a thousand words" therefore in order to show how an asymmetrical airfoil is more "efficient" than a symmetrical type, I decided to give supportive diagrams in the form of NACA graphs for both the symmetrical (NACA 0012) and the asymmetrical (NACA 65,3-618) airfoils. The two supportive graphs we will be using are the ones plotting the "Section Lift coefficient, cl versus Section Angle of Attack a_o," and the plot of the "Section Drag Coefficient, cd versus the Section Lift Coefficient, cl" for these airfoils.

Although I used these two specific airfoils out of the many possibilities I believe they are representative of their respective profiles. The experimental data plotted covers a wide range of Reynolds numbers and ultimately what we want to determine is which airfoil generates the most lift with the least drag over the airfoils' range of angle of attack. Summarized below are the results of this comparison, with the highlighted graphs presented next. Please note that airfoil efficiency is only part of the selection process when specifying the airfoil type to be used for our rotary-winged craft. Since this subject is much more complex to discuss at the present time it will be covered in detail in the Engineering Chapter under Subsection 8.2.2 titled Airfoil Analysis– Symmetrical or Asymmetrical.

In Graph 4.3.1-1, titled "NACA 0012 Lift Coefficient versus Section Angle of Attack" we see that for an arbitrary 8-degree angle of attack the corresponding lift coefficient is 0.8. Then using this 0.8 lift coefficient as a starting point in Graph 4.3.1-2, "NACA 0012 Section Drag Coefficient versus Section Lift Coefficient" we find the corresponding drag coefficient to be 0.008. Comparing these numbers with the asymmetrical airfoil; from Graph 4.3.1-3, NACA 65,3-618 for an 8 degree angle of attack the corresponding lift coefficient is 0.9 and in Graph 4.3.1-4, NACA 65,3-618 using the 0.9 lift coefficient as a starting point, the drag coefficient is 0.005. Conclusion: for a given angle of attack (8 degrees) not only does the asymmetrical wing section (NACA65, 3-618) presented here have a greater lift coefficient of 0.9 versus 0.8 for the symmetrical profile (NACA0012) but it also generates a smaller drag coefficient 0.005 versus 0.008. We have conclusively shown that the asymmetrical airfoil is more efficient not only producing more lift but also less drag than the referenced symmetrical blade.

Graph 4.3.1-1 NACA 0012 Lift Coefficient versus Section Angle of Attack

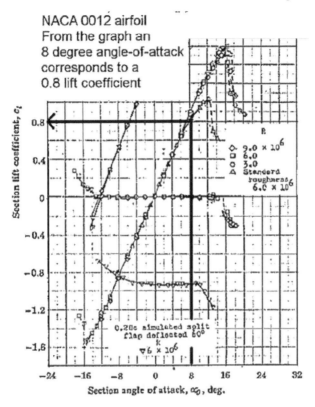

Graph 4.3.1-2 Section Drag Coefficient versus Section Lift Coefficient

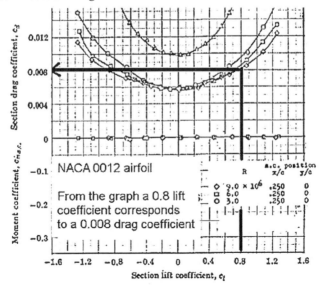

Graph 4.3.1-3 NACA 65,3-618 Lift Coeff. versus Section Angle of Attack

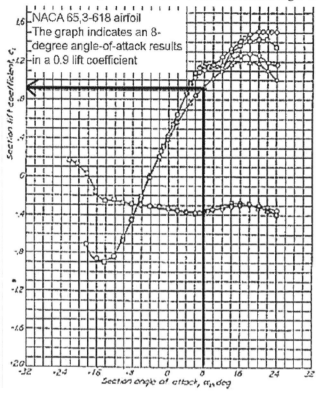

Graph 4.3.1-4 Section Drag Coefficient versus Section Lift Coefficient

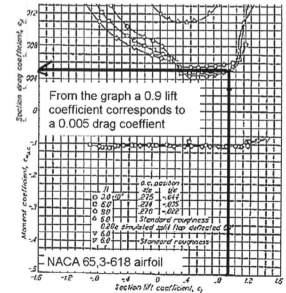

4.3.2 Angle of Attack versus Stall

Graph 4.3.2-1 NACA 65,3-61 Angle of Attack versus Stall:

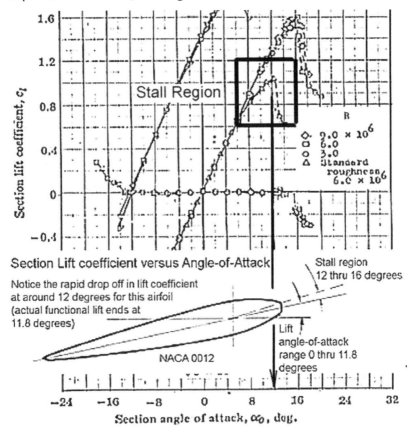

In the previous subsection we evaluated two airfoil shapes at an arbitrary Angle of Attack of 8 degrees. Now lets take a closer look at the above graph, given a constant main rotor rpm. As the blade Angle of Attack is increased Lift increases. Of course there is a limit to the amount of Lift that can be attained by increasing the Angle of Attack. Starting at 12 to 16 degrees depending on the airfoil shape, the airflow can no longer continue it's laminar flow over the top surface and the smooth airflow streamlines start to disconnect from the contact surface creating turbulence. This turbulent flow randomizes the flow velocity patterns such that the lower dynamic pressure starts to increase locally, effectively decreasing Lift. Once this critical Angle of Attack is reached any further angle increase creates a stall whereby Lift drops off quickly.

Stall is a function of both Airspeed and Angle of Attack. When that critical Angle of Attack / Airspeed condition occurs in a fixed-wing aircraft Stall takes place over the entire wingspan. In rotary wing aircraft, Stall is localized at the ends of the rotor blade where its tangential air speed is greatest. In "Hover", Stall, happens with equal consequence at the rotor tips. In "Forward Flight" Stall affects the outer portion of the retreating blade on the left side of the helicopter where the angle of attack is maximized. Blade stall causes roughness in both the helicopter and the flight controls. Obviously, for a pilot to eliminate a blade stall condition one must either decrease rotor tip speed by either slowing down (reducing forward speed) or diminish the current Angle of Attack (less main rotor pitch) or both.

4.4 The Main Rotor:

The significant point concerning the design of the helicopter rotor is maintaining directional flight without losing vertical stability. This topic was discussed at some length in the section on helicopter history where Juan de la Cierva was the first to incorporate a flapping hinge to offset rotor blade forward flight Lift imbalance in his aircraft design. To elaborate more on this topic lets briefly discuss the dynamics of the flapping hinge. For example our main rotor hub has the capability to not only have Angle of Attack rotation but also enables the individual blades to pivot upward. Under normal hover operating conditions, at the design rotational speed of the rotor, the centrifugal force developed on each blade of a 21-foot diameter rotor (this rotor diameter is close to the diameter of the helicopter we will build) could be over 6,000 pounds directed in the horizontal plane. Then as Lift is increased the combination of the horizontal centrifugal force, in conjunction with the vertical Lift force, causes the rotor blades to cone upward as a function of structural integrity. This coning of the rotor blades is constant throughout the complete hub rotation of 360 degrees. Lets go to forward flight. Obviously, with rotating blades establishing Lift the blade moving into the forward flight direction (advancing blade) will experience greater airflow resulting in greater Lift versus the blade moving with the flight direction (retreating blade). This Lift imbalance between advancing and retreating blades can generate a fuselage tipping moment if not addressed. Use of a flapping hinge allows the advancing blade to pivot up thereby reducing the effective lift area, and vice versa for the retreating blade. The end result is Symmetry of Lift between advancing and retreating portions of the disc area.

All in all there are only three basic types of main rotor assemblies used for helicopters today. They are 1.) "Articulated": this rotor allows each blade two degrees of freedom in the plane of motion. It can "Flap" (rotate up or down about the pivot point) and / or "Drag" (pivot forward or aft). This blade's motion can be individual or collective, and the blades can Feather. 2.) "Semi-Rigid": (commonly referred to as the seesaw type) This rotor utilizes a central hub for mounting both blades. Feathering and Flapping of both blades is accomplished together as a unit. 3.) "Rigid": As the name implies the blades, hub and rotor drive shaft are fixed (Rigid) to one another allowing only for blade feathering. In chapter 8.3 "Defining the Helicopter's Rotor Hub Type" I present the case for the type rotor we will use.

Chapter 5-Basic Flight Aerodynamics:

This chapter addresses how the four aerodynamic forces acting on an airborne helicopter affect flight maneuvers.

5.1 Powered Rotor

"Powered Rotor or Powered Flight" is the term used when the helicopter's engine provides the energy for the helicopter to hover, climb, and maneuver or for directional flight. It is specifically noted because flight dynamics change when a helicopter is in an "Autorotation" mode where engine power is not available. In Powered Flight the Total Lift and any Thrust forces developed by the rotating main rotor system are perpendicular to the plane of rotation (or tip-path plane).

5.1.1 Hovering

Diagram 5.1.1-1 Hovering Helicopter

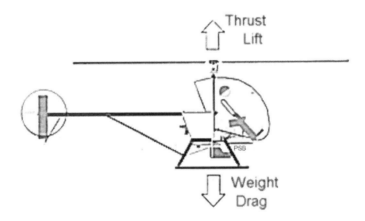

The basic scenario considered for this discussion is with the helicopter "Hovering" a few feet above the ground at zero airspeed in a stationary position with no cross winds. Hovering occurs when the Lift and Thrust, acting vertically up equal the Weight and Drag acting vertically down. Increasing altitude from this position is accomplished by raising the collective while maintaining constant engine RPM control. (Note: In the Engineering Section it is calculated that the AIRSPORT-254 will start to Hover when the main rotor rotation is maintained at ~ 438 rpm and the Collective Lever pitch angle position reaches 8 degrees on a 70 degree F day at sea level.)

While Hovering Drag is generated on the rotating main and tail rotor blades in two forms - Induced Drag, the primary drag force and Profile Drag. As mentioned before this Drag on the main Rotor blades generates a torque equal and opposite to the torque applied to the rotor system to keep the rotor blades rotating. As the main rotor rotates counterclockwise the helicopter itself wants to pivot clockwise about the main rotor shaft centerline. This unwanted motion is countered-acted by the Tail Rotor that develops an opposing torque to keep the helicopter pointed in a specific direction while Hovering. Anytime more power is applied to the Main Rotor, the Tail Rotor must generate more counteracting thrust.

I have included the following paragraph taken from the "Aeronautical Information Manual (AIM)" covering FAA policy regarding "Safety of Flight" to give a clearer perspective of the effects of a hovering helicopter. Under section 7-3-7 "Helicopters" the paragraph, reads:

"In a slow hover-taxi or stationary hover near the surface, helicopter main rotor(s) generate downwash producing high velocity outwash vortices to a distance approximately three times the diameter of the rotor. When rotor downwash hits the surface, the resulting outwash vortices have behavioral characteristics similar to wing tip vortices produced by fixed wing aircraft. The vortex circulation is outward, upward, around and away from the main rotor(s) in all directions. Pilots of small aircraft should avoid operating within three rotor diameters of any helicopter in a slow hover taxi or stationary hover. In forward flight, departing or landing helicopters produce a pair of strong, high-speed trailing vortices similar to wing tip vortices of larger fixed wing aircraft. Pilots of small aircraft should use caution when operating behind or crossing behind landing and departing helicopters."

Although the above statement omits all the theoretical backup of how these rotor vortices are generated, it does indicate how much air turbulence a Hovering helicopter generates.

5.1.1.1 Drift

"Drift" is a direct result of the summation of forces acting on a Hovering helicopter. When the Tail Rotor produces the force necessary to keep the helicopter pointed in one-direction its other effect is to move the entire helicopter in the direction the force is being applied. Looking from above, this direction is to the right of the helicopter's pointed direction and the actual movement called a Translation Tendency or more simply stated as Drifting. The magnitude of the Drift is a function of Tail Rotor thrust.

To minimize this effect these two mechanical features can be incorporated in the helicopter design:

• The main rotor mast may be angled such that the Tip-Path-Plane of the rotor blades produces a small force to oppose the Drift.

• The linkage is adjusted on the Joystick (cyclic control) such that when it is centered the rotor disc is angled to accomplish the same result.

Once either of these two features is built into the design, the helicopter assumes a slightly altered vertical position while Hovering. This slight rotation from centerline results in the left side of the fuselage being lower with the left skid closer to the ground than the right.

Diagram 5.1.1.1-1

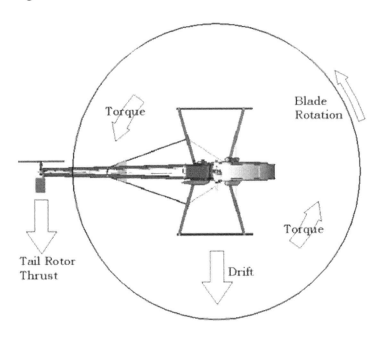

5.1.1.2 Pendulum motion

Diagram 5.1.1.2-1 Helicopter Pendular Oscillation:

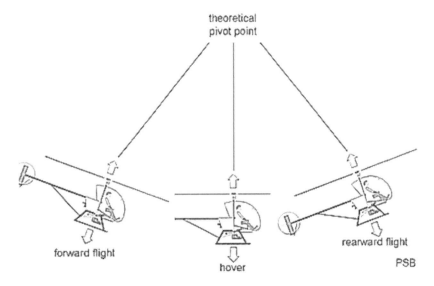

A simple pendulum is a mass supported by a string. When set in motion it is able to oscillate either laterally or longitudinally with only air resistance to dampen and eventually stop the motion. Similarly a single rotor helicopter with its entire fuselage weight suspended from the Main Rotor acts the same way as a pendulum when set in motion. This undesirable swing in a helicopter is caused when the flight control motion is over controlled or erratic. To prevent this effect from occurring flight control operation should be smooth.

5.1.1.3 Coning

Diagram 5.1.1.3-1

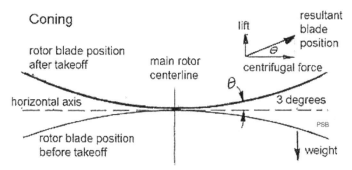

The term "Coning" refers to the deflected shape of the rotor blade during airborne operation. At rest the cantilevered rotor blade is acted upon by gravity and is deflected toward ground with the actual deflection of the blade at the free end dependent on its structural stiffness. After startup as rotor RPM increases a centrifugal force is generated on the rotor blade mass which acts to force the blade radial out in the horizontal direction and perpendicular to the rotor mass centerline. As Angle of Attack is increased to enable Hovering, Lift is generated along the chord length with the largest proportion produced at the free end of the rotor blade where tangential velocity is the greatest. At Hover the centrifugal force is extremely large with respect to either the Rotor Blade weight or the Lift produced. For example for the AIRSPORT-254 the Lift force generated by each blade is slightly more than 250 lbs, and the blade weight approximately 13 lbs, but the centrifugal force generated is ~ 6,000 lbs. The resultant force on the blade (where 6000 lbs is the horizontal force and 250 lbs is the vertical force) deflects the blade upward approximately 3 degrees. This dynamic Rotor Blade condition is called Coning.

It should be noted that the centrifugal force generation by the rotor, at the design operational rpm, is a major factor in determining the structural requirements for both the Main Rotor hub assembly and the Rotor Blades.

5.1.1.4 Coriolis Effect

"Coriolis Effect" (Law of Conservation of Angular Momentum) is the name given to the phenomena of rotating bodies increasing angular speed as the center of the rotating mass moves closer to the axis of rotation. The result of this principle is what enables a figure skater to increase the speed of a spin when their extended arms are retracted closer to the pivot point.

On most helicopters as with the AIRSPORT-254 the Main Rotor system although designed to operate in a plane perpendicular to the main rotor shaft centerline does in fact provide for a limited amount of pivoting about the connection point. This allows the helicopter time to adjust to a sudden Lift imbalance caused by, among other factors, a non-uniform gust of air flowing over the rotor blades. In this flight condition the Rotor Blade with the most Lift rises momentarily during the rotation or "Flaps up." Concurrent with "Flapping up" that Rotor Blade's Center of Mass moves closer to the axis of rotation and it accelerates to conserve angular momentum. Similarly, when the Coned rotor blade "Flaps down" it decelerates. In both cases this Flapping motion is limited by main rotor system design and the energy absorbed at the end points is either by dampers or handled by the blade structure. Flapping motion is also limited to an envelope of travel only above a plane perpendicular to the axis of rotation and through the rotor hub basically due to the effects of Coning.

Diagram 5.1.1.4-1

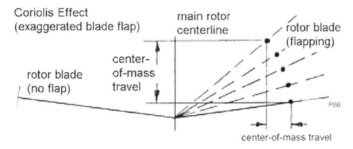

Diagram 5.1.1.4-2 Blade Flapping Motion

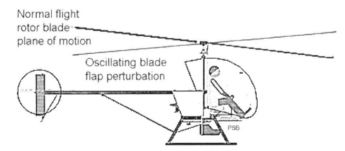

This instantaneous acceleration and deceleration of the Rotor Blade mass increases the internal stress on the blade structure itself and has to be accounted for in the design of the blade. On fully articulated rotor systems and those two bladed rotors where the blades are inline with the rotor hub the Coriolis Effect is pronounced since the change in distance from the blade center of mass to the axis of rotation is measurable. In contrast on a two-bladed main rotor system that has the blades "under-slung" (below the rotor hub connection pivot point) the Coriolis Effect is minimized since the blade Center of Mass remains approximately the same distance from the rotor centerline during positive and negative Flapping. This "under-slung" main rotor configuration is the one incorporated on the AIRSPORT-254 rotor. A more detailed discussion on the topic is presented in "Chapter 8.3 Defining the Helicopter's Rotor Hub Type".

5.1.1.5 Ground Effect

Diagram 5.1.1.5-1 In Ground Effect (IGE)

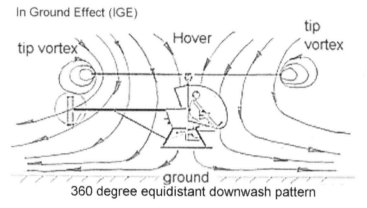

360 degree equidistant downwash pattern

Diagram 5.1.1.5-2 Out of Ground Effect (OGE)

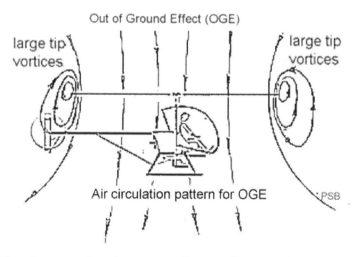

Hovering approximately one rotor diameter distance off the ground produces a phenomenon known as "Ground Effect". This effect increases the

amount of Lift for a set rotor blade angle. This happens for two reasons:
• The frictional interaction between the airstream and the ground results in a reduced induced airflow, which increases Lift.
• Blade tip vortices are minimized because more airflow is directed radially outward allowing more of the blade surface area to produce Lift.

This particular effect is maximized over a hard smooth level surface with no crosswind. The effect is minimized on other surfaces such as tall grass, water, walls, rough and uneven surfaces where airflow is increased and as a result the effect limited.

The benefits of "In Ground Effect (IGE)" are lost as altitude is increased in the hovering helicopter. In this region called "Out of Ground Effect (OGE)" Drag increases due to additional Induced airflow, increased pitch and larger blade tip vortices thereby requiring more power to sustain Hover.

5.1.1.6 Gyroscopic Precession

Diagram 5.1.1.6-1 Gyroscopic Precession, Phase-lag and the Gyroscope

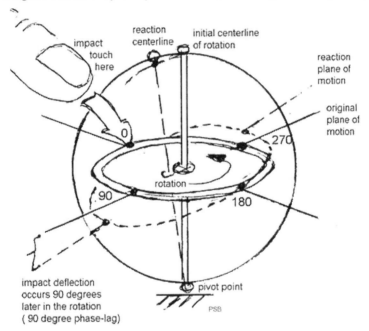

If you ever played with a gyroscope you quickly realize touching it while it is spinning at top speed may cause temporary imbalance but not necessarily loss of total control or rotational momentum. The rotating Main Rotor acts the same way. "Gyroscopic Precession" applies to the way the rotational motion of the object acts when subjected to a disturbing force. The resultant action or deflection of this force occurs approximately 90 degrees from where the force was applied in the angular direction of the rotation.

Phase-lag: is defined as that position in a Rotor Blade's rotation when the full effect of a pitch change is realized. The actual Phase-Lag angle varies based upon the type rotor (ridged, semi-ridged, articulated). For those rotors with flap hinges with only a slight off-set the phase lag is typically between 80 and 90 degrees while a semi-ridged rotor phase-lag range is 75 to 80 degrees.

In the science of gyroscopes one of the Laws states that when a gyroscope has a couple applied to it, the gyroscope reaction will "precess or tilt" in the plane at right angles to the plane containing the couple. It is relevant because on helicopter rotor designs where blades pivot and flap at the geometric center, the rotor behaves similar to a gyroscope and the result of this natural phenomenon of Gyroscopic Precession is technically called a "90 degree phase-lag".

Diagram 5.1.1.6-2 Gyroscopic Precession, phase-lag and the Helicopter: note the main difference in phase-lag between the gyroscope and the helicopter is the location of the new axis pivot point.

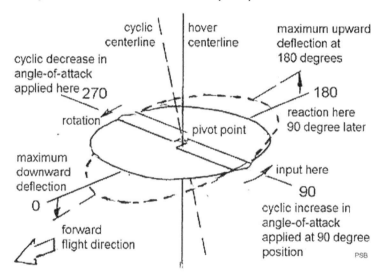

In a two-bladed rotor system the Cyclic control lever (joystick) generates the unbalancing force necessary to alter the tip-path plane of the rotor disc in order to maneuver the helicopter. When the Cyclic control lever is moved in a particular direction it mechanically increases the Angle of Attack and the Lift of the rotor blade as it passes by that directional point in the plane-of-rotation. Correspondingly, at this rotational position the other rotor blade has an equivalent decrease in its Angle of Attack, reducing Lift at its location in the plane-of-rotation. The instantaneous effect of this cyclic control movement is the blade with the increased Angle of Attack "Flaps up" and the opposite blade with decreased angle "Flaps down". Maximum deflection of this input force occurs 90 degrees later in the blades angular rotation where the Tip-Path Plane is tilted forward as it would in a gyroscope.

Once this new plane of rotation is established it will remain in effect as long as the resultant force on the rotating blades does not change. This implies that although individual blades will undergo cyclic Angle of Attack change with subsequent aerodynamic Lift variations (in addition to "blade flap oscillating accelerations as a function of blade inertia and dynamic imbalances) the total force acting on the helicopter remains constant.

This physical behavior can be expected with a semi-ridged articulated rotor (teeter/ see-saw type where both blades are mounted to a hub and flaps and feathers as a unit). The mechanical system used for this type rotor is designed to accommodate this 90 degree phase-lag by having the control input to the rotor feathering lever (that's the lever actually located on rotor blade which controls its pitch angle rotation) 90 degrees from the control input to the swashplate. The rotor blade reaction to this input puts the helicopter forces in the desired flight direction.

The CAD drawing below shows how the mechanical linkage is setup from control joystick to blade feathering lever. As you can see even though the linkage from the joystick is setup so that pushing the stick forward to go forward seems logical, when the control rod pulls the swashplate down the linkage to the rotor blade feathering lever is at an orientation on the swashplate 90 degrees behind the input point.

CAD Dwg 5.1.1.6-1

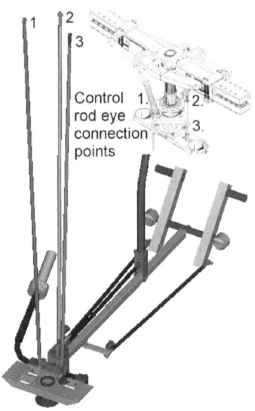

1   2
    3

Control   1.   2.
rod eye
connection   3.
points

5.1.2 Vertical Flight

Diagram 5.1.2-1 Vertical Flight: during hover the lift / thrust force equals the opposite weight / drag force but in vertical flight the magnitude of the lift / thrust force is greater.

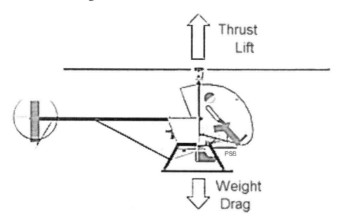

"Vertical" flight takes place when Lift and Thrust (vertical direction) exceeds Weight and Drag. It is initiated by bringing the main rotor RPM to its design rotational speed and increasing the Collective until the Angle of Attack on the rotor blades produces the necessary upward force to achieve ascent. Descent in a no wind condition occurs when the Collective is lowered decreasing the Angle of Attack so that Lift and Thrust force is less than Weight and Drag.

5.1.3 Forward Flight:

Diagram 5.1.3-1 Forward Flight (for steady forward flight A = B

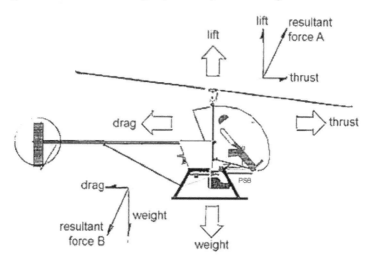

Diagram 5.1.3-2 Definition of Rotor Disc Incidence: angle A = rotor disc incidence

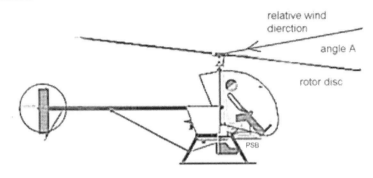

Diagram 5.1.3-3 Definition of Rotor Disc Attitude

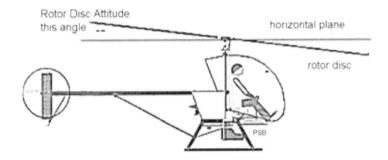

Once the main rotor RPM is at set point and the Collective Lever is positioned to establish Vertical flight, "Forward" flight happens when the Cycle Control Lever (joystick) is moved forward. This causes the Tip-Path Plane to tilt forward generating a Thrust force in the horizontal enabling helicopter forward movement. As said many times before restricting this motion is Drag comprised of wind resistance and inertial resistance.

During a steady-state constant velocity straight and level Forward flight in a "no" atmospheric wind condition, Thrust is equal to Drag and Lift equal to Weight. Implicit in that statement is the resultant force A made up by the horizontal Thrust vector and the vertical Lift vector is equal and opposite the resultant force B comprised of the vertical Weight vector and the Drag vector. Any time one of these forces is dominant over the other the helicopter will move in that direction. For example if the Lift force is greater than the Weight the helicopter will gain altitude. If both Lift and Thrust are dominant the helicopter will gain altitude and accelerate.

After lift-off but prior to steady-state flight the helicopter rotor blades experience some transitional adjustments after the Cyclic control lever is moved, these transitional adjustments can be felt by the pilot.

Diagram 5.1.3-4 Steady-state Hover lift vector shown

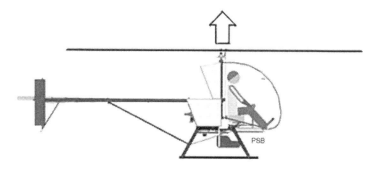

Diagram 5.1.3-5 Hover to Forward Flight Momentary Flight Dynamics: note the momentary lift vector directional change with corresponding loss of altitude due to initial forward cyclic (joystick) movement.

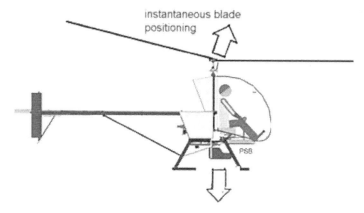

Shown above, in diagram 5.1.3-4, the Lift vector for the Hovering helicopter is vertical. Then as cyclic stick is moved to initiate Forward flight, the individual Rotor Blade's circular path is momentarily imbalanced as the rotor is tilted forward to match the revised swashplate angular orientation. Shown in the above diagram is an exaggeration of a Rotor Blade in instantaneous motion whose time to realign itself into proper angular direction is dependent on its inertia (or reluctance to move to that direction).

During this rotor disc perturbation a portion of the Hover power required to generate Lift is tapped off to provide forward thrust. Unless power is added to make up for the loss there is a momentary loss of altitude. Then as the helicopter accelerates forward and more airflow passes thru the rotor, disc efficiency increases and additional Lift generated. At some point, a forward speed is reached where rotor efficiency is maximized and the power required to maintain altitude at a minimum, after which more power is required to handle increased drag.

5.1.3.1 Translational Lift

Diagram 5.1.3.1-1 Effective Translational Lift (ETL)

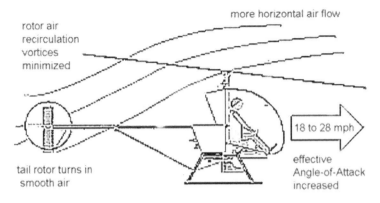

In the above statement on Forward flight it is stated how rotor efficiency is improved with increased airspeed. Technically this is called "Effective Translational Lift (ETL)" and it begins to occur at around 16 to 24 knots when the increased flow of air displaces the Rotor Blade Vortices with relatively undisturbed air. The dynamics of this is more blade surface area available to produce Lift. Additionally airflow becomes more horizontal thereby reducing Induced flow and Drag that in turn effectively increases the Angle of Attack for a set blade Pitch Angle and increases Lift. When ETL begins the helicopter experiences momentary dynamic rotor blade vibration as the rotor disc efficiency increases and the additional Lift is generated.

The result of this less turbulent airflow past the helicopter during ETL is that the Tail Rotor becomes more efficient also and it too generates more Thrust. Since the torque developed by the Main Rotor hasn't changed during ETL this additional Thrust by the Tail Rotor causes the helicopter to Yaw left. Reducing the Tail Rotor pitch with the right foot pedal re-establishes directional control.

It was stated Translational Lift occurs during forward helicopter airspeeds of around 16 to 24 knots, but it is also generated during stationary Hover if atmospheric wind (traveling at approximately the same speed) crosses the rotor disc. This fact is useful to incorporate during takeoff when additional performance is desired.

5.1.3.2 Induced Flow

We have already discussed Relative Wind and Average Relative Wind in relation to Induced Drag. Now lets talk about the air flow involved. "Induced Flow" (Downwash) is that large volume of air forced vertically down by the rotating blades during the production of Lift. "Rotational Relative Wind" develops when the rotor blades rotate. Its resultant direction is parallel and opposite the rotor's rotational plane and is tangential (perpendicular) to it at the leading edge. These two flows, Induced Flow (vertical flow) and

Rotational Relative Wind, (horizontal flow) combine algebraically to form what is referred to as the "Resultant Relative Wind". This Resultant Relative Wind centerline intersects with the Airfoil Chord line to form the all-important Angle of Attack used to determine Lift and Thrust. Induced Flow can significantly impact rotor efficiency by changing the Resultant Relative Wind angle. Increased Induced Flow increases the Resultant Relative Wind angle that in turn reduces the Lift producing Angle of Attack.

Diagram 5.1.3.2-1 Induced Flow

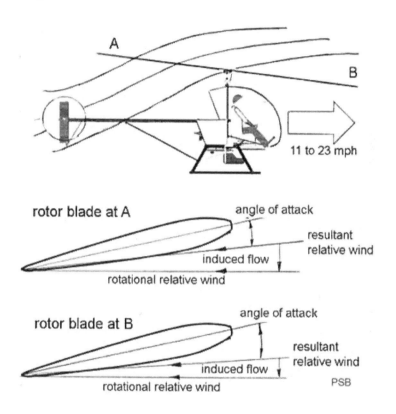

5.1.3.3 Transverse Flow Effects

During helicopter Forward Flight acceleration, as shown in the diagram in the preceding section, Induced Flow approaches zero in the Forward Rotor Disc Area and in the Rear Disc Area it gains in magnitude. The result of this is "increased Lift" produced in the Forward Disc Area where little Induced Flow exists and "decreased Lift" in the Aft Rotor Disc section due to the increased Induced Flow. This rotor blade Lift "imbalance" causes the forward blade to "flap up" and the aft blade to "flap down." Following the principles defined under Gyroscopic Precession the maximum "deflection" to this force is angularly rotated 90 degrees counterclockwise. When the helicopter accelerates to a combined forward velocity plus headwind (if any) of approximately 20 knots the helicopter rolls slightly to the right.

5.1.3.4 Dissymmetry of Lift

Diagram 5.1.3.4-1 Dissymmetry of Lift in Forward Flight

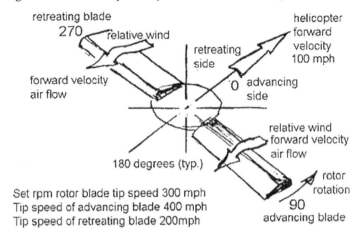

Set rpm rotor blade tip speed 300 mph
Tip speed of advancing blade 400 mph
Tip speed of retreating blade 200mph

Diagram 5.1.3.4-2 Blade Dynamics as a Function of Rotation (forward flight dissymmetry of lift)

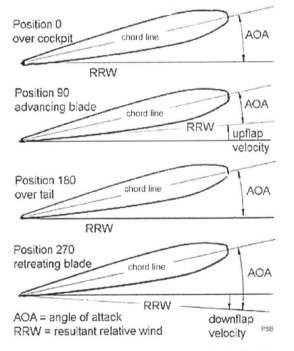

It is apparent by now that high-speed atmospheric air flowing over an airfoil with a certain Angle of Attack produces Lift. It is also a fact that in a rotating system at any one reference point on the rotor disc circumference, one rotor blade is Advancing towards that point and the other blade Retreating. Lift is symmetrical around the rotor disc if there are no additional airflows crossing the rotor disc plane of motion since the air flowing over the blade is solely based upon the rotational speed of the driven blade. An example of this is Hovering over a stationary point with no crosswinds. It is more likely that crosswinds are present and / or the helicopter is in motion. In that situation the air flowing over the Advancing blade is added to that already produced by the blade's rotational velocity and results in an incremental "increase" in Lift. On the Retreating blade this airflow reduces the impact velocity on the airfoil and "reduces" Lift. This Lift differential caused by the Advancing / Retreating rotational motion of the rotor blades into an external airflow is called "Dissymmetry of Lift."

Dissymmetry of Lift if not addressed would cause a forward moving helicopter to roll to the left. Video footage taken of Sikorsky's first flight in his newly designed helicopter did just that and crashed within a short distance after initiating forward motion. On the AIRSPORT-254 the blade tip speed is ~337 mph and when the helicopter reaches a speed of 63 mph (50 knots, maximum allowable limit for an ultralight) the Advancing blade tip speed is 400 mph while the Retreating blade tip speed is only 274 mph. This tip speed differential of 126 mph and the resultant Dissymmetry of Lift is significant and must be eliminated or minimized in order to achieve forward flight stability. On semi-rigid rotor systems, of which the AIRSPORT-254 can be included, an under-slung teetering hinge arrangement is used to mount the two opposing rotor blades. During Dissymmetry of Lift this design allows the Advancing blade to Flap up and the Retreating blade to flap down. When the Advancing blade reaches that rotational angle where airflow across the airfoil reaches its peak, the Flap up velocity is maximized. At that point the Angle of Attack is "minimized" and the Lift generated by the blade is proportionally "reduced." On the Retreating blade the inverse effect happens and at maximum down-flap velocity a proportional increase in Lift is experienced. This designed in capability for the rotor blades to flap thereby reducing Lift on the Advancing blade experiencing maximum airflow speed and increasing Lift on the Retreating blade with minimum airflow speed compensates for Dissymmetry of Lift.

After initial forward Cyclic input the Tip-path plane of rotation of the rotor disc is tilted down towards the front of the helicopter. Once helicopter speed increases and the Dissymmetry of Lift begins to occur the rotor blades start to Flap. The Advancing blade reaches its peak up-flap deflection over the "front" of the helicopter and maximum Retreating blade down-flap deflection over the tail. Under this condition the Tip-path plane of rotation tilts to the "rear" and is called Blowback. The result of this "rearward shift" in the Tip-path plane reduces the forward force component of the total Thrust generated and the helicopter "slows" down. To maintain speed "additional" forward Cyclic input is required.

There is one additional dynamic sequence realized during blade Flap. It was stated earlier that when Flapping occurs the retreating blade's Relative

Wind is equal to the driven blade's rotational velocity minus the wind influx across the blade. This increases the Angle of Attack. There is a critical point where the high Angle of Attack and low Relative Wind speed no longer produces sufficient Lift to neutralize the Dissymmetry of Lift forces and the result is increased vibration, pitching up and rolling left of the airframe. This limit to helicopter airspeed is called "Retreating Blade Stall". On the AIRSPORT-254 that is limited to a maximum speed of 50 knots, Retreating Blade Stall under normal flying conditions would not occur. On factory built FAA certified helicopters Retreating Blade Stall is avoided by not exceeding what is designated as VNE (NEVER-EXCEED Speed) and is normally redlined on the airspeed gauge and noted on a cowl mounted placard.

The above information highlights the functional dynamics for a two bladed under-slung teetering main rotor. The more complex main rotor designs, those having three or more blades, are typically fully articulated, and utilize a separate horizontal hinge (flapping hinge) and velocity dampers for each blade. Any Dissymmetry of Lift is addressed locally by individual blade Flap to neutralize it. Flight dynamics for this type of main rotor system is beyond the scope of the ultralight helicopter builder.

5.1.4 Sideward Flight

Diagram 5.1.4-1 Sideward Flight (for steady flight resultant force A = B)

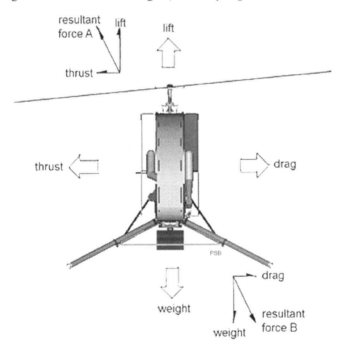

"Sideward" flight follows the same flight dynamics as Forward flight where the Lift force is vertical up and the Thrust force horizontal and the Tip-path plane of rotation perpendicular to the resultant Lift-Thrust vector. The difference is that for Sideward flight the Thrust force is no longer forward but now in the direction that the Cyclic (Joystick) is pointed. Sideward speed as a "function of applied power" suffers, because Drag forces are greater due to the additional airframe wind resistance when moving in that direction. Note that opposing the Lift-Thrust vector is the resultant Weight-Drag force vector.

5.1.5 Rearward Flight

Diagram 5.1.5-1 Rearward Flight (steady rearward flight exists if resultant force A = B)

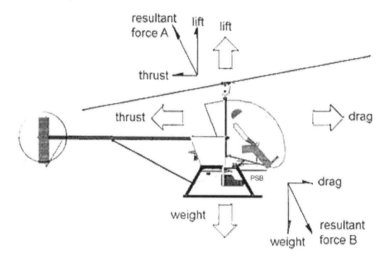

Moving the Cyclic control aft tilts the Tip-path plane of rotation rearward. Helicopter speed in this direction is a function of the magnitude of the resultant Lift-Thrust vector. Like Forward and Sideward flight, the resultant Weight-Drag force vector will oppose motion.

5.1.6 Turning Flight

"Turning" Flight applies to "banked" turns made while in Forward flight. When a helicopter is in forward motion the rotor disc and the Lift-Thrust vector are tilted in the same direction and the Weight-Drag force vector opposite that direction. When the Cyclic lever is moved sideward from its existing position to initiate a "banked" turn the rotor disc shifts in that direction and some of the vertical Lift force is diverted to the horizontal direction to counter act the inertial force resisting that motion. This horizontal component of the original vertical Lift is called the Centripetal force and the resultant vector of these two components called Resultant Lift. Increasing the Bank angle increases the Centripetal force and the rate-of-turn. Since some of the original vertical Lift force is being diverted to counteract the helicopter's "banking" inertia, altitude decreases unless there is an increase in the Angle of Attack. The faster the rate-of-turn the more Angle of Attack is required to maintain altitude. Total power in this flight condition must increase in order to satisfy this new Resultant Lift requirement and the additional thrust force.

Diagram 5.1.6-1 Turning Flight and Inertia

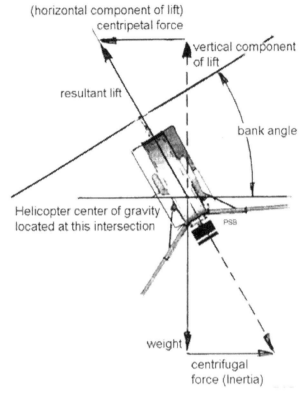

5.2 Unpowered Rotor

"Autorotation" is a term used to define a built in safety on all certified helicopters and the AIRSPORT-254 whereby initial rotor disc inertia combined with the potential energy associated with its altitude can enable a helicopter to land safely during engine failure. During powered flight large volumes of atmospheric air above the spinning rotor disc are forced downward. During an emergency Autorotation maneuver, when the helicopter begins to respond to this zero Lift condition, airflow enters the rotor disc from below. Subsequently the potential energy of the helicopter's altitude is converted into wind kinetic energy that now drives the rotor blades as it begins its rapid descent. It should be noted that a helicopter, including the AIRSPORT-254, is designed with an "over-running clutch" in the main rotor drive-train which allows the main rotor shaft to continue rotation in spite of a stopped engine / power-plant.

Diagram 5.2-1 Normal Flight versus Autorotation Airflow thru Rotor: Note pilot must make collective Angle of Attack change within 2 seconds or helicopter will just drop from the sky.

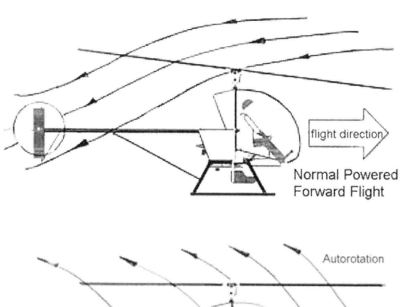

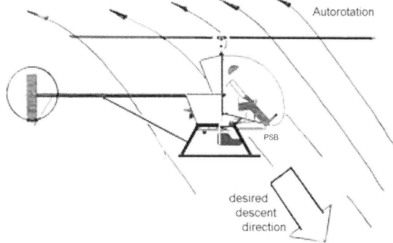

5.2.1 Vertical Flight Autorotation:

The "Vertical" type of Autorotation is the simplest to explain because the Dissymmetry of Lift dynamics do not apply. When the helicopter starts its rapid descent the aerodynamic forces causing blade rotation are uniform and independent of airfoil rotational position in the plane of rotation.

5.2.2 Forward Flight Autorotation:

Moving in the "Forward" direction while in the Autorotation state Lift is produced by an upward airflow, as is the case with a vertical descent in still air. The difference is the forward airspeed produces more Lift on the right side of the helicopter where the counterclockwise rotating blade enters the oncoming air rather than retreats from it. The net effect of this is the Driven and Driving regions shift outboard towards the right side of the rotor hub. On the "retreating side" the relative wind produced "increases" the Angle of Attack whereby more of the blade is in Stall, while on the "advancing side" there is a "reduction" of Angle of Attack and more of the blade is in the Driven region. Additionally, a reversed airflow starts at the blade root reducing some of the Driving region effective area on the retreating side.

Diagram 5.2.2-1

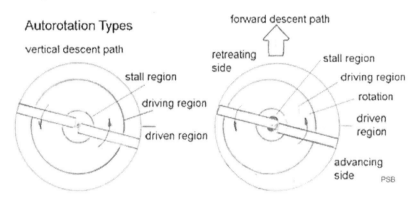

The aerodynamic forces developed during Vertical autorotation can be divided into three distinct zones those being - the "Driven" region; A (30% of the disc), the "Driving" region; C (45%) and the "Stall" region; E (25%).

Diagram 5.2.2-2 Vertical and Forward Descent Autorotation

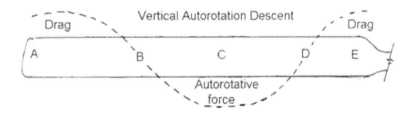

This diagram indicates those zones as they apply to the critical points on the airfoil. The force vectors in each region are different in magnitude because they are a function of the Relative wind generated at that point on the Rotor Blade. This is due to the fact that though rotor RPM is constant the blade "tangential speed decreases" as the distance from the hub to the point the force is acting on the airfoil decreases therefore "reducing" Relative wind generation. To compensate for less Relative wind being produced as tangential speed falls off most constant chord helicopter rotor blades have a "built in linear twist (around 6 degrees)" running the entire blade length to increase Lift efficiency. This design progressively increases the Angle of Attack as Relative wind drops off. This benefits Autorotation because the Angle of Attack is more positive in the Driving region than the Driven region. As can be expected the forces acting on the airfoil when wind powers the rotor blade are different in magnitude and direction than when power plant driven from the main rotor shaft.

The "Driven" region is located closest to the free end of the rotor blade and referred to as the propeller region. Its effect encompasses about 30% of the rotor disk area, but the actual Driven area with respect to the other regions changes in size as a function of blade Pitch angle, rotor RPM, and Rate of Descent. This region produces Lift but because the total force is behind the Axis of Rotation the Drag forces dominate and the net effect causes rotor RPM deceleration.

Two equilibrium points exist on the rotor blade where the total aerodynamic force aligns with the Axis of Rotation. One is between the Driven and Driving region, the other between the Driving and Stall region. At these two points Lift and Drag neutralize each other and rotor RPM is not affected.

The "Driving" region that is also called the Autorotative region produces its total aerodynamic force forward of the Axis of Rotation and at a slight angle. The net effect of this angled force tends to accelerate blade RPM while providing Lift. This Driving region area size changes as airfoil dynamics change such as Pitch angle, Rate of Descent, and rotor RPM. Normally, this region is located between the 25% point on the span and the 70% point.

After engine failure it is imperative to immediately readjust the Angle of Attack of the rotor blades to maximize the potential energy of this newly directed upward airflow, and to retain as much as possible of the existing inertial energy of the rotating rotor. Maintaining constant rotor RPM requires the collective to be manipulated so that the Driving region acceleration forces are equal to the deceleration forces generated by the Driven and Stall regions.

Chapter 6 Flight Controls

The main purpose of this manual is to provide as much information as possible so the potential design/builder would not have to do as much investigative research as I did to gain sufficient knowledge on the various helicopter subsystems. "Flight Control" is one of those subsystems where you should read the military specification to get a real feel for the subject. I have included this document in the appendix but I think it important for you to read it now in order to gain a better appreciation of what is required of helicopter control from a pilot's standpoint.

The Airsport-254 helicopter is a conventional design. The main rotor blade rotation when viewed from above is counterclockwise, and flight requires four basic manual controls. The necessary controls are - Collective, Cyclic (joystick), Anti-torque Pedals and Throttle.

CAD Dwg 6-1

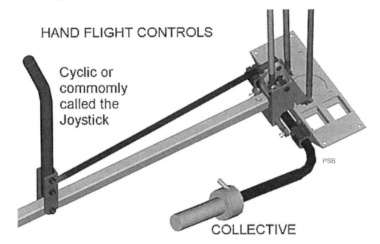

HAND FLIGHT CONTROLS

Cyclic or commomly called the Joystick

COLLECTIVE

PSB

6.1 The Collective:

On the AIRSPORT-254 as in most helicopters the "Collective" pitch control lever is mounted to the left of the pilot. It is used to change the pitch angle of the rotor blades equally and at the same time or as the name implies "Collectively." On the AIRSPORT-254 this is accomplished manually through a mechanical system utilizing levers, mechanical cams, control rods, a swashplate, and other linkage connected to the main rotor assembly. Since this lever is mechanically connected to the rotor blade via the aforementioned parts any movement of the lever will directly affect blade pitch angle within its operational range. Raising the Collective increases the blade angle up to an 11.2 degree "design" maximum. Lowering the Collective decreases the pitch angle to a desired fixed positive value.

In the AIRSPORT-254 located to the left of the pilot a slide mechanism tied to the lever movement is calibrated to indicate the collective pitch angle selected. Additionally, an adjustable friction disc mounted on the lever support rod restrains its position from accidental movement or system vibration. Helicopters in general have some method of pitch angle indication as well as a collective lever-damping device. Some are purely mechanical as the ones described for this helicopter while others may use various electronic angle sensors in conjunction with electro-mechanical or hydraulic Collective lever damping devices. It all depends upon the design specifications.

Even though the Collective is used to increase the Angle of Attack equally on all blades, the motion of the Joystick on the Swashplate is still functional, which means if the Joystick is positioned for forward flight then an increase in Collective means the helicopter if adequately powered should continue Forward flight and also gain altitude.

6.2 The Throttle

The Throttle on a helicopter is equivalent to a gas pedal on a car. If you are driving your car without the aid of cruise control as you approach a hill you must depress the gas pedal further to maintain your speed and the inverse is true when going down the grade as you must let off the gas pedal to maintain speed. In the first case the additional gas supplied enables more power to be developed by the engine up to its horsepower limit at that specific RPM. Should the power required for the car to go up the hill be more than the engine can develop the car will loose speed. The throttle operates the same way but in helicopter flight rotor RPM is the parameter being controlled and not vehicle speed. It is the position of the Collective pitch control lever that varies the power requirements. The Throttle control is located at the end of the Collective lever. Twisting the throttle grip towards you decreases fuel flow, rotating in the opposite direction increases fuel flow. The Throttle controls power and may or may not change engine RPM just as flooring the gas pedal in a loaded car going up a steep grade doesn't necessarily insure maintaining speed, never mind increasing it.

6.2.1. Throttle and Collective Synchronization

Raising the Collective increases the rotor blade Angle of Attack and increases Lift as well as additional Drag forces. The incremental increase in Drag over the length of the Airfoil increases the torque requirement to maintain the current RPM. The engine must supply more power. This is accomplished by increasing the Throttle fuel flow. It should be noted that engine horsepower Hp (ft-lbs) = RPM x Torque / 5250. For simplicity lets say that the torque value used in this equation is a function of the combustion force on the piston times the distance between the piston-connecting-rod-crankshaft-journal-bearing and the crankshaft centerline. Conversely, the torque value developed for the main rotor hub is a function of the blade drag force times the distance from the main rotor centerline to the rotor blade position the drag force acts. This new power requirement or "Load" is measured by manifold pressure on piston engine helicopters and by a torque meter for those powered by turbines.

The convention established for piston driven helicopters is that the primary control for manifold pressure is the Collective and for RPM it is the Throttle. Any movement of either control impacts the other as a secondary response. The control to use first is determined by viewing respectively the Tachometer (RPM gauge) and the Manifold Pressure Gauge then taking the required action.

Just as a car going up a steep grade requires more power to maintain a set speed other factors such as head winds, cross winds, dips in the road or even sweeping curves dictate constant engine power corrections. The same is true for helicopter rotor RPM control. The dexterity and precision of throttle control is readily apparent to anyone who has pushed the limit riding a production dirt bike over wooded hilly terrain. If you haven't, then when you get a chance go see a factory sponsored motor-cross race. Any time the thing you are trying to control weighs close to what you weigh, dirt bikes, motorcycles, standup jet ski, go karts, ultralight helicopters etc. you will definitely work the throttle.

The main rotor system requires more engine torque as the Collective is increased. That's a steady-state given, but a helicopter main rotor in operation is anything but a steady-state operation. We have already discussed some of the dynamics of it all with geometric precession, flapping etc, not to mention structural flexing. When these things occur localized airflow over the blade changes, which impacts Lift, Drag and ultimately the instantaneous Torque. When the Collective is increased or the Joystick (cyclic control lever) moved a lot more happens than just the mechanical changing of the Angle of Attack. What actually happens is a momentary disconnect between applied Torque (engine) versus required Torque (rotor) and the term "Hunting" given to the "Throttle response verse RPM perturbation (lack of synchronization)."

6.2.2 Engine Control

6.2.2.1 Manual: The most basic controlled response to helicopter power requirements comes from the pilot. Any changes to main or tail rotor blade Angle of Attack, wind conditions, or helicopter speed changes the power necessary to maintain a desired flight path. Manual manipulation of the Throttle to regulate fuel flow is required to keep rotor RPM constant.

6.2.2.2 Mechanical: The simplest mechanical device used for engine power management is the Correlator. It consists of a straightforward mechanical link from the Collective to the Throttle. It is designed to adjust fuel flow to produce more or less power in direct relationship to Collective movement. Noting that Collective movement changes the engine load requirements as the Angle of Attack varies. The Correlator is an open loop system in that it adds (or subtracts) power depending on Collective position, but the actual RPM desired must be manually adjusted by the pilot via additional Throttle fine-tuning.

6.2.2 2.3 Automatic: The automatic means of maintaining constant RPM control as flight power demands vary is accomplished by the Governor. In this closed loop system electronic sensors send rotor RPM data to a computer. The computer varies engine power as required via a Throttle mounted actuation device until actual RPM equals set point RPM. This device is used on all turbine-powered helicopters. The AIRSPORT-254 and some piston engine helicopters should employ the device. It is extremely useful as a pilot aid on small lightweight helicopters where quick changes to Relative Wind conditions have more impact on the lightweight airframe and generate larger power requirement aberrations.

6.3 Cyclic Lever (Joystick)

The Cyclic (Joystick) controls the degree of vertical tilt of the Rotor Disc and its radial position with respect to the airframe centerline. Any Joystick direction the pilot moves it to, within its 360 degree field-of-motion, the

Rotor disc corresponds and the degree-of-tilt it assumes is predicated on the magnitude of the Joystick motion in that direction.

When the Joystick moves, that mechanical motion is transferred from the Motion Control Unit control rods to the lower section of the Swashplate. This lower section swivels on a spherical bearing that is aligned with the main rotor shaft centerline and translates the Joystick motion into the desired plane-of-motion the Rotor disc is to follow. The upper housing connects to the lower section via a radial bearing. Internal to this bearing bore a slotted sleeve is press fit. This entire assembly then slips over the main rotor shaft and a shoulder bolt passing thru the slotted sleeve and shaft keep it in relative position until the control rods are attached. The swashplate as described can duplicate every cyclic and collective motion.

Each rotor blade is connected to this upper Swashplate section by means of an upper control rod. Due to the dynamics of Gyroscopic precession in order to have the Rotor disc follow the Swashplate's established plane-of-motion, the upper control rod must be mechanically designed to decrease the Rotor blade pitch angle approximately 90 degrees before reaching the desired Swashplate plane-of-motion and increase the pitch angle approximately 90 degrees after the desired direction is reached..

6.4 The Foot / Anti-torque Pedals

Cad Dwg 6.4-1

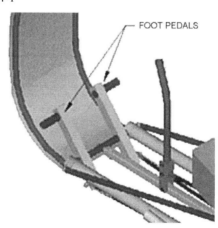

Cockpit Foot-pedals increase or decrease the force generated by the Tail Rotor by uniformly varying the pitch angle on the tail rotor blades similar to what the Collective motion does on the Main Rotor blades. This force acting over the distance between the "vertical" Main Rotor shaft centerline and the "horizontal" centerline of the Tail Rotor" generates the torque (anti-torque) required to counteract Main Rotor torque, and enable the pilot to control helicopter Yaw position. Yaw control is the ability of a helicopter to rotate either clockwise or counter-clockwise about its vertical centerline. During Hovering these Yaw turns are referred to as pedal-turns. During Forward flight the Cyclic (Joystick) is used for directional changes to the flight-path but the Foot-pedals are still required to maintain heading control during takeoffs and approaches during strong crosswinds and any other incidents of Main Rotor torque variations.

6.4.1 Heading

Helicopter flight demands constant power adjustments to the Main Rotor. In order to maintain a constant Heading as the Main Rotor torque varies the Tail Rotor force must also change to cancel that incremental torque differential. The positions of the Foot-pedals are designed such that in the "Neutral" position (where both foot pedals are aligned with each other) the pitch angle set produces sufficient force such that the torque developed is approximately that required to cancel Main Rotor torque and maintain the current cockpit orientation.

When the Foot-pedals are in the "Positive" position (the right pedal forward of "neutral" and the left pedal aft of neutral) Tail Rotor blade pitch angle is reduced. In this case less Main Rotor counteracting torque is generated to keep the helicopter's current orientation so it begins to rotate counterclockwise about it's vertical centerline (Yaw right).

If Foot-pedals are in the "Negative" position (the left pedal forward of "neutral" and the right pedal aft of neutral) Tail Rotor blade pitch angle is increased. In this case more Main Rotor counteracting torque is generated than necessary to keep the helicopter's current orientation so it begins to rotate clockwise about it's vertical centerline (Yaw left).

The dynamics involved with helicopter control are complicated as well as interconnected. An input to any of the flight controls has direct impact on the others. In describing the functionality of these flight controls it is important to realize that what occurs is based upon the initial conditions set before the controls are manipulated. These conditions include things such as cruise power being utilized, the steady state speed, crosswinds etc. For Foot-pedals response to differential Main Rotor torque, a generalization can be made whereby more right pedal displacement is utilized in situations where less power than normal is being used by the Main Rotor and conversely when rotor power demanded is greater than normal more left pedal displacement is necessary.

In addition the Tail Rotor must have some built in negative pitch angle capability because this is utilized to counteract a tendency for the helicopter to Yaw left during Autorotation where the drag resulting from the drive train produces this effect.

PART 2 ENGINEERING

I start with the preliminary design. Preliminary design is the process of defining the core parameters of the helicopter to meet a given set of performance or mission specifications, or more specifically in this case to comply with the guidelines set forth for ultralight / experimental aircraft. Basically, the preliminary design analysis, from the macro point of view, involves sizing the helicopter rotor and power required as a function of gross weight. Specific parameters such as vertical thrust, disc loading, rotor radius, main rotor tip speed and solidity are selected on the basis of the targeted helicopter gross weight. The work continues until analytically we have a good idea of what our helicopter will do when built.

In previous sections I attempted to give some logic and additional explanation as to the evolution of the complex subject of helicopter aerodynamics. In the following chapters we will complete our design parameters and determine the actual components we will need for our project using mathematical calculations to verify our choices.

The following terms are particularly important to review:

Disc: the imaginary surface area swept by the rotation of the rotor blades

Inertia: an object's resistance when attempting to accelerate it in a straight line

Disc loading: the helicopter weight to total main rotor disc area ratio calculated as helicopter Weight divided by rotor disc area. Disc loading changes as a helicopter accelerates or maneuvers. Increases in disc loading requires more power to sustain constant rotor RPM

Rotor RPM: the rotational speed of the main rotor expressed in revolutions per minute or radians per second, our calculations use radians per minute. One radian = 360 degrees / 2 pie = 57 degrees

Rotor disc area = pie x r^2 where r equals the distance from the centerline of the main rotor to the extreme end of the rotor blade, and pie = 3.14.

Total rotor blade area: equals the rotor blade chord times the blade length times the number of rotor blades fixed to the main rotor.

Solidity ratio: total rotor blade area divided by rotor disc area. This ratio enables measurement of rotor system thrust potential.

Chapter 7. Helicopter Design Relationships

For beginning our work we need to consider two areas of importance that being "weight per horsepower" and "disc loading". As I said earlier I never had any intention of reinventing the wheel, so what better way to get a feel for these two design parameters than to look at those relationships on some of the more successful helicopters. This chapter not only states the numerical range for both "weight per horsepower" and "disc loading" but also gives all the helicopter profiles used to develop each range boundary.

To this end I have listed below two important statistics for all the historically significant helicopters stated in section 0.6 "Brief History of the Helicopter". Those statistics being "gross weight per horsepower" and "disc loading." The first set of data indicates how much power was required to lift a certain helicopter weight and the second data set states how that helicopter weight was distributed over the rotor disc area.

A quick perusal of this list of all type configuration helicopters of this period (1936-1949) indicates anywhere from 7 to 15 pounds could be lifted for each horsepower supplied and the rotor disc loads ranged from 1 to 6 lbs. per square foot with the majority being in the 2 to 3 lbs. per square foot envelope.

*Focke (Germany, 1936) tandem: 160 hp / 950 kg
Gross weight per hp = 13, disc loading = 2.5 lbs / sq. ft
*Flettner (Germany, 1940) synchropter: 140 hp/ 1000kg
Gross weight per hp = 15.7, disc loading = ~approximately 1.2 lbs / sq. ft.
*Pullin (Britain, 1938) tandem: 205hp/ 1070kg
Gross weight per hp = 11.4, disc loading = 2.4 lbs / sq. ft.
*Bratukhin (USSR, 1940) tandem: 2-350hp/ 2300kg
 Gross weight per hp = 7.2, disc loading = 6.1 lbs / sq. ft.
*Germany (1941) their Focke-Achgelis Fa-223, tandem: 1000hp/ 4300kg
Gross weight per hp = 9.4, disc loading = 3.8 lbs / sq. ft
Sikorsky (United States, 1941) single: 185hp/ 1100kg
Gross weight per hp = 13, disc loading = 2.13 lb / sq. ft
Bell (United States, 1943) single: 178hp/ 950kg
Gross weight per hp = 11.7, disc loading = 2.16 lbs / sq. ft.
* Piasecki (United States, 1945) tandem: 600hp/ 3100kg
Gross weight per hp = 11.3, disc loading = 2.5 lbs / sq. ft.
*Breguet (France, 1946) coaxial: 240hp/ 1300kg
Gross weight per hp = 11.9, disc loading = 2.3 lbs / sq. ft.
*Hiller (United States, 1948) single: 178hp/ 950kg
 Gross weight per hp = 11.7, disc loading = 2.1 lbs / sq. ft.
*Mil (USSR, 1949) single: 570hp/ 2250kg
Gross weight per hp = 8.6, disc loading = 2.99 lbs / sq. ft
Karman (United States, 1954) single first gas turbine powered helicopter

   Moving forward in time lets look at the "gross weight per horsepower" and
"disc loading" of very successful modern helicopters the Bell model 204,
205 and 209. I started with the Bell model 204 (HU-1B also named the UH-
1B which was an inter-forces designation) because while serving as a C-130
aircraft crew member in Vietnam never a day went by during 1967 thru 1969
that I didn't see that helicopter model flying in our airspace. So many
helicopters were used in Vietnam that it simply cemented their importance in
war operations, especially when the true combat "Cobra" (Bell model 209)
entered the war in 1969. I can remember the first time I saw one. First you
heard it, then as you strained to see where it was you noticed what looked
like a guy in a flying chair coming towards you at 200 mph. Then with all
weaponry in your face it was overwhelming.
   Sketch 7-1

Stated below are the technical summaries of these model 200 series helicopters:

Bell model 204, HU-1A-United States, operational in 1959; utility/ transport craft; one 860-horsepower turbine engine; rotor blade diameter, 44 feet (13.40 meters); length, 44feet 7inch (13.59 meters); height, 12feet 8inches (3.87 meters); empty weight, 4,520 pounds (2,050 kilograms); gross weight, 8,500 pounds (3,856 kilograms); maximum speed, 148 miles (238 kilometers) per hour; range, about 382 miles (615 kilometers); hovering ceiling IGE, 10,600 feet (3,230 meters); service ceiling, 11,483 feet (3500 meters); capacity, pilots of 2+ 8 troops. Gross weight per horsepower = 9.9lbs/hp; disc loading = 5.6lbs/sq. ft.

Bell model 205,UH-1H-United States, operational in 1963; general purpose craft; one 1,400-horsepower turbine engine; rotor blade diameter, 48 feet (14.63 meters); length, 57 feet (17.40 meters); height, 14feet 6inches (4.42 meters); empty weight, 4,800 pounds (2,177 kilograms); gross weight, 9,500 pounds (4,310kilograms); maximum speed, 138 miles (222 kilometers) per hour; range, about 360 miles (510 kilometers); hovering ceiling IGE 13,600 feet (4,145 meters); service ceiling, 21,980 feet (6,700 meters); capacity, pilots 2 + 14 troops. Gross weight per horsepower = 6.8lbs/hp; disc loading = 5.3lbs/sq. ft.

Sketch 7-2

Huey Cobra, Bell model 209,AH-1T-United States, operational in 1969; two seat combat craft; one 1,800-horsepower turbine engine; rotor blade diameter, 44 feet (13.41 meters); length, 57feet 11inches (16.14 meters); height, 13feet 6inches (4.12 meters); empty weight, 6,415 pounds (2,910 kilograms); gross weight, 10,000 pounds (4,535kilograms); maximum speed, 205 miles (330 kilometers) per hour; range, about 342 miles (550 kilometers); hovering ceiling IGE 12,470 feet (3,800 meters); service ceiling, 10,550 feet (3,215 meters); capacity, 2; armaments, one 20mm cannon, four under wing attachments for rockets and machine guns. Gross weight per horsepower = 5.6 lbs/hp; disc loading = 6.6 lbs/sq. ft.

It should be noted that even though I only list the above helicopters just about everything that still flew in the armed forces inventory, both helicopters and winged aircraft, flew missions in Vietnam, an interesting subject from a historical viewpoint but beyond the scope of this book.

The last two sections in this review give the technical information on a popular civilian helicopter and finally a look at three small helicopters similar in size and scope to the one we want to build.

The civilian helicopter below is the Bell model 400 Twin-Ranger:
Sketch 7-3

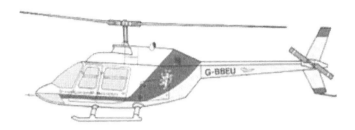

Technical data:

Introduced in 1984; light transport craft; two 420-horsepower turbine engines; rotor blade diameter, 35 feet (10.6 meters); empty weight, 3,075 pounds (1400 kilograms); gross weight, 5500 pounds (2,495 kilograms); maximum speed about 172 miles (278 kilometers) per hour; range, about 450 miles (760 kilometers); ceiling, 19,000 feet (5791 meters); capacity, 2 pilots, 5 passengers. Gross weight per horsepower = 6.5 lbs/hp; disc loading = 5.7 lbs/sq. ft.

Finally, lets look at some small helicopters such as the Hiller YROE-1 Rotorcycle of which the Marines purchased five for evaluation in 1958 and the UltraSport series of helicopter produced in the 1990's.
Sketch 7-4

Hiller YROE-1 Rotorcycle, 1958; ultra light-weight craft; one 43 horsepower engine; rotor blade diameter, 18feet 6inches (5.63 meters); fuselage length 6 feet 11inches (2.10meters); empty weight, 300 pounds (136 kilograms); gross weight, 555 pounds (252 kilograms); maximum speed about 66 miles (106 kilometers) per hour; range, about 166 miles (267 kilometers); service ceiling, 12,000 feet (3,660 meters); capacity, 1 pilot; Gross weight per horsepower = 12.8 lbs/hp; disc loading = 2.0 lbs/sq.ft.

Sketch 7-5

UltraSport 254, ultra light-weight craft; one 55 horsepower engine; rotor blade diameter, 21feet / 6.7 inch chord; tail rotor diameter, 2.6 feet / 2 inch chord; fuselage length (blades folded) 19feet 2inches (5.84 meters); empty weight, 252 pounds (115 kilograms); gross weight, 525 pounds (239 kilograms); maximum speed about 63 miles (101 kilometers) per hour; endurance, about 1.25 hours; service ceiling, 12,000 feet (3,660 meters); capacity, 1 pilot; Gross weight per horsepower = 9.55 lbs/hp; disc loading = 1.5 lbs/sq. ft.

UltraSport 331, single seat experimental craft; one 65 horsepower engine; rotor blade diameter, 21feet / 6.7 inch chord; tail rotor diameter, 2.6 feet / 2 inch chord; fuselage length (blades folded) 19 feet 2inches (5.84 meters); empty weight, 330 pounds (150 kilograms); gross weight, 650 pounds (295 kilograms); maximum speed 104 miles (167 kilometers) per hour; endurance, about 2.5 hours; service ceiling, 12,000 feet (3,660 meters); capacity, 1 pilot; Gross weight per horsepower = 10 lbs/hp; disc loading = 1.88 lbs/sq. ft.

UltraSport 496, two seat experimental craft; one 115 horsepower engine; rotor blade diameter, 23 feet / 6.7 inch chord; tail rotor diameter, 2.6 feet / 2 inch chord; fuselage length 19 feet 9inches (6.02 meters); empty weight, 540 pounds (245 kilograms); gross weight, 1130 pounds (513 kilograms); maximum speed about 104 miles (167 kilometers) per hour; endurance, about 2.5 hours; service ceiling, 12,000 feet (3,660 meters); capacity, 2 pilot; Gross weight per horsepower = 9.83 lbs/hp; disc loading = 2.72 lbs/sq. ft.

In summary gross weight to horsepower and the disc loading for all helicopters referenced is as follows:

- The historical helicopters:
"Gross wt/hp" range, 7.6 to 15 lbs/hp; "disc loading" range, 2 to 3 lbs/sq. ft.
- The military Bell model 200 series helicopters:
"Gross wt/hp" range, 5.6 to 9.8 lbs/hp; "disc loading" range, 5.3-to 6.6 lbs/sq. ft.
- The lightweight helicopters:
"Gross wt/hp" range, 9.6 to 12.8 lbs/hp; "disc loading" ranges, 1.5-to 2.7 lbs/sq. ft.

Chapter 8. The Calculations

The first question in tackling this helicopter-engineering project is where do you begin? I have already stated in Chapter 7 the generalized design relationships for existing helicopters so lets see how close our helicopter compares. In reality it really gets down to the weight you want to move and how fast you want to move it in some vectored direction; whether it is up, down, forward or sideways or more succinctly stated "what type of performance do you want from your helicopter"? Keeping in mind as performance increases so does fuel consumption and that reduces available flight time. "Performance!" what exactly does that mean? For a car a typical performance measure is how fast it can travel a quarter mile. For a good performing car that number is less than 14 seconds. In fact going back to the early 1960's cars such as the XKE Jag and the Ford Cobra were in that category.

A helicopter's performance could be the "rate of climb." For instance an 1100 ft per minute number maybe acceptable for civilian helicopters but some substantially greater "rate of climb" is required for military craft. Another measure could be what is the "maximum altitude" at which the helicopter can operate, realizing that engine power decreases as air density decreases. "Performance", in my definition, is that amount of engine power left over (after getting off the ground to support a hovering state of flight) for use in accelerating in every direction; vertical, translational and angular (banking). Of course this is an over simplified definition because helicopter power requirements vary with speed and that subject will be addressed later in the chapter but I think you get the idea.

If this helicopter is to be classified as an Ultralight its dry weight cannot exceed 254 pounds, excluding safety gear. Its maximum forward speed cannot exceed 50 knots and it is limited to an onboard fuel capacity of only 5 gallons. The Airsport 254 formalized gross weight is rather straightforward. We start with the 254-pound weight constraint placed on the dry weight of an Ultralight helicopter and since only five gallons of fuel maybe carried at takeoff, the gasoline @ 7.4 pounds/ gallon adds 37 pounds to the gross weight. Since Ultralight aircraft are limited to one-person operation a pilot weight of 170 pounds (per FAA) is added. Finally to round out the design an additional 59 to 64 pounds is added to cover safety equipment and instrumentation. The total of these amounts is from 520 to 525 pounds .

8.1 Step 1: Our Helicopter's Design Parameters
Here are some generalizations to be considered:

Rotor rotational speed:
RPM is directly calculated from the rotor radius and established design tip speed, for good autorotation characteristics rotational speed should be high. High rotational speeds reduce the power transmission size due to lower torque transfer.

Rotor blade size:

Rotor blade loading stall limitations establish the blade area (Solidity "S") required. As the blade chord length increases so does the weight of the blade (and subsequently the rotor weight) and the Profile power. The smallest chord that provides an acceptable stall margin is utilized.

For a given gross weight, the disk loading determines the rotor radius. The disk loading is a major factor in determining the power required, particularly the induced power in hover. The disk loading also influences the rotor downwash and the autorotation descent rate. The rotor tip speed is selected largely as a compromise between the effects of compressibility and stall. A high tip speed (tip speeds that approach the speed of sound) increases the advancing-tip Mach number, leading to high profile power, blade loads, vibration, and noise. A low tip speed increases the angle of attack of the retreating blade until limiting profile power, control loads, and vibration due to stall are encountered (blade stall starts at angles greater than 11 degrees and results in loss of lift due to air flow turbulence). There will be only a limited range of acceptable tip speeds, which becomes smaller as the helicopter velocity increases. For a given rotor radius, the tip speed also determines the rotational speed. The rotational speed should be high for good autorotation characteristics and for low torque (and hence low transmission weight). The blade area or solidity is determined by the stall limitations on the rotor blade loading. The limits placed by stall on the blade operating lift coefficient Ct and subsequently (Ct / rotor solidity) works best if the value of [(rotor rotational speed x rotor radius)^2 x Blade Area] is minimized for a given helicopter gross weight. The rotor weight and profile power increase with blade chord, so the smallest blade area that maintains an adequate stall margin is used.

Parameters such as blade twist and plan-form, number of blades, and airfoil section are chosen to optimize the aerodynamic performance of the rotor. The choice will be a compromise for the various operating conditions that must be considered.

To elaborate further on the subject of helicopter weight if we exceed the dry weight of 254 pounds we are no longer classified as an ultralight aircraft and thus are required to have a pilot's license to operate. This defeats the purpose of our objective of being to be able to fly our helicopter without a license. Our helicopter development procedure will deviate from the norm in that we will start with our final weight, as 254 pounds not derive it. Like most helicopters, ours will base our total weight at the rotor centerline to eliminate all destabilizing rotational moments.

The norm for helicopter weight analysis for a new aircraft that has not reached the detailed design stage, is that the component weight estimates can only be obtained by interpolating and extrapolating the trends observed in the weight data for existing vehicles. Preliminary design analyses generally use analytical expressions based on correlation of such weight data.

The fundamental difficulty with such an approach is the reliability of the trends, particularly when it is necessary to extrapolate far beyond existing designs. If this limitation is kept in mind, the formulas expressing empirical weight trends may be successfully employed in preliminary design.

Component weight formulas are typically obtained by correlating weight data from existing designs as a straight line with some parameter Cwt. on a log-log scale, which leads to expressions of the form W = (c1)k(c2) (where c1 and c2 are empirical constants). The parameter Cwt will be a function of those quantities that have a primary influence on the component weight.

As an example, for the helicopter rotor weight, Cwt would depend on at least the rotor radius, tip speed, and blade area. Determining the form of the parameter Cwt. requires a combination of analysis, empirical correlation, and guesswork. There is no unique correlation expression, or even a best one. Consequently there are numerous component weight formulas in use for preliminary design analyses.

The basic difference between our approach and the "norm" is that we will design a specific subcomponent to an actual weight versus scaling up or down an existing component and use whatever weight is generated from the scaling process. Be well aware, from an engineering perspective our approach is way more complex and tedious. Don't expect any first approach to actually satisfy our self-imposed weight limitation criteria.

Having stated the general concept of ratios that are significant in defining our class helicopter lets review the design data of successful helicopters discussed to date so that information can be applied to our situation. Generally speaking the data summarized below gives us a basic idea of those helicopters characteristics.

The following relationships are typical in helicopters built today.

Gross Weight / Engine Horsepower = WT/P=7 to 10#/HP and P~WT^2/3
Gross Weight / Rotor Disc Area range WT/A = 1.5 to 2.5# / ft^2
    WT/A=5 to 10#/ ft^2 disc loading (for military helicopters)
Rotor Disc Area / Engine Horsepower range is from 4.5 to 5 ft^2 / HP
Aspect ratio (span to chord) is typically in the range of 18:1
Rotor Tip Speed for a conservative design is approximately 0.5 cs (where cs
        is the speed of sound at sea level and 70 degrees F equal to 340 m/sec
        (760 mph or 1115.5 ft/sec))
Tail Rotors rotation diameters typically are 1/6 to 1/7 the Main Rotor
        diameter, have a rotational speed somewhere between 1,200 to 2000
        rpm.

    Based upon the above lets further generate the design envelope for our helicopter.

Calculations: General Guidelines for a 525# Helicopter

1.   Engine Horsepower Range Calculation: in this case we have two formulas to use, the first is the generalized formula where Engine Power (P) = P~ WT^2/3 where WT is the gross weight and the second based upon the ratio of Gross Weight to Engine Horsepower (P) over the range stated in the preceding paragraph where WT/P varies from 7~ 10 lbs / hp.

    That being said lets setup the calculations:

Given: WT= 525 lbs,
Formula: P~ 525^2/3; and   P(_) = WT / (lbs /hp)
Solve for P7 the low part of the range and P10 the upper end of the range:
  P7 = 525 lbs. / (7 lbs / hp);
  P10 = 525 lbs. / (10 lbs / hp);
Results:
  P = 65.1 hp, P7 = 75 hp, and P10 = 52.5 hp
Therefore the possible horsepower required ranges from 52.5 to 75

2.   Rotor Disc Area Range Calculation: in similar fashion we will calculate
the values for the Gross Weight to Rotor Disc Area (DA) where the WT/DA
range is equal to 1.5 to 2.5 lbs/sq. ft. to obtain the rotor area and diameter
range we are searching for:
Given: WT = 525 lbs, DA = rotor disc area, D = rotor diameter, R = rotor
radius = D/2
Formula:
  DA1.5 = 525 lbs. / (1.5 lbs./sq.ft.); where D1.5 = (DA1.5 / 0.875)^1/2
  DA2.5 = 525 lbs. / (2.5 lbs./sq.ft.): where D2.5 = (DA2.5 / 0.875)^1/2
  R1.5 = D1.5 / 2
  R2.5 = D2.5 / 2
Solving for DA1.5, D1.5, DA2.5, D2.5, R1.5 and R2.5 we obtain:
Results:
  DA1.5 = 350 sq. ft, D1.5 = 21.11 ft, R1.5 = 10.555 ft.
  DA2.5 = 210 sq. ft, D2.5 = 16.352 ft, R2.5 = 8.176 ft.
Therefore the rotor disc area ranges from:
  210 sq. ft. (radius 8.176 ft.) to 350 sq. ft. (radius 10.555 ft.)

3.   Disc Area to Horsepower Range Calculation: from the above
calculations we have a Disc Area range and a Engine Power range to work
with. Now lets find out how our numbers of DA to P stack up with the DA/P
range given for all helicopters as 4.5 to 5 sq. ft./hp.
Given:
  DA (disc area) range is 210 to 350 sq. ft.
  P (power) range is 52.5 to 75 hp
Solve for all possible DA/P range ratios:
  210 sq.ft./52.5 hp = 4.0 sq.ft./hp;
  210 sq.ft./75 hp = 2.8 sq.ft./hp
  350 sq.ft./52.5 hp = 6.7 sq.ft./hp;
   350 sq. ft./75 hp =4.7 sq.ft./hp
Results:
  Disc Area to Horsepower range is from 2.8 to 6.7 sq.ft./hp

4.   Rotor Blade Chord (width) Range Calculation:
Given: the typical aspect ratio (span/chord) is 18:1 and from calculation #2
  the radius ranges from 8.2 to 10.56 ft.
Formula: Chord = Span (radius)/18

Solve for the rotor blade chord range where
  Chord1.5 = R1.5/18 = .586 ft. (7.04in)
  Chord2.5 = R2.5/18 = .454ft (5.45in.)
Results:
  Chord length range varies between 5.5 to 7 inches

5.   Main Rotor RPM (revolutions per minute) Range Calculation: the rotor blade speed at the blade tip for a conservative design is approximately 0.5 cs (where cs is the speed of sound at sea level and 70 degrees F is equal to 340 m/sec. (760 mph or 1115.5 ft./sec.)), and from basic geometry the distance traveled at the tip of the rotor blade in one revolution is equal to the rotor diameter times pie (~3.14).
Given: Rotor blade tip speed equals one half the speed of sound
Formula
  RPM = 0.5x (1115.5ft./sec.)(60sec./min.) / (3.14 x D)
Solve: for rpm using the diameters determined in calculation #2, where D ranged from 16.352 to 21.11 ft
  RPM1.5= 0.5x (1115.5ft./sec.)(60sec./min.) / (3.14 x 21.11ft/rev)= 504.9 rpm
  RPM2.5= 0.5x (1115.5ft/sec)(60sec/min) / (3.14 x 16.352ft/rev)= 651.7 rpm
Results:
  Main Rotor RPM range is 505 to 652 rev./min.

6.   Tail Rotor Range Calculation: Tail Rotors diameters typically are 1/6 to 1/7 the Main Rotor diameter, have a rotational speed somewhere between 1,200 to 2000 rpm. Using the main rotor diameters from calculation #2, which range find the range of possible tail rotor diameters.
Given:
            D1.5 = 21.11ft.
            D2.5 = 16.35ft.
Formulas:
  TR1.5= D1.5/6= 3.5ft;
  TR1.5= D1.5/7= 3.0ft.
  TR2.5= D2.5/6= 2.7ft.
  TR2.5= D2.5/7= 2.3ft.
Results:
  Tail Rotor diameter ranges from 2.3 to 3.5 ft.

Summarizing the generalized data for our Ultralight Helicopter:
- **Hp range 52.5 to 75**
- **Disc Area Range: 210 sq.ft. (radius 8.2') to 350 sq.ft. (radius 10.6')**
- **Disc Area to Horsepower range is from 2.8 to 6.7 sq.ft. /hp**
- **Chord length range varies between 5.5 to 7 inches**
- **Main Rotor RPM range is 505 to 652 rev./min.**
- **Tail Rotor diameter ranges from 2.3 to 3.5 ft.**

This information gives us the overall design parameters for our helicopter based upon real helicopter build history. Its main significance is to insure that our actual calculated design falls within these guidelines. If it does not, there is a good chance we have a problem somewhere either in our flight testing or in the calculations themselves. With that said, now its time to go to the books and get started crunching numbers.

8.2 Step 2: Define the Rotor Blade Profile

At this point it is critical to know this blade's exact characteristic equations as these define the Lift and Drag relationship that occurs during each rotation of the rotor blade as a function of blade angle. Of course knowing that Lift relationship is rather important since ultimately you have to get off the ground. What is subtler is the Drag term associated with that Lift. Remembering that Drag impacts the power required, the more Drag per revolution the more torque required and the larger the engine needed. The objective in selecting the NACA rotor blade planform (cross sectional profile or shape) is to reduce Drag with respect to the Lift in order to use a smaller engine, which in turn reduces overall weight.

Now that the overall design envelope for our helicopter has been generalized lets discuss the specifics of the cross sectional shape (commonly referred to as the profile or planform) to use for the rotor blades themselves. Going back to Airfoil Theory basically it follows that at each point along the airfoil there is a mathematical relationship that defines the specific velocity and pressure. These equations also give the velocity / pressure profile as the blade Attack of Angle is varied. Resulting from this mathematical capability of being able to define the Lift performance of an airfoil shape, selecting a rotor blade design for a specific flight characteristic is possible.

In addition to what has been discussed the complete list of topics we must review in this rotor blade section is as follows: Lift, Center of Pressure, Tip Velocity, Blade Planform, Blade Twist, and Angle of Attack operational range. In reality if the shape we want doesn't provide the Lift characteristics desired what good is it. This criterion for our blade must be satisfied first. Next must be determined the amount of power required for the rotor shaft to rotate the blades to generate this desired Lift. That calculation requires knowledge of how much Drag on the blade is produced while performing its Lift function.

When the rotor blade rotates to generate Lift, the total Lift produced is actually a summation of all the upward incremental forces generated from the varying velocity pressures produced along the entire rotor blade surface. The smallest forces are generated near the Hub and increase as the radius and / or angular velocity increases. Typically about 2/3's the distance from the Hub is where the Lift Force acts if the summation of all Lift forces along the blade were concentrated. The same can be said about the Drag force.

The "Drag force" multiplied by the" radial distance from the Hub from which it acts" equals the" torque per blade" necessary to turn the rotor shaft. In the simplest terms "total rotor shaft power" equals the "number of rotor blades" multiplied by the "blade torque" times the "rotor RPM" divided by 5250 to give the power necessary in HP. HP= (number of rotor blades) x (torque per blade) x (rotor RPM) / (5250)

A quick look at the difference between an aircraft wing and a helicopter rotor blade operation is useful. In a basic airplane wing, the full length of the leading edge more or less contacts ambient air at the same velocity. A single wing profile can be optimized to provide a rather straightforward Lift / span relationship to satisfy specific design criteria. This applies even though the chord length (the width of the blade) may change near the fuselage for structural reasons.

In contrast a helicopter blade's dynamics are vastly more complex. Its leading edge angular velocity increases as a function of the radial distance from the hub's center of rotation to the blade's leading edge tip. At the blade tip (as a result of the detrimental effects on blade lift performance when tip velocities start to approach the speed of sound) rotor RPM must be limited to a value that performs best at that point for the blade profile and that profiles functional range of "Angle of Attack." Moving closer to the Hub the rotor blade leading edge velocity decreases. This means that whatever planform and Angle of Attack that was optimized at the leading edge tip is no longer valid. Ideally at each leading edge point along the blade radial distance from the hub both the planform and Angle of Attack would have to be optimized for that angular velocity. High performance helicopters do in fact vary the twist, taper and planform of the rotor blade to maximize mission related objectives, ranging from having high speed flight to longer endurance.

Another aspect that makes a helicopter blade more complex is the fact it provides the propulsion for directed flight. This must be accomplished by continuously feathering the blade to a rotational angle whose maximum amplitude corresponds to joystick positioning with every revolution.

Diagram 8.2-1 A Rotor Blade Feathering about it's Neutral Axis

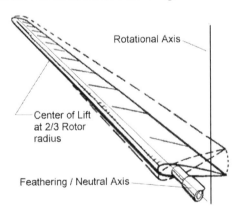

The degree of feathering is determined by the pilot's control of joystick operation. The faster he moves the stick forward the faster the acceleration in that direction restricted only by the helicopter design limits. As can be

expected, this added cyclical blade rotational motion also adds its share of dynamic problems as to how the blade is structurally configured to minimize among other factors, inertial loads.

8.2.1 Rotor Dynamics:

Our earlier discussion of the airfoil was limited to how various airfoil shapes performed in a wind tunnel, how NACA experimental data was collected and graphed and how those graphs were used to rate airfoil efficiency. At that time we determined that for various Reynolds numbers wing efficiency clearly supports an asymmetrical shape for airplanes, and although not mentioned but implied was the fact that the wing was fixed with respect to flight direction.

Continuing our objective to pin down the airfoil NACA type to use in our helicopter we need to review the dynamics of rotor operation in generating lift, take another look at the NACA graphs in Chapter 4.3, make a selection, then determine if that type airfoil can be built with our limited resources. Following this airfoil selection, we will use our NACA airfoil's defining equations to analyze the rotor power required for our helicopters range of operation such as hover, rate of climb, forward speed, etc. As the point-of-blade-contact with the air decreases as you proceed from the rotor-blade-tip to the blade-root so does the air velocity traveling over it. The airfoil we select must not only be efficient at the high Reynolds numbers at the blade tip but also at the low speeds encountered at the root (note that drag increases as the square of the speed).

We start out with this next block of work knowing that the unsymmetrical airfoil is the choice for fixed wing aircraft so why do helicopters tend to use symmetrical profiles? Lets find out by looking once more at the dynamics of helicopter forward flight. Our baseline will be the profile shown below of a symmetrical rotor blade whose Center of Gravity, Center of Pressure and Feathering Axis all align at the quarter chord point.

Diagram 8.2.1-1, Rotor Status Shown: Zero Angle of Attack; Rotor RPM at Lift-off Speed

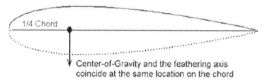

1/4 Chord

Center-of-Gravity and the feathering axis coincide at the same location on the chord

Now lets look at the four-dimensional reference frame used to define the helicopter rotor in forward flight. In the next diagram you see your typical X-Y-Z coordinate system to define a point in three-dimensional space but in addition to this co-ordinate system you must add a time "t" value to find when that specific point was at that location (your forth dimension). This fact complicates helicopter flight and you wind up with terms like "settling in your own downwash", dead-man's curve, tip vortices etc. Basically things go wrong when a rotor blade tip hits the air already churned up by a previous blade rotation. That's where the time "t" factor applies. At 450 rpm, a two-blade rotor will have one of the blades hit the same spot 15 times a second if hovering over a stationary spot.

Diagram 8.2.1-2, Rotor Blade Angle of Attack versus Azimuth Angle

ANGLE-OF-ATTACK VERSUS ANGULAR POSITION

Case 1.HOVER or PURE VERTICAL ASCENT

In reference to our designed helicopter hover should occur when the collective lever sets the angle-of-attack to 8 degrees. An increase to that angle will begin a vertical ascent

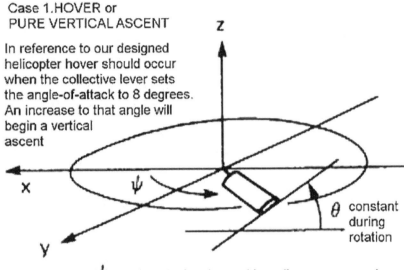

ψ = rotor rotational speed in radians per second.

θ = angle-of-attack (degrees)

PSB

Case 2. FORWARD FLIGHT

For example:
Angle-of-attack at azimuth angle
0 is 8 degrees,
at 90 it is 10 degrees,
at 180 it is 8 degrees
and 270 it is 6 degrees

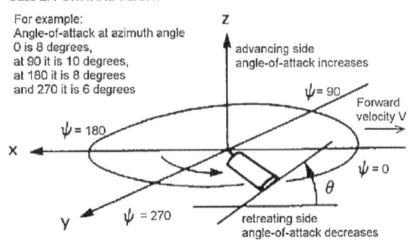

ψ = rotor rotational speed in radians per second.

θ = angle-of-attack (degrees)

PSB

Fifteen times a second we have a rotor blade feathering about its axis. In case 2 of the diagram the feathering motion shown has the range of Angle of Attack noted at the four quadrants of azimuth angle. Those angles being 6, 8 and 10 degrees. I will use those angles as a basis for our comparative analysis of symmetrical versus unsymmetrical profiles to determine what happens when feathering these two types of rotor blades.

8.2.2 Airfoil Analysis- Symmetrical or Asymmetrical

Before I begin let me remind you that in section 4.3.1 these two types of airfoil cross-sections were discussed in great detail and the argument presented supported the use of an Asymmetrical airfoil. The conclusion at the end of that comparison was: for a given angle-of-attack (8 degrees) not only does the asymmetrical wing section (NACA65, 3-618) presented here have a greater lift coefficient of 0.9 versus 0.8 for the symmetrical profile (NACA0012) but it also generates a smaller drag coefficient 0.005 versus 0.008. I also mentioned there was more to airfoil selection of a rotating helicopter blade than just the Lift versus Drag consideration. In this section we get to discuss the next topic, "moments." A "moment" as you already know is a "force" multiplied by a "distance." The force in this scenario is the airfoil Lift generated at the Center-of-Pressure. The distance, typically referred to as the Moment Arm, is measured from the Center-of-Pressure to the feathering axis. What is desirable here is that during blade rotation about the main rotor centerline in which feathering motion takes place no moments are created. Any moments created here generate unwanted disturbing force pulses to the controls. With that said lets start the discussion.

Once stabilized, forward flight requires rotor blades to pivot over a range of Angle of Attacks cyclically about their feathering axis. Lets determine which airfoil shape, symmetrical or asymmetrical, provides the most stable design. The moments corresponding to the Angle of Attack as described in the preceding diagram 6, 8 and 10 degrees will now be evaluated.

• Symmetrical Airfoil

Diagram 8.2-4 is a representative snapshot of a cross section of a symmetrical helicopter rotor blade with an Angle of Attack of 8 degrees at azimuth angles 0, and 180. For the sake of simplicity the theory goes that all the dynamic forces over the surface area in question can be summed as a pressure point acting on an imaginary point on the cord centerline. This point in the vernacular is referred to as the "Center-of-Pressure" and if this point aligns with the feathering axis no moments are created at that point and no forces are transmitted back to the control system which is ideal.

Diagram 8.2.2-1: Center-of-Pressure Location versus 8 degree Angle-of-Attack (azimuth angles 0 and 180 from diagram 8.2.1-1);

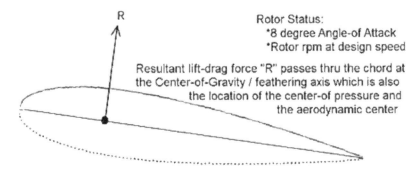

R

Rotor Status:
*8 degree Angle-of Attack
*Rotor rpm at design speed

Resultant lift-drag force "R" passes thru the chord at the Center-of-Gravity / feathering axis which is also the location of the center-of pressure and the aerodynamic center

Center-of-Gravity and feathering axis still coincide on the chord

As we progress along the rotation, Diagram 8.2.2-2 shows the cross section of the rotor blade at the azimuth angle of 90 degrees, the point at which the retreating blade encounters the minimum airflow across its surface and therefore the Angle of Attack increased. The important point here is, in spite of the additional 2 degrees the Center-of-Pressure has remained at the same point as before, on the feathering axis.

Diagram 8.2.2-2: Center-of-Pressure Location at a 10 degree Angle-of-Attack (azimuth angle 90 from diagram 8.2-2)

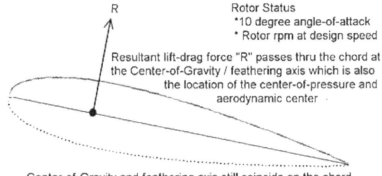

R

Rotor Status
*10 degree angle-of-attack
* Rotor rpm at design speed

Resultant lift-drag force "R" passes thru the chord at the Center-of-Gravity / feathering axis which is also the location of the center-of-pressure and aerodynamic center

Center-of-Gravity and feathering axis still coincide on the chord

As we continue along the rotation (passing azimuth position 180 shown in diagram 8.2.1-1), Diagram 8.2.2-3 shows the cross section of the rotor blade at the azimuth angle of 270 degrees, the point at which the advancing blade encounters the maximum airflow across its surface and therefore the Angle of Attack decreased two degrees. Once again, in spite of the reduction of 2 degrees, the Center-of-Pressure has remained at the same point as before, on the feathering axis.

Diagram 8.2.2-3: Center-of-Pressure at a 6 degree Angle-of-Attack

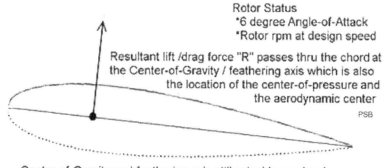

Rotor Status
*6 degree Angle-of-Attack
*Rotor rpm at design speed

Resultant lift /drag force "R" passes thru the chord at
the Center-of-Gravity / feathering axis which is also
the location of the center-of-pressure and
the aerodynamic center
PSB

Center-of-Gravity and feathering axis still coincide on chord

In summary these three symmetrical airfoil diagrams show the ideal case where periodic Angle of Attack change does not move the Center-of-Pressure location with respect to the feathering axis where the aerodynamic center coincides. Shown below is the experimental moment data profile for a symmetrical airfoil.

Graph 8.2.2-1; Symmetrical Airfoil Moment Coefficient

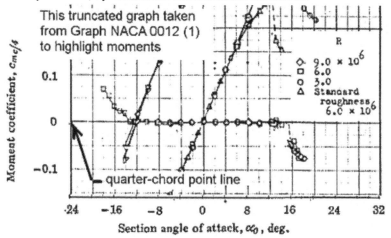

The first fact one notices on this graph is that all the moment coefficient points lay on the horizontal line labeled zero. The second observation is that they remain on that line regardless of Reynolds number (the R in the tabulation chart stands for Reynolds number). Of course what is really conveyed by this graph is the fact that the moment coefficient is zero over the operating Angle of Attack we are interested in. To explain this further, the moment coefficient $C_{mc/4}$ (on vertical axis) represents the "section pitching moment about the quarter-chord point. When that is zero there is no "Disturbing Force" being pulsed back thru the control linkage and that is the desirable effect.

• Asymmetrical Airfoil

Now lets take a look at the dynamics of a generic asymmetrical airfoil (NACA65, 3-618) over the same range of Angle of Attack used above. In this discussion, the three Angle of Attack scenarios 6, 8 and 10 degrees are all superimposed on diagram 8.2.2-4. In this case as the Angle of Attack increases, the resultant lift-drag force increases and when this resultant force passes thru the aerodynamic center (notice that the aerodynamic center is above the airfoil and not on the feathering axis as was the symmetric case) the corresponding Center-of-Pressure shifts progressively forward then retreats rearward as the Angle of Attack decreases. This condition where the Center-of-Pressure line-of-action does not pass thru the feathering axis generates a "pitching moment". The greater the distance from the feathering axis this resultant lift-drag force acts the greater the pitching moment. During the rotation when the Center-of-Pressure location is aft of the feathering axis the moment generated "pitches the rotor-disc up", then as the rotor blade rotation progresses and the Center of Pressure moves forward of the feathering axis "the rotor disc pitches down". This cyclic rotor disc spatial pitch oscillation results in blade flap and undesirable cyclic forces to the control linkage.

Diagram 8.2.2-4: Asymmetrical Airfoil Center-of-Pressure Composite

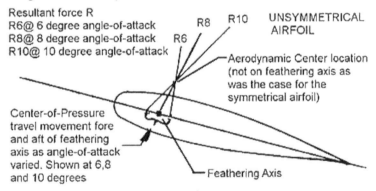

Graph 8.2.2-2 Unsymmetrical Airfoil Moment Coefficients

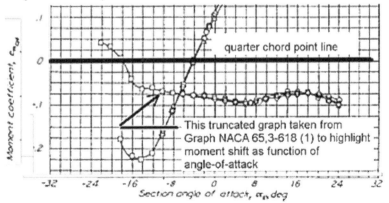

Putting some visuals to the above discussion we have graph 8.2.2-2 where the magnitude of the pitching moment is shown. The control system must neutralize the airfoil pitch forces delivered to the "joystick" in order to maintain system stability and prevent catastrophic damage to the rotor blade. The same resultant force that creates this varying pitching moment transfers that moment via the rotor blade feathering axis pivot shaft to the control linkage and ultimately to the pilots cyclic stick where the feedback is acted on. In those helicopters having a single two bladed rotor this translates to a two-pulse per revolution control system input. Just for the sake of discussion lets analyze this control feedback force from a physics point of view.

In physics a moment (torque) is equal to a "distance" multiplied by a "force" usually stated in the English system as either inch-pounds or foot-pounds depending on the magnitude of the moment (torque) being measured. Power can be described as the moment delivered over a time frame. For example one horsepower is equal to 550 foot-pounds per second. In our asymmetrical airfoil the momentary "distance" is that measured between the actual blade centralized lifting point location and the blade pivot point (feathering axis). The "force" is the actual momentary lifting force of the blade at that centralized location. Multiplication of the described "distance" and "force" gives the instantaneous pitching moment as a function of blade angle of attack feathering position. The greater the distance between the applied lifting force and the feathering pivot point the greater the moment and ultimately the force to your hand while on the cyclic control stick.

What is it going to be, symmetrical or asymmetrical? That is the question. There are analytical ways to answer this but for the time being lets skip the theoretical development of airfoil theory for now (which is what we would need for velocity pressure comparisons) and use the experimental NACA data graphs for these two airfoils. On the plus side an asymmetrical shape allows for greater helicopter forward speeds and is more efficient. After those positives, everything about using an asymmetrical airfoil for our application is down hill. Based upon the factual data presented above who needs to deal with the inherent cyclic moments associated with an asymmetrical planform. Having a 55 knots, (63.3 mph) speed cap for an ultralight class aircraft definitely offsets any need to have an airfoil shape that allows greater speed. In addition to the above just about every small helicopter of the past has used symmetrical airfoils. That's what we will use. Next is to define the actual rotor blade profile.

8.2.3 Airfoil Selection- The Decision

Early on I stated our helicopter would draw heavily from proven designs and what better source of airfoil data to use than that from NACA (topic discussed in subchapter 4.1 "The Airfoil"). Before getting into specific details it would behoove us to understand how the NACA numerical designation is set up. The cambered wing section for each family of NACA blade profiles is developed by mathematically establishing the "x", "y" co-ordinates of the blade geometry as a function of section thickness and percentage of chord length. There is a four-digit, five-digit, and modified four-digit, and five-digit classification for wing sections.

Shown below is the four series NACA 0012 airfoil with its shape defining equations stated below (note for the NACA 0012, t = 0.12), the actual plot follows:

$\pm$ y = (t / 0.20) (0.29690x^0.5 - 0.12600x -0.35160x^2 + 0.28430x^3- 0.10150x^4)

Where "y" represents the distance above and below the chord line, "t" = the maximum thickness expressed as a fraction of the chord and "x" the distance along the chord line starting from the leading edge. The graph shown below +y, as a function of x, is represented as U(x) and –y as L(x) and t equals 0.12, The Leading Edge radius ri is: ri = 1.1019t^2

Graph 8.2.3-1, The NACA 0012 Airfoil

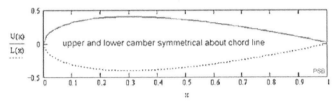

As it turns out, the four-digit section, after the math was completed, was very similar to the efficient Gottingen 398 and the Clark Y wings of World War I. The five-digit section addressed extreme forward positions of maximum camber. For the modified versions the design consists of systematic changes to the thickness distribution and the leading edge radius. NACA also worked to develop a series of wing sections based upon a desired pressure distribution. The 1-series was the first family of high-critical-speed low-drag wing sections. The 2 thru 5-series didn't pan out because the analytical approximation of drag when wing surfaces were rough was not valid therefore their use was discontinued. The 6-series was developed to generate maximum lift and acceptable drag over rough surfaces when operating at critical Mach numbers and the 7-series concentrated on developing a greater amount of laminar flow on the lower surface than the upper.

The four digit series XXXX, which is of most interest to the Airsport 254 rotor design, has the first integer state the maximum mean-line ordinate value in percent of the chord length. The second integer gives the point of maximum camber from the leading edge in tenths of the chord. These two numbers taken together represent the mean line (a line midway between the upper and lower contours of an airfoil profile). The last two integers give the section thickness as a function chord percent. Therefore, an NACA 2412 wing section has 2% camber 4 tenths the chord distance from the leading edge and has a thickness equal to 12% of the chord. For the popular helicopter NACA 0012 profile, the first two digits being zeros indicate there is no camber to the profile therefore the upper and lower wing surfaces are symmetrical.

This explanation of NACA wing section development is extremely brief.

For additional background information I recommend reading more on the subject in the "Theory of Wing Sections" referenced in the bibliography.

We now have to make the decision for the blade planform to be used for our helicopter. The issues to be considered in selecting our airfoil are highlighted below:

• First and foremost the airfoil shape we decide to incorporate in our design must have extensive history in helicopter applications and a substantial amount of research data to support that use.

• The airfoil should have an exceptional lift-drag ratio from blade root to tip over the operational velocity range and Angle of Attack.

• The pitching moment should be minimized in order to promote rotor disc stability implying less aberrant blade motion such as excessive flapping, feathering, lagging, leading and reduced cyclic stick pulsation forces.

• The airfoil must be able to be designed into a structurally sound rotor blade satisfying all flight conditions. The Lift force distribution as a function of rotor RPM and blade Angle of Attack should be such that internal blade localized stresses from root to tip are within safe design limits.

• The designed rotor blade must be able to be purchased or be built in-house without elaborate tooling/ machinery.

When all the smoke cleared researching this topic, it became obvious that the airfoil profile to design our rotor blade around was in fact the NACA 0012. This perfectly symmetrical double convex shape was used on many of the earlier helicopters and as a result volumes of analytical and experimental data was available. In fact this airfoil was used on the 608 hp Sikorsky S-55 helicopter that had a gross weight of over 7000 pounds. This airfoil also had one of the best lift-drag ratios over the operational velocity range and Angle of Attack. Additionally, because of its symmetry its aerodynamic center is always positioned on the median line such that if the feathering axis is designed to coincide with the pivot axis then the pitching moment is virtually eliminated. Its aerodynamic Angle of Attack range 0 ~ 11.8 degrees (0.206 radians) is acceptable (above this value the blade stalls). As far as the structural and construction issues, this has yet to be discussed other than the fact that some of these early helicopters made their rotor blades from wood.

Summary: Based upon experimental data the rotor blade cross sectional planform used will have the NACA 0012 profile, with an aluminum outer surface fabricated from 7075-T6 certified (material shipped with Quality Control documentation) aircraft grade aluminum. Airfoil theory will be the basis for the internal structure design.

8.2.4 The NACA 0012 Air Flow Dynamics:

This particular discussion has to do with air dynamics as it relates to the NACA 0012 profile. Fortunately we have a lot of data on this particular helicopter airfoil shape, both experimental and analytical. I have already used the NACA0012 experimental curves for determining "Lift" versus

"Angle of Attack" and the "Pitching Moment" versus "Angle of Attack." With reference to the analytical, in the Airfoil Theory Section 2.2.3 the formulae developed to generate a "velocity profile as a function of chord length" was systematically presented and the plot for this airfoil given.

My logic for mainly using the experimental data curves in my airfoil selection assessment was easy because everything I needed for comparison was right there in the book. Coming up with a more elegant solution directly from formula did in fact interest me. That's why the presentation of the Airfoil Theory section was included in this book. In the end, both theoretical and experimental results should be the same if all factors are considered. Taking the shortcut and just using the experimental data seemed the fastest way to keep this project moving forward nevertheless the theoretical is required for determining pressure distributions "at" various angle of attacks.

Lets briefly take a look at the analytically developed "velocity profile curve with corresponding data tabulation". Look at the right column of the tabulated data where we have "delta v_a" (which is the increment of local velocity over the surface of the wing associated with angle of attack) and "V" = (velocity of the free stream). If you average out all these incremental "delta v_a" over the chord you can obtain the average differential velocity which in turn is directly proportional to average pressure and ultimately the total Lift of that wing section. If you discard leading and trailing edge effects the amount of incremental Lift generate forward of the 25% chord point and that aft of it are approximately equal for this NACA0012 profile. This is the justification you were looking for when it was stated during the symmetrical versus asymmetrical discussion that the "Center-of-Pressure" forces on the airfoil during Angle of Attack rotation were considered focused at the quarter chord point. The actual graph below (stated earlier as Graph 2.2.2.5-1 & Figure 2.2.4-1) shows the air velocity distribution for this airfoil in a no lift condition, where v = local velocity over the surface.

This graph shows how this velocity pressure is distributed over the chord surface. This is significant when we determine the type of internal structure necessary to handle the design loads and the method required tying it all together. Revisited again are Graph 2.2.2.5-1 and Figure 2.2.4-1

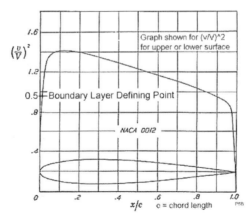

x (per cent c)	y (per cent c)	$(v/V)^2$	v/V	$\Delta v_a/V$
0	0	0	0	1.988
0.5	.....	0.640	0.800	1.475
1.25	1.894	1.010	1.005	1.199
2.5	2.615	1.241	1.114	0.934
5.0	3.555	1.378	1.174	0.685
7.5	4.200	1.402	1.184	0.558
10	4.683	1.411	1.188	0.479
15	5.345	1.411	1.188	0.381
20	5.737	1.399	1.183	0.319
25	5.941	1.378	1.174	0.273
30	6.002	1.350	1.162	0.239
40	5.803	1.288	1.135	0.187
50	5.294	1.228	1.108	0.149
60	4.563	1.166	1.080	0.118
70	3.664	1.109	1.053	0.092
80	2.623	1.044	1.022	0.068
90	1.448	0.956	0.978	0.044
95	0.807	0.906	0.952	0.029
100	0.126	0	0	0
L.E. radius: 1.58 per cent c				

NACA 0012 Basic Thickness Form

8.2.5 NACA0012 Blade Drag Coefficient Calculation:

At this point we have selected the rotor blade NACA0012 profile and along with it received a plethora of related experimental data and analytical theory. All that's missing is the equation that tells us the Drag generated as the blade Angle of Attack (alpha) varied. The operational dynamic characteristics of blade Drag are sometimes given as a mathematical mean. On the other hand, since we have so much information on this airfoil shape we can easily obtain the Drag relationship as an actual function of Angle of Attack in equation form referred to as the "Drag Polar". Actually we need this mathematical relationship because almost all the forth coming equations use it either implicitly or explicitly.

Every wing profile has its own unique equation for calculating the magnitude of Drag associated with a specific blade angle and at that angle there corresponds a coefficient of Lift (see experimental NACA0012 graph 4.3.1-2). Although not apparent now this equation is important when we start analyzing flight power requirements. In those formulae it is our intent to know exact numbers so it makes sense to use the drag equation representative of the blade profile we intend to use specifically that for the NACA 0012. In order to take all the real world three-dimension velocity distribution effects discussed previously into account we refer to the work done by Bailey whose expression for the NACA0012 airfoil "Drag Coefficient as a function of Angle of Attack, (alpha), cdo(alpha)" is:

cdo(alpha) = 0.0084 – 0.0102alpha + 0.384 alpha^2, (alpha in radians)

That being said lets find out what happens as the drag coefficient changes with angle by plotting the equation.

The following is how this is accomplished using Mathcad:

MATHCAD REVIEW: The formula cdo (a) requires the angle "a" to be in radians. The first step is to define the angle range over which we want to plot. The NACA0012 airfoil has an allowable "Angle of Attack" operational upper limit of 11.8 degrees (0.206 radians), so our calculations will start at 0 radians, end at 0.020 and the equation will be recalculated in increments of 0.02 radians. This incrementing portion is called the variable range. Next for reader clarity, in the tabulated data section, I include the degree equivalent next to the radian quantity along with the incremented angle "a". The final equation is the NACA 0012 drag coefficient formula cdo (a). Mathcad looks at the variable "a" in the parenthesis, assigns the first increment in the range to it (in this case zero) then substitutes that value in every point in the right hand expression where "a" is used and calculates the value. This process is then repeated with the next sequential "a" value (0.02) then (0.04), (0.06) … until the ending range value of 0.20 is reached.

Note: The calculations will limit the angle of attack to 0.20 radians (11.5 degrees) just to error on the conservative side.

Calculation 8.2.5-1

A. Defined Variables:

cdo(α) = drag polar (coefficient of drag as a function of angle-of-attack)

α = angle-of-attack (in radians) to get this value in degrees multiply α by (360/2π)

B. Formula for the NACA0012 profile to be calculated along with the degree equivalent:

$$degrees(\alpha) := \alpha \cdot \frac{360}{2\pi}$$

$$cdo(\alpha) := 0.0084 - 0.0102\alpha + 0.384\alpha^2$$

C. Function range:

$$\alpha := 0, 0.02 .. 0.2$$

Shown above is the Mathcad setup for this equation and reflects the methodology to be used thru out the remaining book. Following the Mathcad setup the data points and graph of the NACA drag coefficient "cdo(a)" versus attack angle "a" are presented.

D. Results Tabulated:

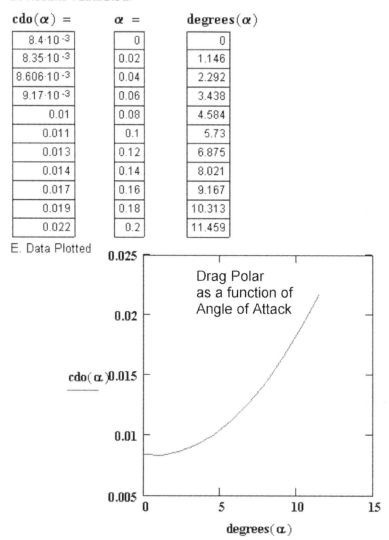

$cdo(\alpha) =$	$\alpha =$	degrees(α)
$8.4 \cdot 10^{-3}$	0	0
$8.35 \cdot 10^{-3}$	0.02	1.146
$8.606 \cdot 10^{-3}$	0.04	2.292
$9.17 \cdot 10^{-3}$	0.06	3.438
0.01	0.08	4.584
0.011	0.1	5.73
0.013	0.12	6.875
0.014	0.14	8.021
0.017	0.16	9.167
0.019	0.18	10.313
0.022	0.2	11.459

E. Data Plotted

Drag Polar as a function of Angle of Attack

degrees(α)

Review of the graph of "Drag Coefficient as a function of Angle of Attack" indicates a parabolic increase in the drag coefficient with increasing angle. In the simple example stated earlier of having your horizontal positioned palm out the window of a forward traveling car at speed, the drag force on your hand also increases as it's attack angle is increased subsequently forcing your hand back toward the rear of the car.

8.2.6 Span-wise Efficiency due to Twist and Taper

Now that the profile has been defined the next question is whether or not some twist to the blade should be added to enhance Lift performance. It stands to reason that if Lift is a function of path length and velocity, then as Leading Edge velocity decreases "path length" should be augmented. This indicates three possibilities:

1. Increase the chord length (widen the blade as you proceed toward the main rotor shaft, keeping the same profile relationship)

2. Increase the Angle of Attack as you progress from the blade tip to the Hub

3. Increase both

Lets look at the numbers associated with these three possibilities:

Twist, 0 degrees;   Taper ratio, 1; Power reduction, 0;
Twist, -8 degrees;   Taper ratio, 1; Power reduction, 2.5%;
Twist, -12 degrees;   Taper ratio, 1; Power reduction, 4.0%;
Twist, 0 degrees;   Taper ratio, 3 ; Power reduction, 2.0%;
Twist, -8 degrees;   Taper ratio, 3 ; Power reduction, 5.5%;
Twist, -12 degrees;   Taper ratio, 3 ; Power reduction, 5.5%;
Twist, ideal degrees; Taper ratio, 1 ; Power reduction, 5.5%;

Review of these numbers reveals the most we could expect to achieve is a 5.5% Lift increase, which requires either ideal twisting or twisting and tapering of the blade. Based upon my experience this is way beyond the capability of most machine / fabrication shops to tackle. In reality due to the cost and complexity of these types of blades they are more often than not exclusive to military or major helicopter manufacturers. We will use a straightforward rotor blade shape with neither twist nor taper. Any attempt to gain additional enhanced performance will be achieved by exercising extreme care in the rotor blade fabrication process. A poorly fabricated rotor blade or one with a rough surface can easily offset all the theoretical efficiencies to be gained by adding twist and taper as stated above.

Based upon the experimental NACA0012 information obtained use of a straight rotor blade will result in a 4% increase in the shaft power requirement. Given the power requirements for our helicopter as stated in section 8.1 being in the range of 50 to 72 the horsepower, those numbers must be increased between to 2 and 2.88 horsepower respectfully. At this point in the design stage lets continue our calculations to determine what the real power numbers will be for our specific helicopter.

Summarizing: The NACA 0012 profile rotor blade will be a non-twisted section having no taper.

8.3 Step 3: Defining the Helicopter's Rotor Hub Type:

Having selected the blade profile it is apparent that the majority of our technical work revolves around the main rotor operation. Logically the next step is the selection of the main rotor type to be used. This main rotor configuration must provide the support to our cantilevered NACA0012 rotor blades, provide sufficient structure to allow cyclic Angle of Attack change, and perform these functions in a dynamically stable manner.

An additional point to discuss is the rotor hub structure requirements. Lets say our helicopter gross weight will be 525 pounds. Since there are two rotor blades each must support a minimum of 262.5 pounds just to Hover. Subsequently if our rotor diameter is 21 ft. rotating at a realistic 440 rpm and each individual rotor blade weighs ~13 pounds, then we apply the formula for centrifugal force, "Fcf" (that's the force trying to tear the rotor blades from the rotor hub when rotating at 440 rpm) where mathematically Fcf = (m x vt) / g x dist), and the terms; m = mass lbm, vt = tangential velocity at the center of mass location (dist) ft./sec, g = conversion factor = 32.2 lbm. x ft. / sec^2 x lbf, and d = 0.5 x radius ft. When the math is completed the calculated centrifugal force just to spin the 13+ pound weight at ~ 440 rpm is over 5500 pounds. That's over two tons per blade trying to either tear loose from its fasteners or cause the rotor hub to self-destruct! This subject will be addressed in detail in the subsystem design section of this book. The point I wanted to make is the rotor hub both dynamically and structurally is the most complex subsystem on our helicopter.

There are three basic types of main rotor assemblies used for helicopters they are:

1."Articulated": this rotor allows each bade two degrees of freedom in the plane of motion, it can "Flap" (rotate up or down about the pivot point) and / or "Drag" (pivot forward or aft). This blade's motion can be individual or collective, additionally the blades can Feather.

2."Semi-Rigid": (commonly referred to as the see-saw type) this rotor utilizes a central hub for both blades to mount to. Feathering and Flapping of the both blades is accomplished together as a unit.

3. "Rigid": As the name implies the blades, hub and rotor drive shaft are fixed (Rigid) to one another allowing only for blade feathering.

The choice is a Semi-Rigid Rotor with an under-slung hub. It was selected for a number of reasons. First of all the "Articulated" rotor seemed too complex a design build project to complete on a limited budget. Also due to the high centrifugal loads on the hub bearings, substantial mass would have to be added making it the heavier of the other types. The main focus when designing a helicopter with a dry weight of only 254 pounds is every effort is concentrated on reducing mass so typically all the heavier options are eliminated. The reader may wonder why the "Rigid" rotor being the most compact was not chosen. The fact is that rotor blades flap due to non-uniform airflow (erratic wind gusts) thru the rotor disc envelope. This results in uneven lift distribution across each of the blades. If the blades could not flap to equalize this condition, as is the case in a rigid system, the moments generated cause the fuselage to pitch or roll or both. Rotor blade flex on this type rotor minimizes this condition but generates higher fluctuating stresses within the blade. Adding additional stress to such an already highly stressed member didn't seem a conservative approach. By default the Semi- Rigid rotor was selected. The overall advantages to this selection are as follows:

a.) Fabrication and maintenance is simplified because individual flapping and dragging hinges are eliminated.

b.) Both blades are mounted in housings supported by a common hub therefore the effects of rotor blade centrifugal forces acting on the hub to rotor shaft bearings are minimized.

c.) Due to the rotor blade mounting described above the blades have inherent rigidity independent of rotational induced centrifugal force.

d.) Positioning the hub pivot point on the rotor drive shaft below the blade feathering axis (called under-slung) minimizes the effects of dynamic unbalance due to the blade mounting geometry.

Summarizing: The Rotor will be an under-slung semi-rigid type.

8.4 Step 4: Defining the Main Rotor Diameter

Before getting into our formulas I have to emphasize once again that there are vastly more complex solution methods out there. Those in-depth processes micro analyze the entire rotor hub and blade system dynamics while operating in simulated real world airflow. Exact relationships defining every position, angle, and angular rotation of every point on a rotor blade surface impacting an air stream including the effects of turbulence, stall, system vibration, instability and airstream flow variations across the rotor blades are included in the analyzed model. Also included are such subtle points as feather bearing friction and the effects of internal bearing component clearances. In order to precisely calculate the rotor power of these elaborately defined models a computer numerical analysis program is normally used by helicopter manufacturers. The computer is necessary due to the complex nature of incorporating all these factors (as a function of blade position in real time) to obtain exact solutions. Numerical computer solutions are way beyond the scope of this book.

Fortunately, we don't have to go the "computer model" number crunching route to obtain solutions. In fact a lot of data has been collected on how this main rotor hub and blade assembly performance deviates from the analytical. It is stated that even using the very complex numerical methods employing "lifting surface theory" or "vortex theory for a finite number of blades" to represent three-dimensional rotor "air flow effects", the results may not be consistently more accurate than using a simple real world conversion factor. There will be discussions in later chapters to delve further into the complexities of enhanced helicopter performance but for now lets just use a blend of theory with some "real world conversion factors" thrown in to simplify our work. This approach is the way this work was performed in the past. The applicable experimental conversion factor numbers used will be explained as we proceed and come from directly from the book on "Helicopter Theory."

We'll continue by restating our objective which is to determine the amount of power required to hover, climb and accelerate our helicopter so an engine can be selected. The starting point of course is at the "main rotor hub and blade assembly".

Going back to Momentum theory lets get some additional definition on the rotor operating parameters relating Rotor Thrust, (T) and the Induced

Velocity, (vh) at the rotor disc. It is stated the Rotor Inflow Ratio (Lambda defined as total air flow thru the disc divided by the rotor speed (radians/sec) x rotor radius) for well designed helicopters ranges from 0.05 to 0.07 with the Induced Velocity, vh between 25 to 35 ft./sec. The actual relationship is Rotor Thrust, (T) is equal to the air mass flux thru the rotor disc (amf) multiplied by the rotor Induced Velocity of that air mass in the far wake (w). Therefore: $T = amf \times w = 2 \times roe \times A \times v^2$, where roe equals air density, "A" is the cross-sectional disc area of flow and "v" the air velocity thru it.

8.4.1 The Equations:

"Rotor disc loading" which is the Thrust, T; divided by the Disc Area, A or T/A is commonly called the "thrust to rotor area ratio" and the

"Power loading" the ratio of the power, P divided by the thrust, T or P/T. This relationship P/T determines the basic characteristics of the helicopter.

In Hover "Momentum Theory" is quite useful to state theoretically the "Induced Power, Pi" per Unit Thrust relationship as:

$$Pi/T = vh = (T/(2A \times roe))1/2$$

Where "roe" is the air density and "vh" the Induced air velocity.

The equation is generated from the physics of fundamental fluid flow. Inspection of this formula reveals that Disc loading (T/A) be low to give a small value of power loading (P/T) which as a result increases Hover efficiency. Looking at the term "$T/(A \times roe)$" (called the effective disc loading) you realize that as air density decreases with altitude and temperature this value increases and correspondingly the power loading. The term "vh" is defined as the induced air velocity flowing down thru the rotor disc. This value must also be small in order to keep the induced power loss low. Although not clearly stated here what must happen is the air, which is accelerated thru the disc, must be accomplished via a small air pressure differential.

"Low disc loading" means the "Solidity" will also be low since the "Solidity, (represented as the Greek letter Sigma)" is the ratio of blade area (Ablade) to rotor area (Arotor), or

Sigma =Ablade / Arotor.

Low disc loading requires high aspect ratio aerodynamically efficient blades that inherently have thin cross sections. The problem with thin cross-sections is at the attachment point to the hub there is insufficient space to add structure to handle the dynamic stresses involved, especially the severe harmonic one cycle loads during forward flight. It is necessary to note that there are limitations to low disc loading.

The "Blade Loading" is the ratio of the thrust to the blade area:

T/Ablade = T/("Sigma"/A),

or represented in coefficient form as the thrust coefficient (Ct); divided by the solidity (Sigma), or Ct / Sigma.

T/Ablade = Ct / Sigma

Given all these defined relationships and the fact that we will not be using a computer analysis software program to tell us what the power requirements will be for our rotor blade / hub assembly how do we modify our basic theoretical formula to obtain power numbers? Putting it another way what is the experimental "fudge" factor to use to get our equations to match real world power numbers? Using the term "fudge" factor doesn't sound very professional so the official aerodynamic term used is called (very elegantly) the "Figure of Merit, (M)". It is the actual term given for measuring rotor hovering efficiency and is defined as the ratio of actual rotor performance to ideal rotor performance. Stated another way it is the minimum theoretical power to hover to the actual power required to hover. Ideally, M=1.

 What are some of the aerodynamic conditions that impact a real rotor? The efficiency losses from a macro point of view are; induced (~60%), profile (~30%) and the non-ideal induced losses (~10%). Looking at it more closely in the real world blade tip loss accounts for a 6~9% reduction of rotor thrust for any specific collective pitch. This reduces the effective lift producing disc area and adds about 3% more to the induced power requirement. Although formulas have been written to predict the Tip Loss factor (B), B typically is simply set to 0.97 (B = 0.97) that compares closely to experimental data.

 Another empirical correction is the Induced Power Factor (k), which adds the rotor losses due to non-uniform flow to the tip losses (B). Supportive experimental data has k related to B as "k = 1.13 / B = 1.17". Looking at a helicopter during hover non-uniform flow increases the induced power by 8- to 12% in addition to 2 to 4% due to tip loss.

 Merging all this together, experimentally it has been determined for a given gross weight the Figure of Merit (M) (for a minimum power loading hovering rotor) is a "constant" and is based upon the empirical power factor k. This relationship to "k" is;

$$M = 2 / (3 k) = 2 / (3 (1.13 / B) = 0.57$$

This means that once the correction factors for non-uniform inflow and tip losses are accounted for we can expect our rotor to be only 57% efficient. In reality current helicopter designs use a value near but a little above this value (M ~ 0.60) at operational gross weight. In my calculations I will use M = 0.55.

 Then again we could use a completely different approach to define the Figure of Merit (M) that being in terms of rotor thrust (T), rotor shaft power (P), rotor disc area (A), air density (roe), the thrust coefficient (Ct) and the Power coefficient (Cp). Where the Figure of Merit =

$$M = \{(T/P)(T / 2 \text{ x "Roe" x A})^{1/2}\} = \{(Ct^{3/2}) / (2^{1/2}) Cp\}.$$

 Another important relationship utilizing the experimentally derived correction factor "k" (the Induced Power Factor) deals with the disc loading for minimum power loading for a real rotor as a function of "Angle of Attack, (a)" and is related as:

$$T/A = [(0.5 \text{ Roe}) (\text{Omega x R})^2] [((\text{Sigma}) (cdo(a) / k)^{2/3}]$$

 Given both the equation for disc loading "T / A" (above) and a value of "M = 0.55", we can now start our preliminary calculations. First up is to determine the value to use for the starting Angle of Attack (a) in our Coefficient of Drag equation "cdo(a)." It is intuitively obvious that no one would design for "Hover" at the extreme end of the Angle of Attack operational range (11.8 degrees) for at least two reasons. First there would be no angle movement left to enable directional flight and secondly at this angle the most drag is required resulting in the most rotor shaft power being required then at any other angle in the range. Lets just use 8 degrees as a starting point in our calculations and fine-tune the results, as we go along. I selected 8 degrees because I liked the round number. At 8 degrees I still had 2 degrees for directional flight and another 1.4 degrees left for a design margin (totals 11.4 degrees or 0.2 radians; I could go up to 11.8 degrees but the 0.2 radians number is easier to use in establishing the calculation range and more conservative).

Before going any further lets look back to section 8.1 Step 1: <u>Our Helicopter's Design Parameters</u> where we defined our helicopter to be within this design envelope:

Table 8.4.1-1. Summarizing the generalized data for our ultralight helicopter:

- **Hp range 52.5 to 75**
- **Disc Area Range: 210 sq. ft. (radius 8.2') to 350 sq. ft. (radius 10.6')**
- **Disc Area to Horsepower range is from 2.8 to 6.7 sq.ft/hp**
- **Chord length range varies between 5.5 to 7 inches**
- **Main Rotor RPM range is 505 to 652 Rotor RPM: 517 to 668**
- **Tail Rotor diameter ranges from 2.3 to 3.5 ft**

Looking back at the "T /A" equation we still don't know what specific "rpm" or "radius" to use, but we do know the limits of each. From the above summary, the rpm range is 505 to 652, and for the radius it is 8.18 to 10.56 ft. Based upon our initial requirements that disc loading be within 1.5 to 2.5 lbf / ft ^2 and tip speed between .3 Mach to .5 Mach the boundary conditions are satisfied by this first guess;

Case 1: Thrust 530 lbf, rotor radius 10 ft, rpm 477.46 (50 Hz), a tip speed of 0.448 Mach and a Disc loading of 1.69. Just to be conservative we will also look at rotor radius varying from 9.5 ft to 10.5 ft and angular speeds between 50~60 Hz.

8.4.2 The Rotor Diameter Calculations:

Going back to our original "T/A" disc loading equation we need to do some algebra. Ideally we want a printout that gives us a "Thrust" value for every "Radius" and "rpm (Hz)" stated (radius 9.5, 10 and 10.5, varying from Hz 50 to 60). Mathcad requires that the equation be expressed in a structure where the left hand side of the equation be in the form "T (Hz, radius)" which translates as the "Thrust value calculated" as a function of every "radius" and every "Hz" in the defined range. Although mentioned previously remember that every term in the equation must be defined prior to stating the equation. The same holds true for the "range" and "calculation increment" for the two variables Hz and radius.

Detailed below is the first pass Mathcad solution to our rotor diameter quest. The result is we have three rotor blade solutions for developing the

desired Thrust:
Calculation 8.4.2-1 Defining the Rotor Parameters- first estimates

Calculate Optimal Rotor Radius

A. Defined Variables:

A = area = πr^2; (ft^2)

NB = number of rotor blades on our helicopter = 2,

cs = speed of sound at 70^0 = 66912 ft / min,

κ = induced power factor = 1.17

ρ = air density @ 70^0 = 0.0752 lb / ft^3

$T(\Omega,r)$ = thrust; lbm ft / sec^2

Ω = rotor angular speed in radians per second;

r = rotor radius; ft

$c(r)$ = r / 18 ; is the chord as a function of radius;

$\sigma(r)$ = rotor disc solidity as a function of radius = NB x c(r) / π r ;

$cdo(\alpha)$ = cdo(0.014); is the NACA 0012 rotor blade drag

coefficient at a desired 8^0 hover angle of attack.

Variables Assigned:

$$c(r) := \frac{r}{18} \qquad cdo(0.14) = 0.014$$

$$A(r) := \pi r^2 \qquad NB := 2$$

$$cs := 66912 \qquad \kappa := 1.17$$

$$\sigma(r) := \frac{NB \cdot c(r)}{\pi \cdot r} \qquad \rho := 0.0752 \cdot \frac{lb}{ft^3}$$

B. Formula to be Calculated:

Now lets determine the optimal rotor radius for minimum power loading during Hover. Using $T/A = (1/2) \rho(\Omega r)^2 (\sigma cdo / \kappa)^{2/3}$ solve for T as a function of various rotor speeds, Ω and rotor radius r. Then from the tabulated data select the best rotor diameter based upon a tip speed not to exceed half the speed of sound. Substituting into this equation and rearranging terms the new equation for Thrust as a function of angular velocity and radius is:

$$T(\Omega,r) := 0.5 \cdot \rho \cdot (\Omega \cdot r)^2 \left(\frac{\frac{NB \cdot c(r)}{\pi \cdot r} \cdot cdo(0.14)}{\kappa} \right)^{\frac{2}{3}} \cdot \pi \cdot r^2$$

Subsequently,the equations for rotor rpm, tip speed (as a percent of the speed of sound) and thrust in pounds force are:

$$\text{rpm}(\Omega) := \frac{60}{2\pi \cdot \text{Hz}} \cdot \Omega \qquad\qquad \text{thrust}(\Omega,r) := \frac{T(\Omega,r)}{\text{lbf}}$$

$$\text{lbf} = 32.174 \frac{\text{lb ft}}{\text{sec}^2}$$

$$\text{TipSpeed}(\Omega,r) := \frac{\dfrac{\text{rpm}(\Omega)}{\text{min}}}{\text{cs} \cdot \dfrac{\text{ft}}{\text{min}}} \cdot \pi \cdot (2 \cdot r)$$

note that T(Ω,r) is in units of (lbm ft / sec^2), To obtain thrust in lbf (pounds force) normally we have to divide by g_c which is in units of (lbm ft / lbf sec^2) unfortunately Mathcad can not do this so the formula shown here gives the value of thrust in lbf but is tabulated without the units given next to the number.

Using the reiterative approach to calculate T as a function of Ω (angular velocity) and r (rotor radius) we will use MathCad to select each rotor radius in the defined range stated below and calculate each value of T for every Ω in the defined range. This calculation will continue for all rotor radii under consideration. Then after all values of T are tabulated we will examine the data to find the radius and rotor rpm that satisfies the desired thrust.

C. Function range:

$$r := 10.5\text{ft}, 10\text{ft} \ .. \ 9.5\text{ft} \qquad\qquad \Omega := 40 \cdot \frac{\text{rad}}{\text{sec}}, 45 \cdot \frac{\text{rad}}{\text{sec}} \ .. \ 55 \cdot \frac{\text{rad}}{\text{sec}}$$

D. Results Tabulated:

r =	A(r) =	c(r) =	σ(r) =
10.5 ft	346.361 ft^2	0.583 ft	0.035
10	314.159	0.556	0.035
9.5	283.529	0.528	0.035

	$T(\Omega,r) =$		TipSpeed(Ω,r)	thrust$(\Omega,r) =$	
r = 10.5 ft	$1.325 \cdot 10^4$	lb ft	0.377	411.975	
	$1.678 \cdot 10^4$	$\dfrac{}{sec^2}$	0.424	521.406	— 429.7rpm
	$2.071 \cdot 10^4$		0.471	643.711	
	$2.506 \cdot 10^4$		0.518	778.891	
r = 10 ft	$1.09 \cdot 10^4$		0.359	338.933	
	$1.38 \cdot 10^4$		0.404	428.962	
	$1.704 \cdot 10^4$		0.448	529.583	— 477.5rpm
	$2.062 \cdot 10^4$		0.493	640.795	
r = 9.5 ft	$8.882 \cdot 10^3$		0.341	276.063	
	$1.124 \cdot 10^4$		0.383	349.392	
	$1.388 \cdot 10^4$		0.426	431.349	
	$1.679 \cdot 10^4$		0.469	521.932	— 525.2rpm

Shown above we have three rotor blade solutions for developing the desired thrust #:
a. 9.5 ft radius. 525.2 rpm, 0.469 cs tip speed, 1.8 disc loading @521.9#
b. 10.0 ft radius. 477.5 rpm, 0.449 cs tip speed, 1.7 disc loading @529.6#
c. 10.5 ft radius. 429.7 rpm, 0.424 cs tip speed, 1.5 disc loading @521.4#

Calculation 8.4.2-2 Determine the "Thrust Coefficient / Solidity" for each of

A. Defined Variables:
C_t = thrust coefficient
A = area = πr^2; ft ^{2}
σ = rotor disc solidity = NB c(r) / πr
ρ = air density @ 70^0 = 0.0752 lb / ft^3
T = thrust; lbm ft / sec^2
Ω = rotor angular speed in radians per second;
r = rotor radius; ft

B. Formula to be Calculated: $C_t := \dfrac{T}{\rho \cdot A \cdot (\Omega \cdot r)^2}$

the three rotor blade cases stated above.
Lets see what the Ct / Sigma values are for these rotor combinations:
a. 9.5 ft radius. 525.2 rpm, 0.469 cs tip speed, 1.8 disc loading @521.9#
b. 10.0 ft radius. 477.5 rpm, 0.449 cs tip speed, 1.7 disc loading @529.6#
c. 10.5 ft radius. 429.7 rpm, 0.424 cs tip speed, 1.5 disc loading @521.4#

Note that the for the r (_), sigma (_), and omega (_) the value in the parenthesis indicates the rotor blade radius under consideration.

In order to use previous Mathcad calculated data we must give revised names for T, A, Ω and r for Mathcad to find the respective new values:

For case a: Variables Assigned

$$r10_5 := 10.5ft \qquad \sigma(10.5) = 0.035 \qquad \Omega10_5 := 45Hz$$

$$A10_5 := 346.361 \cdot ft^2$$

Formula to be
Calculated:
$$Ct10_5 := \frac{T(45Hz, 10.5ft)}{\rho \cdot A10_5 \cdot (\Omega10_5 \cdot r10_5)^2}$$

For case b: Variables Assigned

$$r10 := 10 \cdot ft \qquad \sigma(10) = 0.035 \qquad \Omega10 := 50Hz$$

$$A10 := 314.159 \cdot ft^2$$

Formula to be Calculated
$$Ct10 := \frac{T(50Hz, 10ft)}{\rho \cdot A10 \cdot (\Omega10 \cdot r10)^2}$$

For case c: Variables Assigned

$$r9_5 := 9.5ft \qquad \sigma(9.5) = 0.035 \qquad \Omega9_5 := 55Hz$$

$$A9_5 := 283.529 \cdot ft^2$$

Formula to be Calculated
$$Ct9_5 := \frac{T(55Hz, 9.5ft)}{\rho \cdot A9_5 \cdot (\Omega9_5 \cdot r9_5)^2}$$

C. Results Tabulated

Typically the range for Ct/σ for a moderate rotor Disc Loading is between 0.06 and 0.08, As shown below all three cases are within the upper limit for Ct/ σ = 0.08. Therefore lets calculate which combination requires the least amount of rotor power.

$$\frac{Ct10_5}{\sigma(10.5)} = 0.082 \qquad \frac{Ct10}{\sigma(10)} = 0.082 \qquad \frac{Ct9_5}{\sigma(9.5)} = 0.082$$

We will now determine which one of the following requires the least amount of rotor shaft horsepower to develop the stated thrust.

a. 9.5 ft radius. 525.2 rpm, 0.469 cs tip speed, 1.8 disc loading @521.9#
b. 10.0 ft radius. 477.5 rpm, 0.449 cs tip speed, 1.7 disc loading @529.6#
c. 10.5 ft radius. 429.7 rpm, 0.424 cs tip speed, 1.5 disc loading @521.4#

Our approach for this calculation is to use the simple Figure of Merit relationship.

Calculation 8.4.2-3 Rotor Power for case a, b & c:

A. Defined Variables:

P = power (note the "hp conversion term" is already built into MathCad

M = figure of merit = 0.55

A = area = πr^2

σ = rotor disc solidity

ρ = air density @ 70^0 = 0.0752 lb/ft^3

T = thrust;

$$M = 0.55 \qquad \rho = 0.075 \frac{lb}{ft^3}$$

B. General Formula to be Calculated:

$$M = \frac{T\sqrt{\dfrac{T}{2\cdot\rho\cdot A}}}{P}$$

Then rearranging terms to solve for P, the shaft horsepower, we get this general formula:

$$P = \frac{T\cdot\sqrt{\dfrac{T}{2\cdot\rho\cdot A}}}{M}$$

For case a: Defined Variables (from previous calculations):T (thrust) is evaluated at Ω equal to 55 Hz and a rotor radius of 9.5 feet

$$T(55Hz, 9.5ft) = 1.679 \times 10^4 \frac{lb\,ft}{sec^2} \qquad A9_5 = 283.529\,ft^2$$

Formula

$$P9_5 = \frac{T(55Hz, 9.5ft)}{M\cdot hp}\sqrt{\frac{T(55Hz, 9.5ft)}{2\cdot\rho\cdot A9_5}}$$

For case b: Defined Variables (from previous calculations):T (thrust) is evaluated at Ω equal to 50 Hz and a rotor radius of 10 feet

$$T(50Hz, 10ft) = 1.704 \times 10^4 \frac{lb\,ft}{sec^2} \qquad A10 = 314.159\,ft^2$$

Formula

$$P10 = \frac{T(50Hz, 10ft)}{M\cdot hp}\cdot\sqrt{\frac{T(50Hz, 10ft)}{2\cdot\rho\cdot A10}}$$

For case c: Defined Variables (from previous calculations):T (thrust) is evaluated at Ω equal to 45 Hz and a rotor radius of 10.5 feet

$$T(45Hz, 10.5ft) = 1.678 \times 10^4 \frac{lb\,ft}{sec^2} \qquad A10_5 = 346.361\,ft^2$$

Formula

$$P10_5 := \frac{T(45Hz, 10.5ft)}{M\,hp} \cdot \sqrt{\frac{T(45Hz, 10.5ft)}{2\,\rho\,A10_5}}$$

C. Results Tabulated:

$$P9_5 = 34.239 \qquad P10 = 33.245 \qquad P10_5 = 30.932$$

Summary: The rotor shaft horsepower calculation reveals that case c. 10.5 ft. rotor radius rotating at 429.7 rpm requires only 30.9 horsepower to generate 521.4# of thrust. We will start our detailed calculations around this configuration.

8.4.3 The New Chord Size Calculation:

Now that our rotor blade length is defined lets take a closer look at the actual blade chord we want to use. The above calculations used a fixed aspect ratio of 18 to 1 that calculates as a 7-inch (0.583 ft.) chord for a 10.5 ft radius blade. Hexcell Corporation produced a 170 mm (6.693 inch) chord rotor blade approximately the same length for a similar size helicopter so I decided to adopt that chord size. Based upon this new chord size the calculation shows that a 10.5 ft radius blade rotating at 45 Hz only delivers 506 pounds of thrust. That's not enough so lets rerun the numbers to find the new rpm that generates the 520 to 525 pounds of thrust given the new chord "c" = 0.558 ft (6.693 inches).

Before we proceed we should note that rotor blade tip speed should not be increased to the point air compressibility becomes a factor. Our design rotor rpm should be such that tip speed should be less than half the speed of sound. Determination of the rotor blade chord length is impacted by stall limitations for a specific rotor blade loading. Since this blade operating lift coefficient is limited by retreating blade stall for a given gross weight, it follows that at that gross weight, "Ct / Sigma" (Trust coefficient divided by the rotor solidity) should have the smallest value of {(Omega x R)^2] (blade area)} which will maintain an adequate stall margin. Not only will this reduce rotor blade weight but also reduce the Profile power requirements.

Calculation 8.4.3-1 Determining the Chord Length

A. Defined Variables:

κ = induced power factor = 1.17

ρ = air density @ 70^0 = 0.0752 lb / ft^3

T = thrust; lbm ft / sec^2

Ω = rotor angular speed in radians per second;

r = rotor radius; ft

c = 0.558 ft

σ = rotor disc solidity = NB x c / π r ;

cdo(α) = cdo(0.014); is the NACA 0012 rotor blade drag coefficient at a desired 8 degree hover angle of attack.

$$\sigma := \frac{NB \cdot 0.558ft}{\pi \cdot 10.5ft}$$

B. Formula

$$T(\Omega 1) := 0.5 \cdot \rho \cdot (\Omega 1 \cdot 10.5ft)^2 \cdot \left(\frac{\sigma \cdot cdo(0.14)}{\kappa} \right)^{\frac{2}{3}} \cdot \pi \cdot (10.5ft)^2$$

$$rpm(\Omega 1) := \frac{60}{2\pi \cdot Hz} \cdot \Omega 1 \qquad\qquad thrust(\Omega 1) := \frac{T(\Omega 1)}{lbf}$$

C. Function range:

$$\Omega 1 := 45Hz, 45.2Hz .. 46Hz$$

D. Results

Tabulated:

thrust($\Omega 1$) =	$\Omega 1$ =		rpm($\Omega 1$) =
506.199	45	Hz	429.718
510.708	45.2		431.628
515.238	45.4		433.538
519.787	45.6		435.448
524.357	45.8		437.358
528.946	46		439.268

Based upon this tabulated data for a 10.5-foot rotor blade having a 170mm chord the rotor should be rotating at 437.4 (45.8 Hz) to develop the 524.4 # of thrust desired.

Calculation 8.4.3-2 Recalculation Of Thrust Coefficient "Ct" Using The Revised Chord And RPM Numbers:

A. Defined Variables:

ρ = air density @ 70^0 = 0.0752 lb / ft^3
T (Ω1) = thrust@ Ω1; lbm ft / sec^2
Ct(Ω1) = is the thrust coefficient @ 437.4 (45.8 Hz);
Ω1 = rotor angular speed in radians per second = 45.8 Hz (437.4);
r = rotor radius = now defined as 10.5 ft
A10.5 = area for 10.5 ft radius rotor = π r^2; ft^2

$$r := 10.5\text{ft} \qquad \text{A}10_5 = 346.361\,\text{ft}^2 \qquad \rho = 0.075\,\frac{\text{lb}}{\text{ft}^3}$$

$$\Omega1 := 45.8\text{Hz} \qquad \sigma = 0.034$$

$$\frac{T(45.8\text{Hz})}{\text{lbf}} = 524.357$$

$$T(45.8\text{Hz}) = 1.687 \times 10^4\,\frac{\text{lb ft}}{\text{sec}^2}$$

B. Formula

$$Ct(\Omega1) := \frac{T(\Omega1)}{\rho \cdot \text{A}10_5 \cdot (\Omega1 \cdot r)^2}$$

C. Function range:

$$\Omega1 := 45.8\text{Hz}$$

D. Results Tabulated:

$$\frac{Ct(\Omega1)}{\sigma} = 0.083 \qquad Ct(\Omega1) = 2.801 \times 10^{-3}$$

Looking at the ratio of Ct/σ = to 0.083 this is above the normal range of 0.06 to 0.08 but acceptable.

The Figure of Merit (M) Subject Revisited:
 As stated previously the Figure of Merit (M) is a measure of Aerodynamic efficiency. Well-designed rotors typically have an M of 0.75 to 0.80. In the conservative design stage the M value typically used is M = 0.55 to 0.60. Since we used the lower Figure of Merit (M) value of 0.55 lets just check our numbers using the other generalized formula stated below for this M and compare values.

Calculation 8.4.3-3 Determining the New Figure of Merit:

A. Defined Variables:
k   = additional losses of a real rotor = 1.17
$\Omega 1$ = rotor angular speed in radians per second = 437.4 (45.8 Hz);
$Ct(\Omega 1)$ = is the thrust coefficient @ 437.4 (45.8 Hz)
Cdo = section drag coefficient, subscript o -@profile 0.0065
Cl_{mean} = mean section lift coef = $6Ct/\sigma$ = 6x0.0034/0.0338= 0.604

λh = rotor inflow ratio, subscript h - @hover = $(C_T/2)^{1/2}$

$$Ct(\Omega 1) = 2.801 \times 10^{-3} \qquad \kappa = 1.165 \qquad cdo(0.14) = 0.014$$

$$\lambda h = \sqrt{\frac{Ct(\Omega 1)}{2}} \qquad Clmean = 6 \cdot \frac{Ct(\Omega 1)}{\sigma}$$

B. Formula

$$M = \cfrac{1}{\left[\kappa + \cfrac{0.75}{\lambda h} \left(\cfrac{cdo(0.14)}{Clmean} \right) \right]}$$

C. Results Tabulated: M = 0.571

Now lets look at how the power requirements differ between the two numerically different Figure-of-Merits 0.55 versus 0.57.

Calculation 8.4.3-4 Power required based upon the above Figure of Merit.

A. Defined Variables:
$thrust(\Omega 1)$ = design lift for hover @ 437.4 (45.8 Hz) = 524.357 lbs
$\Omega 1$ = rotor angular speed in radians per second = 45.8 Hz; (437.4)
M = Figure-of-Merit = 0.55 (original assumption); = 0.57 (as calculated above)
A10.5 = disc area for 10.5 foot radius rotor
P = rotor horsepower

B. Formula

$$P = \frac{T}{38M} \cdot \sqrt{\frac{T}{A}} \quad \text{becomes} \quad P = \frac{thrust(\Omega 1)}{38M} \cdot \sqrt{\frac{thrust(\Omega 1) \cdot ft^2}{A10_5}}$$

C. Function range:

$$thrust(\Omega 1) = 524.357 \qquad\qquad M = 0.571$$

D. Results Tabulated:

$$P = 29.711$$

Using M = 0.57 for the Airsport 254 approximately 30 hp is required for hover. Based upon our previous calculations using M = 0.55, 30.95 hp was needed. These values as far as I'm concerned check out favorably.

8.4.4 Helicopter Drag Calculation

Let's compare "helicopter drag versus frontal area" (f / A), and " hub drag versus frontal area" (fhub / A) where "A" is the drag area of helicopter fuselage. Using actual data from existing designs (f / A) ranges from 0.010 to 0.015 for current production helicopters and (fhub / A) ranges from 0.0025 to 0.0050. It is stated that the rotor hub can account for up to 50% of the total parasite drag area. Older design helicopters had (f / A) = 0.025.

CAD Dwg 8.4.4-1

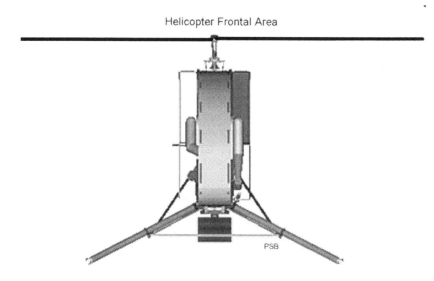

Helicopter Frontal Area

The following is the formula for helicopter drag, where in the summation, Cdi is the drag coefficient for the corresponding Si (component frontal area).

$$f = \sum_{i} CDi \cdot (SI)^2$$

In this equation we know the frontal area (Si) of the helicopter based upon my preliminary sketches to be 8.667ft^2 and for the coefficient of drag (Cdi) I used a value typically used in automobile aerodynamics of 0.035. Since rotor hub drag accounts for up to 50% of the total parasite drag area. I added that to my f/A calculation.

Calculation 8.4.4-1 Calculation of Airsport 254 Fuselage Drag

A. Defined Variables: $Cd := 0.35$

$$Si := \left(\frac{18 \cdot 20 + 12 \cdot 54 + 2 \cdot 2.5 \cdot 48}{144} \right) \cdot ft^2$$

Note this is a summation of all frontal areas with the exception of the hub which is addressed in the formula below

$$A10_5 = 346.361 ft^2$$

$$Si = 8.667 ft^2$$

B. Formula to be Calculated:

$$fdrag := Cd \cdot Si + 0.5 \cdot (Cd \cdot Si)$$

This second term is rotor hub contribution to drag

C. Results Tabulated for f /A:

$$\left(\frac{fdrag}{A10_5} \right) = 0.013$$

Our new f/A summation value becomes 0.013, which is within the typical range stated above as from 0.010 to 0.015. To err on the conservative side lets use f/A = 0.025 in the continuing calculations.

8.5 Step 5: The Helicopter and Energy Balance:

   With the blades determined, the power plant is sized by a rotor power performance analysis that consists primarily of a calculation of the thrust required for the specified objective. At the present time very sophisticated computer programs perform these analysis for the major helicopter manufacturers, however since this methodology is not available to the average person, I basically relied on established helicopter guidelines and then supported the generalized design using the "Energy Balance Method" for performing the analysis. We now come to the core of our power calculations. Very complex stuff indeed! As stated before much of Airfoil Theory was beyond my original plan for this book. Somewhere along the line when I say something is complex it sure helps convey the message when one can visibly see what I am referring to. It is beyond the scope of this book to delve into the derivations and experimental results to obtain the formulas used in this next engineering exercise. The simple fact is a substantial effort was made on my part to sort thru the various theories before I decided to use the Energy Balance Equations in finalizing our helicopter's power requirements. Fortunately the two books, "Helicopter Theory" and "Theory of Wing Sections" contained all of the mathematics needed.

   The "Energy Balance Equations", are defined for each individual energy loss involved with rotor operation. Each loss is named and defined by dynamic function. Factors such as air density, rotor radius, rpm, blade shape drag characteristics, and blade chord are some of the quantities used in formulating these equations. The actual Energy Balance "Method" simply requires totaling all the rotor losses to ultimately get to a helicopter power requirement. The net result is dependent on what you want to find out. For example, if maximum altitude at hover is desired obviously the equations have to be a function of air density since both air density and engine power output decrease with increasing altitude. Subjects such as this, along with maximum potential speed, rate of climb, etc. will be reviewed but for now lets just focus our effort on finding out "Helicopter Power as a function of Forward Speed (during flight at sea level)" for this helicopter. What will not be included is any disturbing force that either distorts the rotor blade shape due to dynamic structural loads or flight condition causing any rotor blade to take an aberrant rotor blade path. Note this type of detail is in fact included in those computer programs mentioned previously.

Looking at the total rotor power: in Hover, it is the summation of the Induced power and Profile power, in directional flight we add to those Parasite power (associated with helicopter drag) and Climb power when ascending. We can now add the four helicopter losses; Induced, Profile, Parasite, and Climb directly to arrive at total power. Another approach would be to total these factors by totalizing the power coefficients. For example let the System power coefficient, Cp = Induced power coefficient + Profile power coefficient + Parasite power coefficient + Climb power coefficient.

Using Power coefficients is the way we will proceed. We begin by restating the definition of each of the four power terms we will be mathematically developing Energy Balance Equations for:

Induced power is the power required to get the air moving thru the rotor or more specifically to add downward momentum to that mass of air enabling Lift to occur. General momentum theory only looks at the macro view of uniform airflow to develop the induced power. What is needed is some way to determine exactly how this Induced power loss is generated at the rotating blade surface as a function of its length. This detail incremental interaction between rotor blade length and a non-uniform air mass inflow distribution is in fact very complex to model. Tackling this expanded dynamic scope the experts turned to a variation of both Blade Element Theory and Momentum Theory. The result was the derivation of the "differential form" of the Momentum Theory equation. This equation incorporated such factors as "blade twist, pitch, drag and chord as a function of radius, rotor tip losses, and non-uniform flow losses in the calculation of rotor thrust and power. In spite of this rigorous attempt to include all the possible variables the effects of the rotor vortex wake was omitted due to the difficulty in defining its complex boundary conditions.

Profile power: If Induced power applies to getting the air mass to move then it seems logical to define Profile power as the energy or power needed to keep the rotor blade in motion to maintain that air momentum. This basically means the power necessary to sustain the required rpm while overcoming the viscous drag on the blade. Once again a "differential form" of the applicable Momentum Theory equation is used. This means incorporating all the same variables as mentioned above such as using the integrated drag coefficient over the entire span of our specific blade profile and due to the sophistication of the dynamic complexities at the rotor blade tip, Prandtl's Tip Loss Factor will be used as a realistic approximate solution. Probably you would intuitively suspect the rotor Profile power to decrease with increased altitude, ambient air temperature or both as air's density decreases and it does as the Drag forces on the blade are less. In order to achieve the same flight characteristics a greater effective disc loading is necessary requiring more Profile power so that when summed the "Total Power" is greater than before.

   Parasite power: These are those power losses incurred while in directional flight, basically attributed to air resistance. As helicopter speed increases and / or progressively higher loading of the rotor blades takes place the increased airflow across the extreme leading edge surface of the blade will reach a point that air compressibility factors and stall will become an issue. To address these Parasite power effects the numerical solution requires the blade angle of attack distribution be a function of forward velocity. Noting that Parasite power increases proportionally to the velocity cubed, V^3.

   Climb power: This is that amount of energy required to move the helicopter from one position to another having a vertical position of greater potential energy.

   Miscellaneous information:
   Our helicopter will be what is called a balanced design. A 'Balanced" helicopter design is one in which the power required to Hover is about equal to that required at maximum speed. Since a rotor "in Hover" represents between 80 to 85% of the total engine power utilized our effort will be to design a Main Rotor that minimizes the power consumed in this mode of operation.
   We have already discussed in detail the inefficiencies of the rotor during my explanation of the terms "Figure of Merit, (M)", "Induced Power Factor (k)" and "Tip Losses, (Beta)." Just for review the power losses of the rotor "in Hover" are:
a) Induced power 60% (the energy dissipated in the rotor wake by imparting a downward momentum to the air from which the Lift reaction on the rotor is obtained)
b) Profile power 30% (the energy dissipated by the viscous drag of the blade)
c) Non-uniform inflow 5~ 7% (applies to a varying ambient airflow entering the rotor disc)
d) Swirl in the wake < 1% (applies to air leaving the control space)
e) Tip losses 2%~4%

What we haven't discussed is the additional power losses associated with the system, so here they are:

f) Engine & Transmission power loss 6% to 9% of total power generated

g) Tail rotor power requirement 7% to 9%

h) Aerodynamic Rotor / Fuselage Drag approx. 2%

System losses reduce available power to the main rotor to 80~ 85% of the rated engine power output.

Looking at "Power distribution as a function of Speed" we find in Hover, zero speed, Induced power dominates. Then as helicopter speed increases Induced power quickly drops off with a slight increase in Profile power.

8.5.1 Energy Balance Equations:

We will now setup the Energy Balance equations. I will present the equations in their integral form first then simplify them using accepted approximations. The first pass will be to develop the equations in the form of the power coefficients as a function of "mu" the rotor advance ratio. The resulting equations must then be transformed to give us rotor power as a function of forward velocity since that's the number we really want to know. The conversion of the power coefficient equations to the more useful form is derived next.

We begin with the explanation of terms:

The power coefficients: Cpi, induced power; Cpo, profile power; Cpp, parasite power and Cpc, climb power.

A. Defined Variables:

λi = induced inflow ratio (v / ΩR)

μ = rotor advance ratio ($V\cos\alpha$ / ΩR)

Ω = rotor rotational speed (rad/sec)

σ = rotor solidity = Nc / πR

ρ = air density

A = rotor disc area (πR^2)

cd = section drag coefficient

Ct = thrust coefficient

D = helicopter drag force

R = rotor radius

ur = radial air velocity of blade section

ut = air velocity of blade section tangent to the disc plane

V = helicopter velocity relative to the air

Vc = climb velocity

W = helicopter weight

α = blade section angle of attack

$\varkappa$ = inflow correction factor = 1.2

N = number of rotor blades

v = rotor induced velocity

T = rotor thrust

c = blade chord

f = helicopter equivalent drag area including hub $(D / 0.5 \rho V)$ = to .075 A

f /A = helicopter drag = 0.025 for older production helicopters

η = is the power train efficiency loss = 0.8

B. Formulae to be Calculated: $Cp = Cpi + Cpo + Cpp + Cpc$ ■

Now lets start simplifying the power coefficient terms: ■

1. $Cpi = \int_{rR}^{B} \lambda i \, dCt$

For Cpi (induced power coefficient in forward flight) stated above we can approximate λi as $Ct / 2\mu$ for all but the lowest flight speeds. Furthermore this assumption is independent of climb or descent speed. If we include the correction factor κ for the additional inflow losses due to nonuniform inflow and tip vortices we obtain: $Cpi = \lambda i \, Ct = \kappa \, Ct^2 / 2\mu$ the corresponding power term is $Pi = \kappa T^2 / \rho 2AV$

For hover Cpih (induced power coefficient where forward speed = 0) instead of using $Cpi = \kappa \, Ct^2 / 2\mu$ we must use the following expression:

$Cpih = \kappa Ct^2 / 2(\mu^2 + \lambda^2)^{0.5}$

2. $Cpo = \int_{0}^{1} \frac{\sigma \cdot cd}{2} \cdot \left(ut^2 + ur^2\right)^{\frac{3}{2}} dr$

For Cpo (profile power loss coefficient) stated above as: (which included losses due to reverse and radial flow inaddition to the forces due to radial drag) could be simplified by using a mean section drag coeficient for rotor advance ratios up to 0.5μ. This equation is:

$Cpo = (\sigma \, cdo / 8)(1 + 4.6 \, \mu^2)$

3.
$$Cpp := \frac{D \cdot V}{\rho \cdot A \cdot (\Omega \cdot R)^3}$$

For Cpp (parasite power loss coefficient) stated above : if we set D equal to an equivalent helicopter drag area "f" then $D = 0.5\rho V^2 f$ and the equation becomes:

$$Cpp = 0.5(V/\Omega R)^3(f/A) \text{ approximately} = 0.5 \mu^3 f/A$$

4.
$$Cpc := \frac{Vc \cdot W}{\rho \cdot A \cdot (\Omega \cdot R)^4}$$

For Cpc (climb power loss coefficient) stated above :
if the climb velocity Vc then the climb inflow ratio λc is equal to Vc / ΩR. Substituting into the original equation we obtain:

$$Cpc = \lambda c(W / \rho A (\Omega R)^3 \text{ approximately} = \lambda c \, Ct$$

Now lets restate power coefficient summation in terms of the original formula:

$$Cp := \int_{rR}^{B} \lambda i \, dCt + \int_{0}^{1} \frac{\sigma \cdot cd}{2} \left(ut^2 + ur^2\right)^{\frac{3}{2}} dr + \frac{D \cdot V}{\rho \cdot A \cdot (\Omega \cdot R)^3} + \frac{Vc \cdot W}{\rho \cdot A \cdot (\Omega \cdot R)^4}$$

This equation is now restated below as a conditional simplified algebraic equivalent in two forms, one for hover, and one valid up to 0.5 μ. (Note that for helicopter speeds above 0.5 μ effects of blade stall and leading edge air compressiblity presents significant losses). For Hover the simplified power coefficient equation is:

$$Cp(\mu) := \frac{\kappa \cdot Ct^2}{2 \cdot \left(\mu^2 + \lambda^2\right)^{0.5}} + \sigma \cdot \frac{cdo}{8}\left(1 + 4.6 \cdot \mu^2\right) + \lambda_c Ct + \frac{1}{2}\left(\frac{f}{A}\right)\cdot \mu^3$$

for forward flight the power coefficient equation is:

$$Cp(\mu) := \kappa \cdot \frac{Ct^2}{2\mu} + \sigma \cdot \frac{cdo}{8}\left(1 + 4.6\mu^2\right) + \lambda_c Ct + \frac{1}{2}\left(\frac{f}{A}\right)\cdot \mu^3$$

Now that we have established the power coefficient equation using the energy balance method. We notice that the equation is stated as a function of μ, the rotor advance ratio, what would be more advantageous to us is to have this equation changed to represent rotor power as a function of forward velocity. We can do this by using the following relationships, then substituting them back into the power coefficient equation.

rotor advance
ratio is defined as: $$\mu := \frac{V \cdot \cos(\alpha)}{\Omega \cdot R}$$

$$Cp(V) := \kappa \cdot \frac{Ct^2}{2\left(\dfrac{V \cdot \cos(\alpha)}{\Omega \cdot R}\right)} +$$

$$\sigma \cdot \frac{cdo}{8}\left[1 + 4.6\left(\frac{V \cdot \cos(\alpha)}{\Omega \cdot R}\right)^2\right] +$$

$$\lambda \cdot Ct + \frac{1}{2}\left(\frac{f}{A}\right)\left(\frac{V \cdot \cos(\alpha)}{\Omega \cdot R}\right)^3$$

The following Power Coefficient equations are setup in two ways one for Hover (Vo), which is applicable from 0 to 50 ft./sec (34 mph), and the second for Forward speeds (V), above 50 ft./sec.

Furthermore by definition Cp equals:

$$Cp := \frac{P}{\rho \cdot A \cdot (\Omega \cdot r)^3}$$

Then:

$$P := Cp \left[\rho \cdot A \cdot (\Omega \cdot r)^3 \right]$$

Substituting for μ to obtain Cp as a function of forward velocity V (ft/sec), we obtain the following relationships for these power coefficients:

Induced power (hover)

$$Pih(V) := \left[\rho \cdot A \cdot (\Omega \cdot r)^3 \right] \cdot \left[\frac{\kappa \cdot Ct^2}{2 \cdot \left[\left(\frac{V \cdot \cos(\alpha)}{\Omega \cdot R} \right)^2 + \lambda^2 \right]^{0.5}} \right]$$

Induced power : (forward velocity)

$$Pi(V) := \left[\rho \cdot A \cdot (\Omega \cdot r)^3 \right] \cdot \left[\kappa \cdot \frac{Ct^2}{2 \left(\frac{V \cdot \cos(\alpha)}{\Omega \cdot R} \right)} \right]$$

Profile power:

$$Po(V) := \left[\rho \cdot A \cdot (\Omega \cdot r)^3 \right] \cdot \left[\sigma \cdot \frac{cdo}{8} \cdot \left[1 + 4.6 \left(\frac{V \cdot \cos(\alpha)}{\Omega \cdot R} \right)^2 \right] \right]$$

Climb power: $Pc(V) := \left[\rho \cdot A \cdot (\Omega \cdot r)^3 \right] \cdot (\lambda \cdot Ct)$

Parasite power:

$$Pp(V) := \left[\rho \cdot A \cdot (\Omega \cdot r)^3 \right] \cdot \left[\frac{1}{2} \cdot \left(\frac{f}{A} \right) \left(\frac{V \cdot \cos(\alpha)}{\Omega \cdot R} \right)^3 \right]$$

$$Phover(Vo) := \frac{Pih(Vo) + Po(Vo) + Pp(Vo)}{\eta}$$

$$Pfwd(V) := \frac{(Pi(V) + Po(V) + Pp(V))}{\eta}$$

$\eta := 0.8$
Drivetrain efficiency factor

8.5.2 Helicopter Power versus Forward Velocity:

In part A, the variables are defined that we will be using. I do this because some variables will be set at different values depending on what we are trying to determine. For example air density for this set of equations is set at a value used for flight close to the ground or more specifically at sea level and at 70 degrees F. Having these variables defined for each calculation section will aid in quicker understanding of what is being calculated. Part B, gives the formula to be calculated in its simplest form. For this initial use of the Energy Balance Method there is an extensive path followed to go from the complex to simplified equations. Take the time to completely understand the development of the "Thrust" equation since it is used consistently. Buried within it is also the "specific blade angle of attack drag formula". Part C, is the function range. For example the "velocity range" to be calculated is from "0 to 63 mph (ultralight limit 55 knots = 63.293 mph = 92.83 ft/sec)." In Part D, the results are tabulated. How many times have you come across a formula with absolutely no idea what the answer should be? In some cases if you are lucky you can purchase an answer key but not always. I have avoided all that inconvenience by giving the results of the calculated formula. In Part E, the data is plotted. Please note these preliminary calculations use a rotor rpm of 45 Hz and not the revised 45.8 Hz.

Calculation 8.5.2-1 Helicopter Power versus Forward Velocity

A. Defined Variables: $\quad \eta := 0.8 \qquad N := 2 \qquad \Omega 1 = 45.8\,\text{Hz}$

$$\mu := V \cdot \frac{\cos(\alpha)}{\Omega \cdot r} \qquad V := 0.5\mu \cdot \Omega \cdot \frac{r}{\cos(\alpha)}$$

$$A := 346.36\text{ft}^2 \qquad f := 0.025\,A \qquad r = 10.5\,\text{ft} \qquad \sigma = 0.034$$

$$c := \frac{170\cdot\text{ft}}{25.4\cdot 12} \qquad \rho = 0.075\,\frac{\text{lb}}{\text{ft}^3} \qquad \lambda := \kappa \cdot \sqrt{\frac{Ct(\Omega 1)}{2}}$$

$$\kappa = 1.165 \qquad \alpha := 0.14 \qquad c = 0.558\,\text{ft} \qquad Vc := 8.33\,\frac{\text{ft}}{\text{sec}}$$

$$cdo(0.14) = 0.014 \qquad\qquad mph(Vo) := Vo \cdot \frac{3600\cdot\text{sec}}{5280\cdot\text{ft}}$$

$$Ct(\Omega 1) = 2.801 \times 10^{-3} \qquad \lambda c := \frac{Vc}{\Omega 1 \cdot r}$$

B. Formula:

During our initial preliminary calculations we set the angle of attack "α" for hover at 8 degrees which corresponds in radians to 0.14. Using this value for α and our selected rotor blade combo of 10.5ft radius rotating at 45Hz (429.7 rpm) we will now calculate the maximum forward speed these two equations are valid for.

$$\text{Phover}(Vo) := \frac{\text{Pih}(Vo) + \text{Po}(Vo) + \text{Pp}(Vo)}{\eta}$$

$$\text{Pfwd}(V) := \frac{(\text{Pi}(V) + \text{Po}(V) + \text{Pp}(V))}{\eta}$$

$$\text{constant} := \frac{\rho}{\text{hp}} \cdot A \cdot (\Omega 1 \cdot r)^3$$

$$\text{constant} = 1.637 \times 10^5$$

$$\text{Pih}(Vo) := (\text{constant}) \cdot \left[\frac{\kappa \cdot Ct(\Omega 1)^2}{2 \cdot \left[\left(\frac{Vo \cdot \cos(\alpha)}{\Omega 1 \cdot r} \right)^2 + \lambda^2 \right]^{0.5}} \right]$$

$$\text{Po}(Vo) := (\text{constant}) \cdot \left[\sigma \cdot \frac{cdo(0.14)}{8} \cdot \left[1 + 4.6 \cdot \left(\frac{Vo \cdot \cos(\alpha)}{\Omega 1 \cdot r} \right)^2 \right] \right]$$

$$\text{Pp}(Vo) := (\text{constant}) \cdot \left[\frac{1}{2} \left(\frac{f}{A} \right) \left(\frac{Vo \cdot \cos(\alpha)}{\Omega 1 \cdot r} \right)^3 \right]$$

$$\text{Pi}(V) := (\text{constant}) \cdot \left[\kappa \cdot \frac{Ct(\Omega 1)^2}{2 \cdot \left(\frac{V \cdot \cos(\alpha)}{\Omega 1 \cdot r} \right)} \right]$$

$$Po(V) := (\text{constant}) \cdot \left[\sigma \cdot \frac{cdo(0.14)}{8} \cdot \left[1 + 4.6 \cdot \left(\frac{V \cdot \cos(\alpha)}{\Omega l \cdot r} \right)^2 \right] \right]$$

$$Pc(V) := (\text{constant}) \cdot \left(\lambda c \cdot Ctl0_S \right)$$

$$Pp(V) := (\text{constant}) \cdot \left[\frac{1}{2} \cdot \left(\frac{f}{A} \right) \cdot \left(\frac{V \cdot \cos(\alpha)}{\Omega l \cdot r} \right)^3 \right]$$

$$Phover(Vo) := \frac{Pih(Vo) + Po(Vo) + Pp(Vo)}{\eta}$$

$$Pfwd(V) := \frac{(Pi(V) + Po(V) + Pp(V))}{\eta}$$

C. Function range:

$$Vo := 0 \, \frac{ft}{\sec}, 10 \, \frac{ft}{\sec} \ldots 50 \, \frac{ft}{\sec} \qquad V := 55 \, \frac{ft}{\sec}, 65 \, \frac{ft}{\sec} \ldots 95 \, \frac{ft}{\sec}$$

D. Results Tabulated:

$mph(Vo) =$	$Phover(Vo)$	$Pih(Vo) =$	$Po(Vo) =$	$Pp(Vo) =$
0	33.992	17.157	10.037	0
6.818	31.985	15.513	10.056	0.018
13.636	28.413	12.472	10.115	0.143
20.455	25.735	9.893	10.213	0.482
27.273	24.399	8.026	10.35	1.143
34.091	24.311	6.69	10.526	2.233

$mph(V) =$	$Pfwd(V) =$	$Pi(V) =$	$Po(V) =$	$Pp(V) =$
37.5	25.257	6.604	10.629	2.972
44.318	26.698	5.588	10.864	4.906
51.136	29.397	4.843	11.138	7.536
57.955	33.369	4.273	11.451	10.971
64.773	38.679	3.824	11.803	15.316

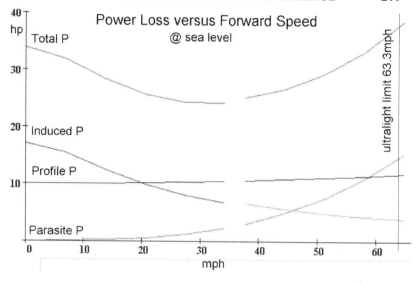

Least amount of rotor power occurs at approximately 30 mph.

8.6 Step 6: Performance Calculations

8.6.1 The Engine

Now that we have the "Energy Balance Equations" in place we add the engine power available and calculate the possible flight capabilities of our helicopter. We need to look at the engine to determine it's power, torque and fuel consumption as a function of engine RPM. Then with that knowledge we can calculate "Range and Endurance" (section 8.6.2), "Maximum Speed" (section 8.6.3), "Rate of Climb" (section 8.6.4), "Engine Horsepower versus Angle of Attack" (section 8.6.5), "Maximum Altitude" (section 8.6.6), and finally "Flight Operation at 100 degrees F".

Before getting to the next series of calculations lets talk engines. At this time in our preliminary work it is critical to know what the weights are of engines this size. Always we must keep in mind that this ultralight class has a 254-pound maximum dry weight limitation. Any engine weighing over 70 pounds would more than likely doom this ultralight project to failure. We need approximately 35 horsepower in order to hover. Consequently, I investigated all engines currently manufactured with that power capability; two stroke, four stroke, diesel, gas turbine and rotary to find the one with the best power to weight ratio. The outcome of this laborious search revealed: gas turbines and rotarys are too expensive. A Rotax FAA certified four stroke is too heavy and also pricey. The two stroke Rotax and Hirth aircraft engines were strong possibilities and were preferred because they were light (both around 70 pounds) and air-cooled. The downside was that Rotax and Hirth retailers, service personnel and owners were fiercely loyal and as a result quick to expound a detrimental viewpoint of the competitor's product.

Ultimately, because of this negativity and their cost of thousands more than I was willing to spend at this point in my helicopter development I decided not to use either of them.

As it turned out I committed to using a 70 pound three cylinder 2-Stroke International engine whose manufacturing facility was only a short distance away. Unfortunately it was a mistake to select this small company as an engine source since they went bankrupt and all the work I put into the helicopter using their engine plans was for naught.

Concurrently, while working on the design I was at a Kawasaki Jet Ski dealership and met Steve, the service manager. He was a true "motor-head" working only to pay for his toys, motorcycles, jet skis, jet boat, jacked up 4 x 4 etc. He got me interested in Kawasaki engines. I wouldn't say we were friends but the more I showed him the analytical stuff on helicopter function the more he would go out of his way to help me with establishing engine component weights. When time permitted we would weigh various parts of engines being worked on in the repair shop or in stock. It soon became apparent the Kawasaki engine could be used to replace the 2-Stroke International engine. Another benefit was the Kawasaki came with an electric start. As luck would have it I was able to purchase two Jet Skis with those 70-pound 650 Kawasaki engines in them for only $500 each. Now that the engines were in my possession the analytical work and the build could continue.

Years have gone by since this engine selection process took place and recently I came across a company called Compact Radial Engines that offer an air-cooled MZ202 engine that could be used to replace the water-cooled Kawasaki. Coincidently, with all this automotive hybrid work in process now it's only a matter of time when a small lightweight powerful 4-stroke FAA certified engine may be available for this helicopter.

That's the future, right now we have the Kawasaki 650 engine and the performance charts we need for the next set of calculations are shown next:

Graph 8.6.1-1

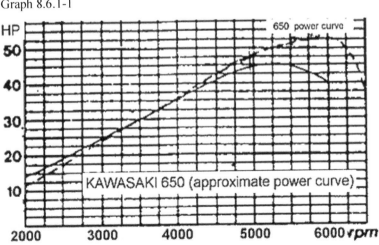

Graph 8.6.1-2

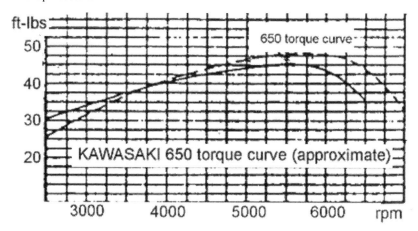

8.6.2 Range and Endurance:

Graph 8.6.2-1

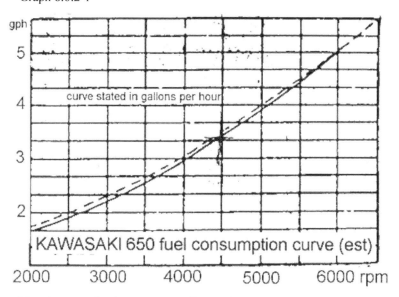

Now is as good a time as any to discuss these two terms since both "range" (like how far a distance can you fly) and "endurance" (even if you fly nowhere, how long can you stay airborne) are both a function of engine fuel consumption. The entire previous "Energy Equation" section generated the relationship between helicopter speed and power. If we know the engine's specific fuel consumption rate to power output, both range and endurance are easily calculated. That is true if you know the specific characteristics of your flight plan including your intended speed, altitude, and gross weight.

In our particular case we are dealing with an ultralight class of aircraft limited to 5 gallons of fuel, a maximum speed of 50 knots, and an engine that uses 5 gallons per hour of fuel at maximum power (52 hp.@ 6000 rpm). You could almost guess and be close to correct that you are not going to spend more than one hour in the air if you intend to have some fuel left in the tank when you land. That by the way is a good idea! Note: Based upon the power numbers generated in the previous section, Hover requires approximately 35 hp. Our calculations for fuel consumption are setup as a "ratio of calculated power versus 50 hp (hp at maximum torque of engine)" times "fuel rate at maximum power (at 5500 rpm)."

Graph 8.6.2-2

Formula

$$gpm(Vo) := \frac{5}{60} \cdot \frac{Phover(Vo)}{50} \qquad gpm(V) := \frac{5}{60} \cdot \frac{Phover(V)}{50}$$

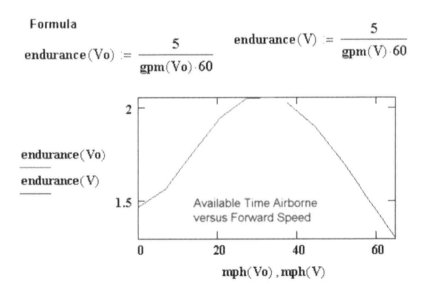

Plotted are both the "engine fuel consumption rate (in gallons per minute)" and our helicopter's "Endurance (in hours)" as a "function of forward speed". As you can readily notice, the lowest fuel consumption rate and the best flight Endurance occurs at a speed of 30 mph, a point of minimal rotor power. Before you get to that point you must first Hover, then accelerate for a time and all that must be accounted for. To obtain an actual Endurance number you would have to add up all your fuel usage of each flight segment in your flight plan. For example to calculate total trip fuel usage, lets say our flight plan includes 15 minutes of initial Hover (0.057 gpm), 40 minutes of flight time at an average speed of 60 mph (0.058 gpm) and 5 minutes of landing / Hover. Total Trip Fuel (TTF) consumption for this simplified case is calculated as follows:

TTF = (15 min)(0.057 gpm) + (50 min)(0.058 gpm) + (5 min)(0.057 gpm) = TTF = 4.04 gallons.

In this case our range would be (60 mph)(50 min.) = 50 miles

Now lets move on to the more sophisticated helicopter analysis of this subject. The following equations have Range and Endurance calculated by integrating the specific Range and Endurance values over the total fuel weight for a planned flight and initial gross weight condition.

Calculation 8.6.2-1

$$\text{Range} := \int \frac{dRange}{dFuel\_wt}\, dFuel\_wt$$

$$\frac{dRange}{dFuel\_wt} := \frac{V}{Power \cdot (Specific\_fuel)}$$

$$\text{Endurance} := \int \frac{dEndurance}{dFuel\_wt}\, dFuel\_wt$$

$$\frac{dEndurance}{dFuel\_wt} := \frac{1}{Power \cdot (Specific\_fuel)}$$

Note: specific Range and Endurance are given by the specific fuel consumption of the engine in # / hp-hr

Basically we do the same work as the example only we integrate the specific range for each flight segment over the total fuel weight given the initial gross weight and flight condition. The endurance integral is similar but uses the specific endurance for each flight segment.

In actual flight even optimal conditions vary and these fuel consumption rates versus power numbers will vary from the estimates for each flight segment. Since power is directly related to gross weight and altitude as fuel is used the helicopter weight to be moved is less. Fuel weights around 7-1/2 pounds per gallon and that decrease in a helicopter's gross weight reduces flight power requirements.

These gross weight change dynamics are important to determine in the design stages of a helicopter's development. It is even more critical to military helicopters required for mission specific duties. From a pilot's perspective the integrals determining a range and endurance for a particular flight plan may be approximated by using the specific range and endurance at the midpoint of the flight where the revised helicopter weight used in the calculations is equal to the initial gross weight minus one-half the total fuel weight.

8.6.3 Maximum Speed

Referring once again to our "Power versus Forward Velocity" graph our 50 hp helicopter can attain a forward velocity of 55 knots (63.3 mph) , the maximum speed allowable for ultralight aircraft. The question is how fast can our helicopter go? Using the generalized book formula where: Vmax = vh (4 / (f / A)) ^(1 / 3) we calculate that speed as 62.08 mph.

Calculation 8.6.3-1 Maximum Speed Calculation

A. Defined Variables:

ρ = air density at 70 degrees; lb / ft^3
A = rotor area; ft^2
f / A = helicopter drag
κ = inflow correction factor
$T(\Omega 1)$ = thrust used in energy equation; lbm ft / sec^2
η = drivetrain efficiency

$$\Omega 1 = 45.8\,\text{Hz} \qquad \eta = 0.8 \qquad \kappa = 1.165$$

$$\frac{f}{A} = 0.025 \qquad \rho = 0.075\,\frac{\text{lb}}{\text{ft}^3} \qquad A = 346.36\,\text{ft}^2$$

B. Formula

$$vh = \kappa \cdot \sqrt{\frac{T(\Omega 1)}{2\rho \cdot A}} \qquad Vmax = \frac{3600\text{sec}}{5280\text{ft}} \cdot vh \cdot \eta \cdot \left(\frac{4}{\dfrac{f}{A}}\right)^{\frac{1}{3}}$$

C. Results Tabulated: $vh = 20.965\,\dfrac{\text{ft}}{\text{sec}}$ $\qquad$ Vmax = 62.08 mph

8.6.4 Rate of Climb

First of all let's establish the fact that there is no theoretical possibility of a perpetual motion device. I mention this because in the real world power is always required to keep any device moving. Even the most efficient device I know of "the pendulum" eventually slows down and for it to maintain certain amplitude of motion constant power must be applied. There is no free lunch. If you only have enough power to hover you are stuck in that mode until you burn off some fuel weight. If by design you have surplus power while Hovering then that power can be used to either gain altitude, accelerate in some desired level flight direction or any combination of both. The point is with a finite amount of power available there are flight limitations as to what you can do. This section takes off from the previous section where "Power versus Forward Speed" was plotted. In that section it was stated that at approximately 30 mph minimal power was required to maintain flight or stated another way at this speed the maximum excess amount of power is available for gaining altitude or accelerating.

The calculations continue using the climb power expression developed in the previous section. Basically this equation is set equal to the net engine power remaining after that required for maintaining steady-state flight and solved for potential climb velocity.

Be aware there is a subtle point concerning the engine. Even though the power available of the Kawasaki 650 is 52 hp @ 6000 rpm we will not use this number in our calculations because at this peak horsepower engine torque is decreasing. Remember it is engine torque that makes everything turn therefore lets just skip the peak horsepower number and use that horsepower where peak torque is developed. This ultimately minimizes main rotor rpm fluctuation. The standard horsepower equation is Torque (ft-lbs) x RPM) / 5250. Given the Kawasaki 650 peak torque of 47.7 ft-lbs occurs at 5500 RPM the horsepower at that point is approximately 50 horsepower. This is the number we will use in these calculations.

Calculation 8.6.4-1

A. Defined Variables:

ρ = air density @ 70^0 = 0.0752 lb/ft^3

$\Omega 1$ = rotor angular speed in radians per second = 437.4 (45.8 Hz);

r = rotor radius = 10.5 ft

A = area; ft^2

λc = Vc (climb velocity) / Ω r; ft / sec

Peng = 50; hp

Pnet(Vo) = Net power available from hover to μ; hp

Pnet(V) = Net power available from μ to top speed; hp

$Ct10_5$ = thrust coefficient for 10.5 ft radius rotor

Phover(Vo) = power required to go from hover to μ calculated in energy equation

Pfwd(V) = power required to go from μ to 50 mph

V500 = 500 ft/min (8.33 ft/sec) was taken from the Hiller Rotorcycle which is simillar in size to this helicopter

B. Formula:

$$Pc(V) := \left[\rho \cdot A \cdot (\Omega \cdot r)^3 \right] \cdot (\lambda c \cdot Ct)$$

Note: set equal to the net power available to obtain climb velocity as a function of forward speed.

In the above equation substitue $Vc / \Omega r$ for λc, then solve for Vc as a function of Vo, V

$$Pnet(Vo) := 50 - Phover(Vo)$$

$$V500 := 8.33 \frac{ft}{sec}$$

$$Pnet2(V) := 50 - Pfwd(V)$$

$$Pc(Vo) := \left[\frac{\rho}{hp} \cdot A \cdot (\Omega 1 \cdot r)^3 \right] \cdot \left(\frac{Vc}{\Omega 1 \cdot r} \cdot Ctl05 \right)$$

$$Pc(V) := \left[\frac{\rho}{hp} \cdot A \cdot (\Omega 1 \cdot r)^3 \right] \cdot \left(\frac{Vc}{\Omega 1 \cdot r} \cdot Ctl05 \right)$$

$$Vc(Vo) := \frac{Pnet(Vo) \cdot (\Omega 1 \cdot r)}{\left[\frac{\rho}{hp} \cdot A \cdot (\Omega 1 \cdot r)^3 \right] \cdot Ctl05}$$

$$Vc2(V) := \frac{Pnet2(V) \cdot (\Omega 1 \cdot r)}{\left[\frac{\rho}{hp} \cdot A \cdot (\Omega 1 \cdot r)^3 \right] \cdot Ctl05}$$

D. Results Tabulated:

$Vc(Vo) =$		$Pnet(Vo) =$	$Vc2(V) =$		$Pnet2(V) =$
16.301	ft	16.008	25.196	ft	24.743
18.345	sec	18.015	23.729	sec	23.302
21.983		21.587	20.981		20.603
24.71		24.265	16.936		16.631
26.07		25.601	11.528		11.321
26.159		25.689			

E. Data Plotted

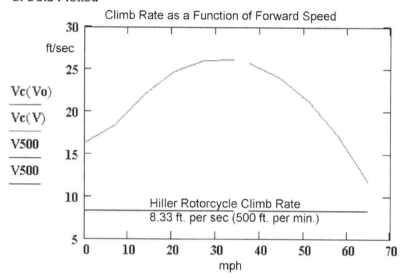

$$\frac{Vc(Vo)}{Vc(V)} \qquad \frac{V500}{V500}$$

8.6.5 *Engine Horsepower versus Angle of Attack*

Upon actual field testing it was apparent that moving the Collective too quickly to increase the Angle of Attack to initiate Hover, instantly bogged down the engine. Needless to say this was a serious consequence. Shown below once again is the engine power band as a function of engine rpm. Notice how the power drops off in either rpm direction once you pass the 5500 to 6000 rpm bandwidth. Now the calculations determine the Angle of Attack where the engine power is insufficient.

Reprinted Graph 8.6.1-2

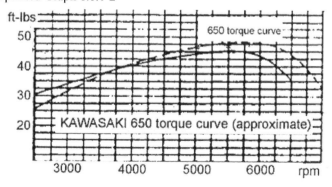

The equations explore two initial conditions, the first where "Ct / sigma" equals 0.08 and the second where "Ct / sigma" equals 0.06. Then the Angle of Attack will be varied up to its limiting value and the resulting hp calculated for each case.

Our proposed power plant develops approximately 50 horsepower at peak torque @ 5500 rpm and our drive train will be designed to support this peak torque availability. If this rpm can be maintained, the engine will have sufficient power to handle any subtle misapplication of collective positioning while increasing the angle of attack, (alpha).

Calculation 8.6.5-1 Angle of Attack Limitation Calculations

A. Defined Variables:

λ = induced inflow ratio (v / ΩR)

Ω = rotor rotational speed (rad/sec)

σ = rotor solidity = Nc / πR

ρ = air density @ 70 degrees; lbm / ft^3

A = rotor disc area (πr^2); ft^2

cd(α) = section drag coefficient as a function of angle of attack

Pih(α) = induced power as a function of angle of attack

Po(α) = profile power as a function of angle of attack

Ct(α) = thrust coefficient as a function of angle of attack

Phover(α) = power to hover as a function of angle of attack

r = rotor radius; ft

V = helicopter velocity relative to the air at hover V = 0

α = blade section angle of attack

κ = inflow correction factor = 1.2

NB = number of rotor blades

c = blade cord; ft

η = is the power train efficiency loss = 0.8

$\kappa = 1.165$ $c = 0.558\,\text{ft}$ $\sigma = 0.034$ $\lambda = 0.044$

$\Omega = 45.8\,\text{Hz}$ $r = 10.5\,\text{ft}$ $\eta = 0.8$

$Ct(\Omega) = 2.801 \times 10^{-3}$ $Vo := \dfrac{0\text{ft}}{\text{sec}}$ $A = 346.36\,\text{ft}^2$

$\rho = 0.075\,\dfrac{\text{lb}}{\text{ft}^3}$

$\text{Pengine}(\alpha) := 50$ $\text{constant} := \dfrac{\rho}{\text{hp}} \cdot A \cdot (\Omega 1 \cdot r)^3$

B. Formula:

Thrust formula $\quad T(\alpha) := 0.5 \cdot \rho \cdot (\Omega \cdot r)^2 \cdot \left(\dfrac{\dfrac{NB \cdot c}{\pi \cdot r} \cdot cdo(\alpha)}{\kappa} \right)^{\frac{2}{3}} \cdot \pi \cdot r^2$

Thrust coefficient formula $\quad Ct(\alpha) := \dfrac{T(\alpha)}{\rho \cdot A \cdot (\Omega \cdot r)^2}$

Hover Power from Energy equation:

$$Phover(Vo) := \frac{Pih(Vo) + Po(Vo) + Pp(Vo)}{\eta}$$

The above hover power equation is setup as a function of forward speed ranging from 0 to 50 ft/sec. Now we want to determine power as a function of pitch angle (or from a pilot's perspective "collective lever position"). This mathematical exercise is to determine if any pitch angle position overwhelms the engine's capability to keep up. To calculate this we must setup this power equation as a function of pitch angle "alpha" over the operational range from 8 to 11.4 degrees (0.2 radians). Then set the velocity equal to zero, which implies hover, and since there is no parasite power loss at zero speed we will leave that term out. As a result of these assumptions we obtain the following formula. Note that the Ct (a) term is now a function of "alpha" in these equations and not "omega".

$$Phover(\alpha) := \frac{(Pih(\alpha) + Po(\alpha))}{\eta}$$

Solving for Pih(α) and Po(α) we get the following:

$$Pih(\alpha) := (constant) \cdot \left[\frac{\kappa \cdot Ct(\alpha)^2}{2 \cdot \left[\left(\dfrac{Vo \cdot cos(\alpha)}{\Omega \cdot r} \right)^2 + \lambda^2 \right]^{0.5}} \right]$$

$$Po(\alpha) := (constant) \cdot \left[\sigma \cdot \frac{cdo(\alpha)}{8} \cdot \left[1 + 4.6 \cdot \left(\frac{Vo \cdot cos(\alpha)}{\Omega \cdot r} \right)^2 \right] \right]$$

Now that these terms have been redefined we can now generate the numbers for the power (as a function of pitch angle) equation:

$$\mathrm{Phover}(\alpha) := \frac{(\mathrm{Pih}(\alpha) + \mathrm{Po}(\alpha))}{\eta} \qquad\qquad \mathrm{angle}(\alpha) := \alpha \cdot \frac{360}{2\pi}$$

C. Function range: $\alpha := 0.14, 0.15 .. 0.20$

D. Results Tabulated:

Phover(α) =	α =	angle(α) =	$\dfrac{T(\alpha)}{g}$ =
34.103	0.14	8.021	525.71 lb
37.007	0.15	8.594	549.888
40.181	0.16	9.167	575.321
43.634	0.17	9.74	601.94
47.373	0.18	10.313	629.681
51.409	0.19	10.886	658.483
55.748	0.2	11.459	688.292

Now that these terms have been redefined we can now generate the graph for power as a function of pitch angle.

E. Data Plotted

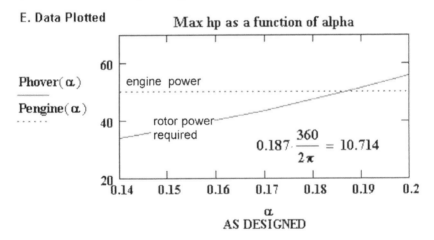

Max hp as a function of alpha

Phover(α)

Pengine(α)

engine power

rotor power required

$$0.187 \cdot \frac{360}{2\pi} = 10.714$$

α
AS DESIGNED

F. Summary

The data calculated and plotted indicates the engine that develops approximately 50 horsepower at peak torque @ 5500 rpm, will not be capable of handling any collective pitch angle above 10.7 degrees. At this pitch angle the upward thrust developed is over 650 pounds, which is substantially more than the 524 pounds required to hover. At this collective position we have two concerns. The first concern is if this collective lever is rapidly jerked to this position the coning angle developed exceeds that built into the main rotor hub momentarily creating substantially higher rotor blade root stress than desired with the potential to damage the blades. Secondly, if this angle is exceeded the engine cannot maintain peak torque. When this occurs not only does engine rpm drop but also engine power, which quickly causes the main rotor rpm to drop. Unless the collective is backed off to a lower blade pitch angle the condition basically stalls the engine.

8.6.6 Maximum Altitude

The helicopter ceiling is defined as the altitude at which the maximum power available is equal to the power required.

Standard Atmosphere Chart
Altitude / Pressure / Density

ft	psia	(lbm/ft ^3)
0	14.696	0.0752
1000	14.175	0.0725
2000	13.664	0.0699*
3000	13.168	0.0674
4000	12.692	0.0649
5000	12.226	0.0626
6000	11.778	0.0603
7000	11.341	0.058
8000	10.914	0.0558
9000	10.501	0.0537
10000	10.108	0.0517

*Note: At 100 F air density is 0.0709 lbm/ft ^3 ~ equal to 2000 ft on this chart.

In the following set of calculations we will determine the amount of horsepower required at a set Angle of Attack as a function of air density.

A. Defined Variables:

Ω = rotor rotational speed (rad/sec)
$\alpha\_\_$ = blade section angle of attack at various angles within range
x = inflow correction factor = 1.2
σ = rotor solidity = Nc / πR
c = blade cord; ft
r = rotor radius; ft
ρ = air density @ various altitudes; lbm / ft^3
A = rotor disc area (πr^2); ft^2
V = helicopter velocity relative to the air at hover V = 0
NB = number of rotor blades
η = is the power train efficiency loss = 0.8

The tabulated data in each of these cases has been truncated to save space and only the numbers germane stated in the "Results Tabulated" section. In that section you find the weight to be lifted (~525#), the power required, and the density used in the calculation. In subsequent calculations as the Angle of Attack is increased and the weight criteria satisfied the density number satisfying the equation will be compared to the Altitude Chart to come up with an altitude corresponding to that density. Note: a14 = 8 degree angle of attack, respectively a17 = 9.74 degrees, a18 = 10.7 degrees, and a20 =11.46 degrees.

Calculation 8.6.6-1 Calculation of Maximum Altitude

A. Defined Variables:

$$\Omega = 45.8\,Hz \qquad \alpha = 0.14\,,0.15\ldots 0.20$$

$\alpha 14 = 0.14$	$\alpha 17 = 0.17$	$\alpha 18 = 0.187$	$\alpha 20 = 0.20$
$\kappa = 1.165$	$\sigma = 0.034$	$c = 0.558\,ft$	$r = 10.5\,ft$
$\lambda = 0.044$	$A = 346.36\,ft^2$	$Vo = \dfrac{0\,ft}{sec}$	

B. Function range of "ρ" is defined locally in the following specific formulae

C1. Formula to be Calculated: Determine helicopter altitude and thrust while in ground effect having a pitch angle:of 8°

Function range

$$\rho = 0.0752\,\frac{lb}{ft^3}\,,0.0751\,\frac{lb}{ft^3}\ldots 0.0517\,\frac{lb}{ft^3} \qquad cons = A\cdot\frac{(\Omega 1\cdot r)^3}{hp}$$

$$T14(\rho) = 0.5\cdot\rho\cdot(\Omega\cdot r)^2\cdot\left(\frac{\dfrac{NB\cdot c}{\pi\cdot r}\cdot cdo(\alpha 14)}{\kappa}\right)^{\frac{2}{3}}\cdot\pi\cdot r^2$$

design thrust at 8 degree pitch angle

thrust coefficient $\quad Ctl4(\rho) = \dfrac{T14(\rho)}{\rho\cdot A\cdot(\Omega\cdot r)^2}$

power for hover portion only

$$Pihl4(\rho) = \rho\cdot cons\cdot\left[\frac{\kappa\cdot Ctl4(\rho)^2}{2\cdot\left[\left(\dfrac{Vo\cdot cos(\alpha 14)}{\Omega\cdot r}\right)^2+\lambda^2\right]^{0.5}}\right]$$

profile power

$$Po14(\rho) := \rho \cdot cons \cdot \left[\sigma \cdot \frac{cdo(\alpha14)}{8} \cdot \left[1 + 4.6 \cdot \left(\frac{Vo \cdot cos(\alpha14)}{\Omega \cdot r} \right)^2 \right] \right]$$

power requirement summation

$$PAlt14(\rho) := \frac{1}{\eta}(Pih14(\rho) + Po14(\rho))$$

D1. Results Tabulated:

$$\frac{T14(\rho)}{g} =$$

$\frac{T14(\rho)}{g}$	$PAlt14(\rho) =$	$\rho =$
525.71 lb	34.103	0.075 lb
525.011	34.058	0.075 $\frac{}{ft^3}$

At an 8-degree pitch angle 525# can be lifted at sea level (air density 0.075 lb / ft^3) using 34 hp.

C2. Formula to be Calculated: Determine helicopter altitude and thrust while in ground effect having a pitch angle: of 9.74°

Function range $\rho := 0.0656 \cdot \frac{lb}{ft^3}, 0.0657 \cdot \frac{lb}{ft^3} \ldots 0.0659 \cdot \frac{lb}{ft^3}$

design thrust at 9.74 degree pitch angle

$$T17(\rho) := 0.5 \cdot \rho \cdot (\Omega \cdot r)^2 \cdot \left(\frac{\frac{NB \cdot c}{\pi \cdot r} \cdot cdo(\alpha17)}{\kappa} \right)^{\frac{2}{3}} \cdot \pi \cdot r^2$$

thrust coefficient $$Ct17(\rho) := \frac{T17(\rho)}{\rho \cdot A \cdot (\Omega \cdot r)^2}$$

$$\text{Pih17}(\rho) := \rho \cdot \text{cons} \cdot \left[\frac{\kappa \cdot \text{Ct17}(\rho)^2}{2 \cdot \left[\left(\frac{Vo \cdot \cos(\alpha 17)}{\Omega \cdot r} \right)^2 + \lambda^2 \right]} \right]^{0.5}$$

power for hover portion only

$$\text{Po17}(\rho) := \rho \cdot \text{cons} \cdot \left[\sigma \cdot \frac{\text{cdo}(\alpha 17)}{8} \cdot \left[1 + 4.6 \left(\frac{Vo \cdot \cos(\alpha 17)}{\Omega \cdot r} \right)^2 \right] \right]$$

profile power

power requirement summation

$$\text{PAlt17}(\rho) := \frac{1}{\eta} (\text{Pih17}(\rho) + \text{Po17}(\rho))$$

$$\frac{\text{T17}(\rho)}{g} = \qquad \text{PAlt17}(\rho) = \rho =$$

D2. Results Tabulated:			
525.097 lb	38.063	0.066 lb	
525.897	38.121	0.066 $\frac{}{\text{ft}^3}$	

At a 9.74-degree pitch angle 525# can be lifted to 3500 ft (air density 0.066 lb / ft^3) using 38 hp.

C3. Formula to be Calculated: Determine helicopter altitude and thrust while in ground effect having a pitch angle:of 10.7°

Function range $\qquad \rho := 0.0608 \cdot \frac{\text{lb}}{\text{ft}^3}, 0.0609 \cdot \frac{\text{lb}}{\text{ft}^3} \cdots 0.061 \cdot \frac{\text{lb}}{\text{ft}^3}$

design thrust at 10.7 degree pitch angle

$$\text{T18}(\rho) := 0.5 \cdot \rho \cdot (\Omega \cdot r)^2 \cdot \left(\frac{\frac{NB \cdot c}{\pi \cdot r} \cdot \text{cdo}(\alpha 18)}{\kappa} \right)^{\frac{2}{3}} \cdot \pi \cdot r^2$$

thrust coefficient $\qquad \text{Ct18}(\rho) := \frac{\text{T18}(\rho)}{\rho \cdot A \cdot (\Omega \cdot r)^2}$

$$Pih18(\rho) := \rho \cdot cons \cdot \left[\frac{\kappa \cdot Ct18(\rho)^2}{2 \cdot \left[\left(\frac{Vo \cdot cos(\alpha 18)}{\Omega \cdot r} \right)^2 + \lambda^2 \right]^{0.5}} \right]$$

power for hover
portion only

$$Po18(\rho) := \rho \cdot cons \cdot \left[\sigma \cdot \frac{cdo(\alpha 18)}{8} \cdot \left[1 + 4.6 \cdot \left(\frac{Vo \cdot cos(\alpha 18)}{\Omega \cdot r} \right)^2 \right] \right]$$

profile power

power requirement
summation

$$PAlt18(\rho) := \frac{1}{\eta}(Pih18(\rho) + Po18(\rho))$$

$$\frac{118(\rho)}{g} = \qquad PAlt18(\rho) = \qquad \rho =$$

D3. Results Tabulated:

$\frac{118(\rho)}{g} =$		$PAlt18(\rho) =$	$\rho =$	
525.317	lb	40.56	0.061	lb
526.181		40.627	0.061	$\frac{}{ft^3}$

At a 10.7-degree pitch angle 525.3# can be lifted to 5500 ft (air density
0.061 lb / ft^3) using 40.6 hp.

**C4. Formula to be Calculated: Determine helicopter altitude and
thrust while in ground effect having a pitch angle:of 11.46°**

Function range $\rho := 0.057 \cdot \frac{lb}{ft^3}, 0.0575 \cdot \frac{lb}{ft^3} \cdot 0.058 \cdot \frac{lb}{ft^3}$

design thrust at 11.46 degree pitch angle

$$T20(\rho) := 0.5 \cdot \rho \cdot (\Omega \cdot r)^2 \cdot \left(\frac{\frac{NB \cdot c}{\pi \cdot r} \cdot cdo(\alpha 20)}{\kappa} \right)^{\frac{2}{3}} \cdot \pi \cdot r^2$$

thrust coefficient $Ct20(\rho) := \frac{T20(\rho)}{\rho \cdot A \cdot (\Omega \cdot r)^2}$

$$\text{Pih20}(\rho) := \rho \cdot \text{cons} \cdot \left[\frac{\kappa \cdot \text{Ct20}(\rho)^2}{2 \cdot \left[\left(\frac{\text{Vo} \cdot \cos(\alpha 20)}{\Omega \cdot r} \right)^2 + \lambda^2 \right]^{0.5}} \right]$$

power for hover
portion only

$$\text{Po20}(\rho) := \rho \cdot \text{cons} \cdot \left[\sigma \cdot \frac{\text{cdo}(\alpha 20)}{8} \cdot \left[1 + 4.6 \cdot \left(\frac{\text{Vo} \cdot \cos(\alpha 20)}{\Omega \cdot r} \right)^2 \right] \right]$$

profile power

power requirement
summation

$$\text{PAlt20}(\rho) := \frac{1}{\eta} (\text{Pih20}(\rho) + \text{Po20}(\rho))$$

$$\frac{\text{T20}(\rho)}{g} = \qquad \text{PAlt20}(\rho) = \rho =$$

D4. Results Tabulated

$\dfrac{\text{T20}(\rho)}{g}$		$\text{PAlt20}(\rho)$	ρ	
521.711	lb	42.256	0.057	lb
526.287		42.626	0.058	$\overline{ft^3}$

At a 11.46 pitch angle 526.3# can be lifted to 7000 ft (air density 0.058 lb / ft^3) requiring 42.6 hp.

Calculation of engine output power at altitude pressures stated below.

A. Defined Variables:

$$\rho 1 := 0.075 \frac{lb}{ft^3} \qquad\qquad \rho 3 := 0.0609 \frac{lb}{ft^3}$$

$$\rho 2 := 0.066 \frac{lb}{ft^3} \qquad\qquad \rho 4 := 0.058 \frac{lb}{ft^3}$$

B. Formula to be Calculated:

$$\text{engineHP1} := 50 \cdot \left(1 - \frac{0.075 \cdot \dfrac{lb}{ft^3} - \rho 1}{0.075 \cdot \dfrac{lb}{ft^3}} \right)$$

$$\text{engineHP2} := 50 \cdot \left(1 - \frac{0.075 \cdot \dfrac{lb}{ft^3} - \rho 2}{0.075 \cdot \dfrac{lb}{ft^3}} \right)$$

$$engineHP3 := 50 \cdot \left(1 - \frac{0.075 \cdot \frac{lb}{ft^3} - \rho 3}{0.075 \cdot \frac{lb}{ft^3}} \right)$$

$$engineHP4 := 50 \cdot \left(1 - \frac{0.075 \cdot \frac{lb}{ft^3} - \rho 4}{0.075 \cdot \frac{lb}{ft^3}} \right)$$

D. Results Tabulated:

engineHP1 = 50	engineHP3 = 40.6
engineHP2 = 44	engineHP4 = 38.667

Now that all this data is calculated Mathcad will plot this information if it is put in a matrix format. In the G matrix defined below the data in the first column is altitude data, the second column the corresponding angle of attack data, the third column the required hp and the last column the engine hp derate for that altitude.

E. Data Plotted

$$G := \begin{pmatrix} 0 & 8 & 34 & 50 \\ 3500 & 9.74 & 38 & 44 \\ 5500 & 10.7 & 40.6 & 40.6 \\ 7000 & 11.46 & 42.6 & 38.67 \end{pmatrix}$$

$$altitude := G^{\langle 0 \rangle}$$

$$pitchangle := G^{\langle 1 \rangle}$$

$$HPreq := G^{\langle 2 \rangle}$$

$$HPavail := G^{\langle 3 \rangle}$$

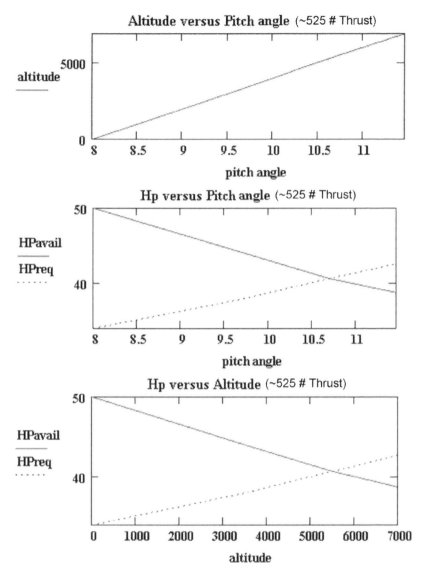

Summary: At an 8 degree pitch angle 523.5# can be lifted in 70 degree air having a density of 0.075 lbs/ft ^{2} (sea level) using 33.9 hp, (50 hp available). At a 9.74 degree pitch angle 524.4# can be lifted in air having a density of 0.066 lbs/ft ^{2} (3500 ft) using 36 hp, (44 hp available). At a 10.7 degree pitch angle 524.7# can be lifted in air having a density of 0.061 lbs/ft ^{2} (5500 ft) using 40.5 hp, (40.7 hp available). Unfortunately at the NACA0012 Angle of Attack limit of 11.46 degrees the engine cannot support this due to lack of power available. At an 11.46 degree pitch angle 524.7# can be lifted in air having a density of 0.058 lbs/ft ^{2} (7000 ft) using 42.5 hp, but our engine at that altitude produces only 38.7 hp.

According to the plotted data we need 42.5 hp to satisfy the power requirements near the upper limit of pitch angle at altitude. Lets work backwards to determine the size engine we should be using in order to enable us to use the full range of Angle of Attack available.

A. Revised Horsepower Calculation

$$\text{engineHP} = \cfrac{42.533}{\left(1 - \cfrac{0.075\,\dfrac{\text{lb}}{\text{ft}^3} - \rho 4}{0.075\,\dfrac{\text{lb}}{\text{ft}^3}}\right)}$$

engineHP = 55

If 55 horsepower is necessary to generate the required 42.53 horsepower at altitude with an 11.46 pitch angle then lets determine what the engine's gross horsepower should be. I realize every engine's performance differs so for simplicity we will use the number ratio of our engine (52 gross horsepower/ 50 hp at rated torque) and simply multiply that by the 55 number obtained above. The result of this straight ratio multiplication gives us an engine with a gross horsepower number of 57.2.

Remember way back when, when the formula for gross helicopter hp was given as:

P = (Gross Weight) ^(2/3); well plugging in the 525# gross weight of our helicopter we obtain P = 65 horsepower, this number is within 12% of our number, and look at all the work we had to do to get here. Personally, when I replace this engine I'd go with the 65 hp engine.

B. Formula to be Calculated: Generalized Power Equation

$$\text{GWT} = 525 \qquad \text{Pest} = \text{GWT}^{\frac{2}{3}}$$

D. Results Tabulated:

Pest = 65 Hp

8.6.7 Flight Operation at 100 degrees F

Parameters:

1. Air density, (Roe), at sea level is 0.075 lb/ft ^3 and air density at 100 F is 0.0709 lb/ft^3

2. Rotor rpm is to stay constant and only the Angle of Attack, (alpha), varied to support the helicopter Gross Weight.

Going back a bit in the calculations I extracted the Altitude versus Pitch Angle matrix. Using this as a starting point I created two matrixes one called "B" which used the first two columns of the Altitude versus Pitch Angle and "A" a Temperature versus Air Density matrix. In the "B" matrix instead of using altitude I used the density equivalent.

E. Data Plotted

$$
G := \begin{pmatrix}
0 & 8 & 34 & 50 \\
3500 & 9.74 & 38 & 44 \\
5500 & 10.7 & 40.6 & 40.6 \\
7000 & 11.46 & 42.6 & 38.67
\end{pmatrix}
$$

$$\text{altitude} := G^{\langle 0 \rangle}$$

$$\text{pitchangle} := G^{\langle 1 \rangle}$$

$$\text{HPreq} := G^{\langle 2 \rangle}$$

$$\text{HPavail} := G^{\langle 3 \rangle}$$

Calculation 8.6.7-1 Flight Operation at 100 degrees F

Using these data matrixes I generated two plots. The first is a plot of Air density versus Temperature. At 100 degrees F the air density is 0.0709 lbs/ft^3 which is approximately equal to flight at 2000 ft. The second plot shows air density as a function of angle of attack. The horizontal line that intersects the curve represents the air density at 100 degrees F. Dropping down from that point to the horizontal axis we obtain an angle of attack of 8.77 degrees.

We now know how ambient air temperature affects the angle of attack of a hovering helicopter. I mention "hovering" because temperature decreases with altitude so to keep things simple the calculations are based upon a sea level starting point.

$$A := \begin{pmatrix} 100 & 0.0709 \\ 80 & 0.0735 \\ 68 & 0.0752 \\ 60 & 0.0763 \\ 40 & 0.0794 \\ 20 & 0.0827 \end{pmatrix} \qquad B := \begin{pmatrix} 0.075 & 8 \\ 0.066 & 9.74 \\ 0.061 & 10.7 \\ 0.058 & 11.46 \end{pmatrix} \qquad C := \begin{pmatrix} 0.0709 \\ 0.0709 \end{pmatrix}$$

$$\text{AirDensity} := A^{\langle 1 \rangle}$$

$$\text{TempA} := A^{\langle 0 \rangle} \qquad \text{Angle} := B^{\langle 1 \rangle} : B^{\langle 0 \rangle}$$

$$\text{Temp100} := C^{\langle 0 \rangle}$$

Angle of Attack

8.6.8 The Collective Angle Required for Hovering:

This calculation is for a constant chord linearly twisted blade. The blade used on our helicopter is straight, therefore our collective angle may be several tenths of a degree more. I used this formula because it was simple and a negative 12 degree twist results in only a 4% power reduction. I felt the results of this calculation were a realistic starting point for field testing.

Calculation 8.6.8-1

A. Defined Variables:

θ_{75} = collective pitch angle at 75% rotor radius based upon blade element theory for a linearly twisted constant chord rotor blade with uniform flow.

a = blade section two-dimensional lift-curve slope. This slope value is obtained from the NACA 0012 graph titled section lift coefficient versus section angle of attack (in radians) and is equal to a = 6.571

λ = rotor inflow ratio $(V \sin a + v) / \Omega r$, or $\kappa (Ct / 2)^{**}0.5$

Cteq = thrust coefficient from energy equation @ 45.8 Hz = 2.801×10^{-3}

A10.5 = disc area for 10.5 ft radius blade = πr^2

σ = rotor disc solidity

ρ = air density @ 70^0 = 0.0752 lb/ft^3

GWT = same as thrust value = 525# name changed for Mathcad clarity;

$\Omega 1$ = rotor angular speed in radians per second from energy equation = 45.8 Hz

vh = rotor induced air velocity

Vhover = at hover forward velocity is 0

r = rotor radius = 10.5 ft

$$\sigma = 0.034$$

$$Cteq := 2.801 \times 10^{-3} \qquad GWT := 525 lb$$

$$a := 6.571$$

$$Vhover := 0 \frac{ft}{sec} \qquad Vhover \cdot \sin(a) = 0 \frac{ft}{sec}$$

$$r = 10.5 ft$$

$$\Omega 1 = 45.8 Hz$$

$$vh := \kappa \cdot \sqrt{\frac{GWT \cdot g}{2 \rho \cdot A10_5}} \qquad \rho := 0.0752 \frac{lb}{ft^3} \qquad A10_5 = 346.361 ft^2$$

Calculation of λ can be performed in two ways as indicated in the definition of terms so I calcuted one as λ and the other as $\lambda 1$ to see if they agreed, and they did as shown below:

$$\lambda := \kappa \cdot \sqrt{\frac{Cteq}{2}} \qquad \text{or} \qquad \lambda 1 := \frac{Vhover \cdot \sin(a) + vh}{\Omega 1 \cdot r}$$

$$\lambda = 0.044 \qquad\qquad \lambda 1 = 0.044$$

B. Formula to be Calculated: Hover Collective Angle at sea level elevation

$$\theta_{75} := \frac{6 \cdot Cteq}{\sigma \cdot a} + \frac{3}{2}\lambda$$

C. Results Tabulated:

$$\theta_{75} = 0.141 \text{ radians} \qquad \theta_{75} \cdot \frac{360}{2\pi} = 8.078 \text{ degrees}$$

The basic sizing of this helicopter is complete, and the general layout can be sketched. The components can be selected now from the size of the rotor and power plant and from the fuel and pilot weight. The component weights are summed to obtain the gross weight of the helicopter, and the procedure is repeated until the gross weight converges. You can't imagine how many lightening holes I added to parts in order to get their weight reduced.

Design optimization for production type helicopters is based on an examination of mission cost parameters (such as direct operating cost, or even gross weight, which controls first cost) or various performance indices (such as range, maximum speed, or noise) as a function of the basic rotor and helicopter parameters. Even rotor size and helicopter size can be considered in the optimization process if the performance analysis and weight estimation are detailed enough to be able to distinguish between the variations. In my case the only factor under consideration was making the ultralight 254 lb base weight.

Persons planning a proposed production type helicopter must also look into other factors influencing the basic design that do not appear directly in the preliminary design analysis. For example, the rotor type could be determined more by its influence on the helicopter handling qualities, aero elastic stability, and maintenance than by its influence on performance and weight. Here again for our helicopter the rotor type selected was the semi-rigid type due mainly for its weight but also its inherent simplicity.

Detail design finalizes the shape of all components of the helicopter. All the individual components are designed to perform their required tasks in accordance with the results of the preliminary design analysis and the weight constraints. The major task is the structural analysis of all components, which requires a realistic estimate of the aerodynamic and inertial loads. Hopefully those estimates will reflect real world operating conditions and a complete calculation of helicopter stresses and stress points can be performed. This stage in the helicopter design brings to bear the full capabilities of the builder to solve. If technical issues or problems arise one could seek outside engineering guidance thru an organization such as EAA.

PART 3-THE BUILDER'S TOOLBOX

At this point in the book we already know a substantial amount of performance detail regarding our helicopter. The next task is to translate this into an actual design. For that we need to know the types of material available and the limitations of the machining equipment we will be using. Once we know the ramifications of the machine shop process of each material we can sit down and devote time to develop the basic helicopter CAD (computer aided drafting) layout. That sounds rather straightforward but the CAD layout is way more than a mere exercise in drafting. The model you build in the computer is exactly the "same size" as the real helicopter! Wow, "a direct one to one relationship", try doing that on paper! In addition by selecting the types of material for the parts you create you can input the material's density and generate an x-y-z centroid for the model. The centroid is the location of the craft's center-of-gravity which on takeoff should be located at the main rotor centerline. As parts are created and added to the model we can obtain a real time output of where the CG is and more importantly where it should be.

Chapter 9- Raw Materials

9.1 Metal Stock

Early on rotor blades were made from wood but due to its lack of stiffness and the propensity to lose dynamic balance as ambient air humidity changed, this material will not even be considered even if it was possible to carve out the desired shape.

Composite will also not be considered for several reasons. The most significant reason for me was the 1990 L111 airline crash, which killed 200 passengers. The cause of the crash was due to the fatigue failure of a composite tail attachment housing. The significance of this failure is the difficulty associated with predicting the life of a composite part. Composites are discussed in depth in section 9.2.3 of this chapter but for this discussion lets just say that they are fiber weaves saturated with a structural resin.

There are several areas of concern using composites regardless of how qualified or large a manufacturer. First the supplier of the resin and the fabric weave must have ISO 9000 quality controls in place in order to insure that the stated physical properties of the material you receive match those of the specifications on the material. All handling, storage, shelf life and time dependent blending instructions must be strictly adhered to which mandates maintaining your own Quality Control Program. A material storage space and a fabrication area must be humidity and temperature controlled and monitored with a 24 hour chart recorder to verify those room variables are being maintained. Detail documentation must follow each part fabricated to enable traceability back to any subsequent fact possibly contributing to a part failure so it could be corrected in the future. This level of control in dealing with composites is not an option but obligatory to achieve the desired tensile strengths published.

Even if every condition above is fulfilled how can you be sure that the part you fabricate does not have any voids in the homogeneous mix? The aerospace industry has formal testing programs to provide feedback to manufacturing to insure that the process or processes producing the part will function as it was designed by Engineering. What is the significant of a void in the mix? It is air. It will expand with heat possibly delaminating the structure and also air contributes zero tensile strength properties to the part for the volume it takes up. As a result the part will not be as strong. If this part were a rotor blade a structural failure would be serious indeed.

Finally even if we did everything perfectly in the blending and the fabrication stage what is the failure criterion? All parts will fail under some condition! The strongest 52100 steel ball bearing will turn into a puddle of molten liquid with heat, so we have to know exactly what operational parameters must we stay within to avoid the 1990 L111 airline crash disaster. Does our part weaken over time as the resin ages? Does the part distort (creep) under sustained stress? If stressed what is the yield point? Does it have one? How are properties affected at elevated temperatures? What effect does impact have or how does the part respond to fluctuating stress? What is the result if all of the above occurs at the same time? Personally working with composites exclusively for such a critical part as a rotor blade is beyond my scope. If I were to entertain use of a composite rotor blade with -12 twist I would seek assistance from the Hexcell corporation. Unfortunately, although they manufacture an excellent state-of-the-art rotor blade, they are a volume builder and at the time of this writing would not commit to an individual request.

If wood is out and composites are out what options remain. This was never an issue as far as I was concerned. I knew early on that I would utilize metal. Specifically this helicopter will use a constant chord rotor blade having a 7075-T6 aircraft grade certified aluminum surface and be designed to operate within the endurance limit of the material. The details of the construction will be developed later.

Obviously there is more to the helicopter than just the rotor blade so each part must be evaluated based upon its design function versus weight. Remember that this is an ultralight aircraft. With that in mind we must know what materials are readily available at reasonable cost and the unique qualities that set them apart from each other.

Before proceeding with the detailed properties of the materials under consideration stated below are terms we need to know.

A. Alcad: finished with a thin surface layer of high purity aluminum

B. Stress Relieving Processes: annealing, normalizing, tempering are "Hot Working" processes used to relieve internal stress, refine grain size and improve machineability of allotropic materials (typically steels) thru the use of controlled heating and cooling. The "Cold Working or Work Hardened" process is used on steels, where the temperatures are below the recrystallization temperature, to produce stronger, harder and more brittle type steel. Hardening of non-allotropic materials such as aluminum, copper, magnesium and nickel content stainless steels is performed by solution heat treatment (also referred to as precipitation hardening or age hardening). The steps involve precipitation, quenching and possibly artificial aging. The temper designations for aluminum which produces the stress values given are as follows:

T3 solution heat-treated, then cold worked

T351 solution heat treated, stress-relieved stretched, then cold worked

T4 solution heat-treated, then naturally aged

T5 artificially aged

T6 solution treated, then artificially aged

C. Welding: Two metals are fused together by localized heat or pressure or both. Typical heating methods are gas (for example oxyacetylene 6300 degrees F), Arc (either AC or DC current 10,000 degrees F), TIG (tungsten inert gas, arc welding in a shielded gas such as Argon), MIG (metal inert gas, similar to TIG but uses a consumable wire feed) which is the welder I use. Typically TIG or MIG is used on welding magnesium, aluminum, stainless steels and certain other types of steel.

9.1.1 Aluminum- Aircraft Grades

General Information:

The first digit of the type aluminum indicates the major alloying element. The aluminum alloys we will be considering: for a 2xxx the major alloy used is copper, 6xxx -manganese and silicon, and 7xxx -zinc.

Letter Code Rating: The weldability code designations: "A" generally weldable by all commercial procedures and methods, "B" weldable under special circumstances which requires testing to obtain the correct procedure, "C" limited weldability because of localized crack sensitivity, reduced corrosion resistance and /or loss of mechanical properties, and "D" suitable welding methods have not been developed to date.

• General Types of Aluminum

1)   1000- This commercial grade is pure aluminum. It is easily formed because it is soft, ductile and work hardens less easily than the alloyed types. It also welds better. Strength properties are not enhanced by heat treatment. It is important in our use as a surface oxidation barrier since it has excellent resistance to corrosion. Tensile Strength- yield minimum 3500 psi, elongation 15-30%

2)   2024- This is a high strength alloy with good strength to weight ratio. It has excellent fatigue resistance and is easily machined. Although the surface can be polished to a fine finish, due to it's low corrosion resistance, it is typically anodized or Alcad. The material is formed in the annealed state and further enhanced thru heat treatment. Weldability: no gas or arc but may be spot, flash or seam welded. Applications: helicopter structural components such as rod ends, brackets, bearing housings, main rotor plate and hub. Here is the technically data for 2024-T351 and ALCAD 2024-T351: Tensile Strength; yield minimum 40-41,000 psi; Corrosion Resistance- 2024-T351-"C", ALCAD 2024-T351-"A"; Cold Workability- "C"; Machineability- "B"; Brazability- "D"; Weldability; Gas- "D"; Arc-

"C"; Spot or Seam- "A".

3)   6061- Of the heat treatable alloys this is the least expensive. Its mechanical and corrosion resistance properties are good and it can be machined in a variety of ways. Workability is good in the annealed condition and even in the T4 condition is capable of some moderately severe forming operations. Artificial aging to the T6 state maximized the material properties. Weldability: It can be furnace brazed and welded by all methods. To enhance corrosion resistance it can be Alcad. Applications: Tie rods, aircraft frame structure, housings. The technical data for 6061-T6: Tensile Strength: yield minimum 35,000 psi; Corrosion Resistance-"A"; Cold Workability- "C"; Machineability- "C"; Brazability- "A"; Weldability: Gas- "A"; Arc- "A"; Spot or Seam- "A".

4)   7075-T6 This is a high strength, high cost alloy with an excellent strength-to-weight ratio. Forming takes place in the annealed condition and properties are enhanced by heat treatment. It can be Alcad to improve corrosion resistance but there is some strength degradation. Weldability: gas or arc not recommended but may be spot, flash or seam welded. Applications: rotor blades. The specifics for 7075-T6: Tensile Strength: yield minimum 58-64,0000 psi; Corrosion Resistance-"A"; Cold Workability-"D"; Machineability-"B"; Brazability-"D"; Weldability; Gas- "D"; Arc-"D"; Spot or Seam- "B". Note that the cold workability of this material is rated "D" which implies it is difficult to form at room temperatures but this is exactly the process we will use to make our rotor blades.

9.1.2 Steel- Aircraft Grades
General Information:
Material Science - Steel: One of the most fascinating and yet often overlooked topics of interest is steel. It's used everywhere from engines to computer internal support frames, from bicycle gears to light bulb bases. Everywhere you look steel is being used to enhance our existence yet very few people know anything about it. To me the subject of steel is amazing. In history, there is a period called the Iron Age (in Europe from 1200 BC to 400 AD), where metal workers fine-tuned their craft to enhance the properties of the iron they produced. Even today metallurgists continue to work towards expanding their knowledge of the innermost secrets of the material. In the beginning Iron ore was worked to make some basic metallic shapes. Later someone decided to add carbon. These mixtures were called steel and that opened a new area of metallurgy.

•   It is steel if less than 2% carbon is in the iron-carbon mix. Within that classification you have:
hypoeutectoid steel: consisting of ferrite (essentially pure iron) and pearlite (mixture of two solids, ferrite and cementite, the hardest form of iron) in an iron alloy containing less than 0.8% carbon,
eutectoid steel: consisting of ferrite and pearlite in an iron alloy containing 0.8% carbon and hypereutectoid steel: consisting of cementite and pearlite in an iron alloy containing greater than 0.8% carbon but less than 2%.

- It is called cast iron if the mix exceeds 2% carbon. This alloy
nomenclature is further divided:
  hypoeutectic cast iron when the internal carbon range is 2.0% to < 4.3%
  eutectic cast iron when 4.3% is present
  hypereutectic cast iron when more than 4.3% carbon is present
  Graph 9.1.2-1 The Classic "Iron-Carbon" Diagram:

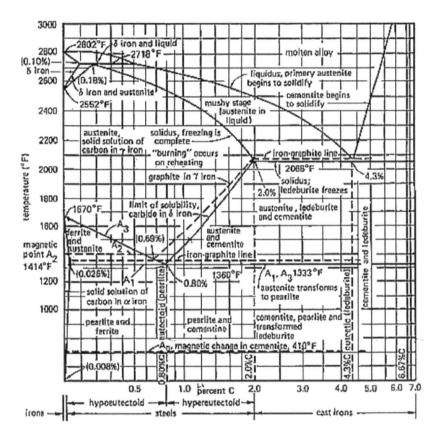

In order to get a better grasp of what those classifications mean it is best to
start at the very beginning of the process when the iron ore is dug out of the
earth. Iron ore is about 50% Gangue (earth and stone mixed with the iron
oxides) and the remainder Hematite (iron oxide Fe_2O_3) and Magnetite (iron
oxide Fe_3O_4). Somewhere along the time-line someone realized that adding
heat, coke and limestone to the iron ore produced molten iron. Today this
process is completed in a blast furnace where high temperature air oxidizes
the coke to generate heat and carbon monoxide. At approximately 600
degrees F the carbon monoxide rises to the top of the furnace and reduces
the iron oxide to FeO. This newly formed FeO settles further down in the
furnace to a region where the temperature range is 1300 to 1500 degrees F.
At this point the surrounding carbon reduces the FeO to iron. The iron

produced continues to flow towards the bottom of the furnace. At a point where the furnace temperature range is 1500 F to 2500 degrees F carbon saturates the iron in both the free carbon and carbide forms and lowers it's melting point. When this occurs it liquefies and travels to the furnace bottom where it is drawn off as pig iron.

Now lets dissect the above Classic Iron-Carbon diagram. As you can see on the x-axis you have the steels, carbon content 2% or less, where you find the hypoeutectoid and hypereutectoid types, and above 2% carbon the cast irons. On the y-axis we have the temperature scale. What intrigues me is the crystalline changes that take place as a function of carbon content and temperature. We're talking about molecular level changes as a function of temperature where the actual atomic structure is transformed impacting not only the molecular layout of the molecule but also its volume, electrical resistance and magnetic properties (the term given to this ability to generate different atomic arrangements of the same crystal is called an "Allotropic" element).

Up to 1670 degrees F you have alpha iron that has a BCC (body center cubic) molecular shape. Between 1670 to 2552 degrees F the structure transforms to gamma iron with a FCC (face centered cubic) molecular shape. Then between 2552 to 2802 degrees F, this delta iron changes back to BCC. Above 2802 degrees F we have liquid. Of course there are lines drawn on the Iron-Carbon diagram to indicate these phase changes and the critical points of each.

Additionally within the alpha, gamma and delta iron changes we have solids developing in the solution with names such as Austenite (in gamma iron), Cementite (hardest form of iron), Pearlite (from decomposition of austenite at 1333 degrees F) and Ferrite (essentially pure iron).

The real importance of the diagram is that what you see on the graph is reversible by just letting the mix cool down at its own rate. Rapid cooling can virtually freeze the picture and results in steel with substantially different properties. Say you want to make a hammer, for sure you need the right steel for it since you wouldn't want it deformed to the point of uselessness the first time you banged in some nails. Specifically you need hard steel, which means it resists plastic deformation.

How do you get hard steel? First of all you need steel with some carbon in it (0.5% produces the hardest steel) so when the temperature reaches some ideal value austenite is produced. After that occurs it is rapidly quenched (cooled). During the quenching operation there is the formation of Martensite (the hard grains) and Bainite (softer grains but good impact strength), which gives you the hardness. To increase toughness the steel is then tempered. This has to be the shortest explanation on how to harden steel ever written! Within these two sentences there lies an infinite amount of possible resultant steel combinations and a myriad of "heat treatment" (cooling / quenching) scenarios. I didn't even touch on the subject of case hardening which deals with steel's localized surface hardening. Additional information on "heat treating" can be found on the Internet.

Adding to the above phase dynamics shown in the Iron-Carbon Diagram, alloys other than simple carbon have been found over the years to further enhance the properties of steel.

Those are as follows:

• Aluminum (Al): is used to regulate austenitic grain size by retarding gain growth and contributes to producing in nitriding steels a uniformly strong case hardening.

• Carbon (C) as discussed above when it is added to iron the combination becomes either steel or cast iron. This component of steel increases it's tensile strength, hardness, and wear resistance but also decreases its toughness, ductility and machineability.

• Chromium (Cr) provides resistance to wear, abrasion, corrosion and scaling at high temperatures, while increasing tensile strength, hardness, toughness, and hardenability.

• Cobalt (Co) enhances the properties of the major alloy components in the more complex steels. It also increases hardness, strength and enables higher quench temperatures.

• Lead (Pb) is virtually insoluble in steel therefore not strictly an alloy element. When added as minute well dispersed particles it improves machineability by reducing friction where the tool cutting edge contacts the piece and facilitates better chip-breaking formations.

• Manganese (Mn) increases hardenability, hardness, and tensile strength. It also enhances resistance to wear, scaling and distortion. It reacts with sulfur to improve forgeability. It is a deoxidizer and degasifier and when used in carburizing it increases the carbon-penetration rate.

• Molybdenum (Mo) enhances the effects of other alloy elements as well as improving machinability, corrosion resistance, increasing hardenability, hardness, strength, and toughness. At elevated temperatures it maintains strength and reduces creep; in hot-worked sheet form red-hardness properties are increased.

• Nickel (N) increases hardness and strength while maintaining its ductility and toughness. In stainless (high chromium) steels it reduces corrosion and scaling at elevated temperatures

• Phosphorus (P) increases strength, machinability, and hardness but also adds cold-shortness (brittleness).

• Silicon (Si) improves tensile and yield strength, increases hardenability, magnetic permeability, and forging range.

• Sulfur (S) in free-cutting steel improves machinability, but requires appropriate amount of manganese or brittleness occurs at red heat. It decreases ductility, impact resistance and weldability

• Tungsten (W) at elevated temperatures has superior hot-working capabilities and greater cutting efficiency. It increases toughness, hardness and strength.

• Vanadium (V) in metal cutting tool steels raises the red hardness properties and because it also retards grain growth, higher quenching temperatures are achieved. It enhances the properties of the major alloy components in the more complex steels, increases strength, hardness and shock resistance.

The steel used in this helicopter is a medium carbon alloy designated as AISI/SAE 4130. This general-purpose alloy group 4130, 4140, 4142, 4145, 4147 and 4150 are widely used due to their low cost. Characteristics of this steel are good hardenability, strength, toughness, wear resistance and ductility. This wide range of strength and toughness is accomplished through various heat treatment techniques. It is one of the most commonly used aircraft steels because of its ease of fabrication, weldability and mild hardenability. In relatively thin sheet sections high strength levels can be attained through heat treatment even when normalized. Increased wear and abrasion resistance is also attainable by nitriding. The structural tubing used for the helicopter conforms to specification MIL-T-6736 Normalized, the round cold finished rod to MIL-S-6758A Normalized and the cold rolled strips to MIL-S-18729C Normalized

What exactly do we mean by alloy steel, medium carbon, 4130, normalized? Lets start with the "alloy" designator. For steel to be considered an alloy type it must have at least one of it's alloy components exceed one or more of the following limits: manganese, 1.65%; silicon, 0.60%: copper, 0.60%, aluminum, boron, chromium up to 3.99%, and or trace amounts of cobalt, columbium, molybdenum, nickel, titanium, vanadium, zirconium or any other alloying element used to obtain a certain effect.

As for the number designator given to the steel, in the AISI numbering system, if steel has no prefix letter then the basic open hearth or oxygen process manufactured it. The basic electric furnace process may be utilized if the steel's phosphorus and sulphur limits are adjusted. If the prefix E is given then the steel was only produced by the basic electric furnace process. In the four- numeral steel series the last two digits indicate the approximate middle of the carbon range. For the steel we are using, 4130, the 30 represent a carbon range of 0.25 to 0.35 percent. In a five-numeral series the last three digits represent carbon content. Deviation to this rule applies to some carbon ranges and the variations that occur with the addition of manganese, sulfur, chromium or other element combinations. That being said for 4130 steels the elemental component breakdown is as follows:

C =0.27 / 0.33; Mn =0.30 / 0.70; Cr =0.30 / 0.70; Mo =0.08 / 0.15

The first two digits indicate the type of alloy elements in the steel according to the following grouping.

13xx	Manganese 1.75%
40xx	Molybdenum 0.20 or 0.25%
41xx	Chromium 0.50, 0.80 or 0.95%; Molybdenum 0.12, 0.20, or 0.30%
43xx	Nickel 1.83%; Chromium 0.50 or 0.80%; Molybdenum 0.25%
44xx	Molybdenum 0.53%
46xx	Nickel 0.85 or 1.83%; Molybdenum 0.20 or 0.25%
47xx	Nickel 1.05%; Chromium 0.45%
48xx	Nickel 3.50%; Molybdenum 0.25%
50xx	Chromium 0.40%

51xx   Chromium 0.80, 0.88, 0.93, 0.95 or 1.00%
5xxxx   Carbon 1.04%; Chromium 1.03 or 1.45%
61xx   Chromium 0.60 or 0.95, Vanadium 0.13 or 0.15%
86xx   Nickel 0.55%, Chromium 0.50%, Molybdenum 0.25%
87xx   Nickel 0.55%, Chromium 0.50%, Molybdenum 0.35%
88xx   Nickel 0.55%, Chromium 0.50%, Molybdenum 0.45%
92xx   Silicon 2.00%

Of course the there are an infinite number of small variations to the above list that can be used in making steel. In an effort to standardize the steel manufacturing industry the types listed above are selected on the basis of satisfying the predominant demands of metallurgists and engineers to support fabrications in generating steel products with the desired properties.

9.2.3 Composites

After reading my initial comments at the beginning of the chapter you know I am not a strong supporter of composites in any critical structural part of my helicopter. Their use has become increasingly popular in both aircraft and auto design where weight reduction is a primary concern. In an effort to make this book complete, here is what I have researched on the subject.

Initially the push to develop stronger lighter more durable substitutes for steel and aluminum was driven by the aerospace industry. For the home aircraft builder, composites started to be utilized in construction around 1970 with the introduction of two kit planes the KR-1 and KR-2 by Ken Rand. Burt Rutan's plane "VariViggen" (which earned him the Stan Dzik design trophy at the Oshkosh air show in 1972) also used composites. Burt continued his effort to convince aircraft builders that the future was composites and did so by conducting numerous seminars and publishing many construction manuals on the subject. The funding for his work was in part, generated by his extremely successful kit plane named the VariEze produced at his Rutan Aircraft Factory. After a string of popular kit plane designs Rutan decided to start another company dedicated to advanced composite design and prototype construction building proof-of-concept aircraft. This company was called SCALED. Of course their were many other scientists and engineers that developed the materials needed to produce these high strength composites Rutan used, but he more or less proved to the aircraft enthusiast that they worked. He also perfected the mold less composite technique to simplify composite construction.

In the 1980's kit plane manufacturers such as Nat Puffer (introduced the Cozy), Quickie Aircraft founded by Tom Jewett and Gene Sheehan (pioneers of the pre-fabricated "fast build" kit), Tom Hamilton (Glasair), Lance Niebauer (Lancair), followed by Ken Wheeler (Express) all extended the composite aircraft build process. Composites are universally recognized as enabling complex designs to be fabricated significantly easier than with conventional materials and at lower total cost (material plus manufacturing).

Listed below are those aerospace composite fabrics available to the builder that replaced the common fiberglass weave for enhanced design application.

Note all costs stated here were as of 2013 and are listed only to give numbers for comparison purposes between these stated composite types. One important point the term "Wet Out" is used throughout these material descriptions. What that means can be compared to trying to soak up water with an oily rag versus a chamois. Obviously, the chamois will be completely saturated within minutes whereas the oily rag virtually water free. In this case we would say the chamois was completely "wetted out" and due to all the water content its color would chance from pale yellow to almost brown. A composite fabric's ability to absorb the epoxy resin is gauged similarly by how well it "Wets Out" the material.

- Composite Materials
  1) S glass:

Characteristics: uses a different chemical formulation from E glass. One ply of S glass can replace several plies of E glass

Workability: good in lay-up composites due to its ability to wet-out with the resin. An additional benefit of using S glass is its change in appearance when saturated by the resin, which is a good visual to the fabricator that the proper wet-out of the material has been achieved.

Properties: S glass is 15% stiffer and 30% stronger than E-glass and retains these properties up to 1500 degrees: This material is considerably tougher than an equivalent E-glass. S glass has half the strength and stiffness of Kevlar and twice the weight.

Availability: typically stocked in 60 inch widths, 0.009 inches thick plain weave (5.8 oz./ yd.)

Cost: $14.45 per lineal yard

2) Bi-directional Kevlar: Du Pont introduced Kevlar 49 (aramid fiber) to the commercial market in 1972

Characteristics: soft to the touch, yellow in color, can be combined with other fabrics to produce custom composites

Workability: hard to work with in hand lay-up process, resins typically used with glass fiber can be used with Kevlar 49 and which resin to use depends on the composite strength desired. When developing custom composites Kevlar 49 does not bond well with polyesters therefore they are not recommended.

Properties: strong, tough, impact resistant, superior vibration damping, light weight. Tensile strength 43,000 psi, modulus of elasticity $19 \times 10^{**}6$ psi, meets all FAA flammability requirements, maintains properties when subjected to water, high humidity, salt water, lubricating oils, or fuel. It also has excellent performance in cryogenic temperatures (-320 F), however it has low compressive strength.

Availability: available as rovings and yarns. There are three common types of fabrics typically stocked by suppliers Kevlar#120 is a light weight fabric 1.8 oz./yd. and only 0.0035"thick, while Kevlar#281 & Kevlar#285 are more substantial cloths differing only in weave pattern (5.0 oz./yd. and 0.010"thick)

Cost: width 38" price per lineal yard, Kevlar#120-$18.80, Kevlar#281-$17.10, Kevlar#285-$20.00

3)   Unidirectional Kevlar: KS-400
Characteristics: combines the benefits of fiberglass S-2 glass with Kevlar 49 generating a unidirectional reinforcing laminate with high tensile strength and high modulus of elasticity along with lightweight.
Workability: good in lay-up composites due to its ability to wet-out with the resin. An additional benefit of using S-2 glass is its change in appearance when saturated by the resin a good indicator to the fabricator that the proper wet-out of the material has been achieved. Impregnating resins should only be epoxies or vinyl esters.
Properties: because of this ability to wet-out the S-2 glass the properties of bonding and impact strength are enhanced compared with pure woven Kevlar 49 even though the amount of the S-2 glass used is relatively small compared with the enhanced properties achieved. Laminates produced are exceptionally stiff and are resistant to delaminating. They can also be produced with greater fiber-resin ratios.
Availability: typically sold in 12" wide rolls
Cost: width 12"price per lineal ft, $3.50
4)   Bi-directional Graphite:
Characteristics: excellent substitute for Kevlar, produced from rayon fibers that are severely stretched while heated which alters the molecular structure.
Workability: cuts easier than Kevlar, standard weaves can be purchased as "pre-pregs" where the graphite is already impregnated with either polyester or epoxy resin, although this appears to save time, which it does for the major aircraft manufactures the equipment and quality control required to properly cure pre-pregs puts the process outside the scope of the home builder.
Properties: as strong as 3 to 4 layers of standard fiberglass, has low density with great strength, tensile strength warp direction 300 lbs./in; fill direction 300 lbs./in.
Availability: typically stocked in three styles Graphite #282 (5.7 oz./yd., 0.007"thick, 42"width), Graphite#284-50 (5.8 oz./yd., 0.007"thick, 50" wide) and Graphite #584 (10.9 oz./yd., 0.013" thick, 42" wide)
Cost: Graphite #282, $16.50 lineal ft; Graphite #284-50, $20.90 lineal ft; Graphite #584, $30.20 lineal ft.
5)   Ceramics:
Characteristics: advanced composite
Workability: can be laminated to a quality similar to that achieved with S glass
Properties: light weight and can handle temperatures up to 3000 degrees, excellent choice for engine compartment firewalls
Availability: difficult to obtain from listed suppliers
Cost: high

In Summary: Graphite fibers are created by extreme stretching and heating of rayon fibers to change their molecular structure. Graphite has very low density (weight/unit volume), is very stiff (high modulus) and very strong (high tensile). S glass uses a different chemical formulation from standard E glass fabrics, and is stronger, tougher and stiffer than E glass. One ply of S glass can replace several plies of E glass, which can result in a stronger and considerably lighter aircraft component. Ceramic fabrics are the latest innovation in advanced composites. These fabrics produce laminates approaching the qualities of S glass plus they can withstand temperatures of almost 3000° F. Ceramic cloth can produce a very lightweight and effective firewall laminate, although at this time the cost is high.

Often the choice of the materials to use for a laminate is difficult because of the required properties. One must consider the advantages of one material over another and its anticipated performance. S glass is about 30% stronger and 15% stiffer than E glass. It has 20-25% of the stiffness of graphite and is as strong, but it is also 30% heavier. S glass though, has only half the strength and stiffness of Kevlar and twice the weight. Kevlar on the other hand, is 40% stronger and 25% lighter than graphite but has only half the stiffness of Graphite. Sometimes, blending different advanced composite fabrics in a laminate can achieve the proper balance of stiffness, strength and weight.

• Epoxy Systems

It is very important for any particular advanced composite cloth to absorb the resin (or as they call it in the industry "to wet out") in order for the material to achieve its' stated strength. Each type of cloth has a specific permeability and only the correct resin match will "wet" it. It is extremely important to verify with the cloth manufacturer if the resin you want to use is an acceptable match.

I used a Kevlar cloth as a filler material for the nose section of the rotor blades and "Poly Epoxy" for the structural epoxy system. Consequently, this is the only epoxy system I will discuss. I'll start off by saying this resin isn't inexpensive and even if you don't open the can or converter it has a limited shelf life so you better intend to use it as soon as it is delivered to your door. That being said Poly Epoxy is a high-performance epoxy resin with unparalleled compressive, tensile, impact and flexural strengths, along with great peel, shear fatigue resistance and fracture behavior. You can either use it with or without molds and it works exceptionally well in vacuum bagging processes. It cures in two phases. The two phases occur all by themselves in the overall curing process and the resulting bond is tougher and stronger than any other.

Poly Epoxy Technical Data

Mechanical Properties:	w/Post Cure	wo/Post Cure
Tensile Strength, psi	9600	8800
Elongation at break, %	7.5	3.6
Tensile Modulus, psi	470,000	460,000
Flexural Strength, psi	19,000	14,500
Flexural Modulus, psi	515,000	500,000
Compressive Strength, psi	32,000	33,000
Shore D Hardness	82	70
Glass Transition Temp. C	72	62
Heat Distortion Temp. C	64	50

Water Immersion Wt. Gain*   2.8   2.9
   * % @140 degrees F, 30 days
Rheology: Mixing ratio: 3 parts Resin to 1 part Converter by Weight; 10 parts Resin to 4 parts Converter by Volume.
Kinetics: Pot life, 100 grams…105 minutes; 1 quart….75 minutes
Mold Open Time….3-4 hours, Tack Free Time….5-6 hours

Chapter 10- Commercial Parts
  This chapter covers the reasoning behind the selection process of the various commercial parts used for this helicopter. I start with fasteners, then move on to an in depth look at power transmission types (timing belts, v-belts, chain, bearings etc.). The information presented here is a general overview of the subject material. Some actual calculations justifying a particular drive type used on the helicopter are found in the Engineering section.

10.1 Fasteners
  Lets put it this way, just use steel aircraft grade castle nuts and bolts of the AN classification type. They are held to high manufacturing standards and the bolts have thru holes enabling cotter pin placement. Why steel instead of aluminum fasteners? The steel ones, size for size, are three times as heavy and all the parts being fastened are aluminum, plus we need over fifty fasteners in the build in this supposedly ultralight design yet they were my choice. The reason they are preferred is that the vibration load on the fastened parts is concentrated at the bolted connection. Since a steel bolt is over five times as strong as the aluminum one, when you're not sure of the stresses, always err on the safe side. To drive that point home just look at the bolt with the sheared off head in the Failure Analysis section.
  Here is one more tip concerning fasteners. As I mentioned earlier in this book, DO NOT DRILL OUT THE MATING PART HOLE TO A LARGER DIAMETER IN ORDER TO MAKE THE BOLT HEAD PERPENDICULAR TO THE SURFACE. The misalignment condition often occurs when installing gusset plates, especially when working to mirror image the right and left hand side of the helicopter. Normally with the allowable bolt hole clearances if all the bolts are positioned first before tightening this alignment problem will be avoided.

10.1.1 Fastener Engineering Data
  The entire bolt / nut hardware used on this helicopter is aircraft rated AN specified. In the strictest terms according to proper QC (Quality Control) this hardware should be stored in separate bins in a dedicated location in your shop. If you were an actual helicopter manufacturer certification information would also be required in order to track any future part failures. Fortunately, these AN bolts have a unique stamping on the bolt head so when you receive them you can easily check the bolt marking to make sure they are what you ordered, nuts however are not marked.

Aircraft rated fasteners are very expensive and after loosing several while trailering the helicopter to the airfield I decided to go exclusively with cotter pins and castle nuts in just about every bolted connection. On the main and tail rotor, safety wire was used instead of cotter pins to insure those bolts didn't vibrate loose or rotate relative to one another. There are other ways to keep nuts from vibrating loose but in the long run I found this castle nut / cotter pin approach to be the simplest. Just for the record from my perspective the next best fastener possibility is using double nuts. This is where you add a second nut to the one already torqued and torque it to the existing nut. You could also use self-locking nuts but I personally like the positive lock a cotter pin gives. Lock-tite may be okay for use on your car but definitely not on your aircraft. One more point concerning castle nuts, the slot on the nut must align with the drilled hole on the bolt so the cotter pin can fit. You torque castle nuts the same way you would any standard nut. If the slot and hole do not line up you may add additional torque until the next slot aligns. This over torquing leeway applies only to castle nuts.

Two more points concerning bolted connections: the nut and bolt head must be flush against the material to be fastened. I personally have snapped a bolt head off a 1/4–20 bolt during the torquing process because of the gap that existed between one side of the bolt head versus the other side. Consequently, if you don't want premature bolt failure fix the alignment problem first instead of trying to make the bolt tightening process do it for you. Next point, no threaded portion of the bolt should be within the material to be fastened. Threads on a bolt are stress risers and are the first part to fail if subjected to a shear load generated by a material shift. The point here is the bolt shank snug fit in the drilled hole keeps the two pieces of material from moving so there is minimal movement at the threaded connection. If the clamping connection is designed correctly the force on the two parts in conjunction with the materials coefficient of friction will make any relative material movement improbable.

The following tabulation gives the rated strength (lbs) of each AN size:

AN Part #	thread	T	Yield Tensile (at root dia.)	Single Shear (at full dia.)
AN3	10-32	NF-3A	1,690	2,125
AN4	1/4-28	UNF-3A	3,130	3,680
AN5	5/16-24	UNF-3A	4,980	5,750
AN6	3/8-24	UNF-3A	7,740	8,280
AN7	7/16-20	UNF-3A	10,430	11,250
AN8	1/2-20	UNF-3A	14,190	14,700
AN9	9/16-18	UNF-3A	18,100	18,700
AN10	5/8-18	UNF-3A	23,080	23,000
AN12	3/4-16	UNF-3A	33,730	33,150
AN14	7/8-14	UNF-3A	46,000	45,050
AN16	1-14	NF-3A	61,870	50,980
AN17	1-12	UNF-3A	61,870	58,900
AN18	1-1/8-12	UNF-3A	78,050	73,750
AN20	1-1/4-12	UNF-3A	99,820	91,050

Stated below are the Torque Values (inch-pounds) for Oil-Free Cadmium-Plated Threads. It should be noted that you should not torque a bolt by turning it. All torque to the fastener should be achieved at the nut. Although it is not prohibited to torque at the bolt end it will induce additional internal bolt stress.

Bolt Size	Nuts	
	Tension-Type MS20365 AN310; 40,000psi	Shear-Type MS20364 AN320; 24,000
8-32	12-15	7-9
10-32	20-25	12-15
1/4 –28	50-70	30-40
5/16-24	100-140	80-85
3/8-24	160-190	95-110
7/16-20	450-500	270-300
1/2-20	480-690	290-410

10.1.2 Nuts & Bolts

The AN types bolts listed above are available in 1/8 inch length increments. The mating nuts for the AN bolts are available as either tension AN310 or shear type AN320 depending on the fastening application.

10.1.3 Washers

Use AN960 type. They are used to provide a smooth bearing surface. The subtle point here is your bolt has a tensile strength of over 100,000 psi and the aluminum pieces you are bolting together have a tensile strength of less than 40,000 psi. Without the washers the aluminum surface is really marked up after several disassembly / retightening operations. Using the washers prevents this deterioration of the aluminum surface but be aware these washers rarely last more than one or two retightening actions and must be replaced, so make sure you order way more than you need. The washers are also important as shims to obtain the correct grip length for a bolt and nut assembly, and to adjust the position of AN310 and AN320 castle nuts with respect to drilled cotter pin holes.

10.1.4 Cotter Pins

Use MS24665. Cotter pins should fit snugly but not so tight they have to be hammered in. One end of the cotter pin should be bent adjacent to the nut and the other towards the center of the bolt. They can be purchased in either cadmium-plated or corrosion-resistant steel.

10.1.5 Rivets

There must be over a thousand types of rivets available for you to use, both standard type and blind. Standard rivets require you to mechanically mushroom the protruding end whereas blind rivets automatically seat the blind end from the insertion side. It seems every time I open a product catalog more are available so for the blind rivets I'm only going to discuss Cherry commercial rivets. The strongest is the Cherry "Q" rivet. It is a structural self-plugging rivet that plugs the entire length of the rivet sleeve providing full shear strength for structural and load bearing applications. It comes in three series, the BS (aluminum rivet with steel mandrel; the 1/8 inch diameter one is tension rated at 325#, this is the one I used), the MS (monel rivet with steel mandrel, 1/8 inch tension rated at 525#) and the CC (stainless steel rivet with stainless steel mandrel, 1/8 inch tension rated at 600#).

The standard rivets used on the tail and main rotor blades are 1/8-inch diameter AN426AD. These come with a flush head having a tensile rating at 38,000 psi and the shear rating 26,000 psi.

10.2 Power Transmission:

DISCLAIMER: be forewarned none of the following power transmission manufacturers will okay the use of their product to be used in your helicopter!!!! Why? Simply put they don't want the liability. Every part I ordered, when asked I said it was for a prototype machine that I was building. I did not specify further.

Power Transmission refers to the way power, in this case engine power, is transferred to the driven device, which on our helicopter is either the main or tail rotor and some fractional amount to the ancillary equipment such as the water pump, tachometer, etc.

Just for example lets look at how that engine / transmission power is transferred to the rear wheels on your typical motorcycle. Transmitting this power from the transmission to the rear wheels is accomplished in one of three ways; chain, belt or shaft.

Lets start with the chain drive. The sprocket chain combination has been around since time itself. Why? It's straightforward simple. Just swapping out the sprockets and repositioning the rear wheel to accommodate the new chain length easily changes final drive ratios. The meshing of chain to sprocket is virtually self-cleaning which is why they're used exclusively in production race dirt bikes. They are relatively inexpensive and easy to maintain with exception to the periodic chain oil bath. The main problem with the chain / sprocket combo is once the chain starts to stretch (exceeds its elastic limit) the chain link cross sectional area decreases and its ability to handle the torque reduced causing the chain to stretch even faster. Commercial equipment manufacturers added take-up idlers to their equipment to address the issue of chain stretch. On motorcycles this was not practical so the alternate solution was to use either a fixed center distance cog belt drive or a shaft. The cog belt is now used on some Harley Davidson touring bikes because they don't stretch, are slightly more efficient (mainly because there is no power loss due to chain link friction), create less running noise and require less maintenance. The down side is you never take one of these bikes mud bogging!

Finally we come to shaft drive. To start it's a complicated setup. You need two right angle gearboxes, one at the transmission the other at the rear wheels and a connecting shaft to tie the two together with universal joints at each end to accommodate rear wheel jounce and maybe a spline connection for any additional rear wheel translation. Needless to say the arrangement has all the advantages of a belt drive and you could theoretically take it mud bogging but for sure it isn't cheap or power efficient. In addition final drive ratios are not changeable.

The motorcycle example gave you a quick overview of what power transmission is all about. Now lets take a look at all the possible power transmitting methods I investigated for this project. Here's the list: "v" belts, synchronous belts "standard and high torque", roller chain, timing chain, and transmissions.

Needless to say when shafts rotate somewhere along the line you're going to need some bearings. During that exploratory work I looked at, needle bearings, ball bearings, roller bearings, spherical bearings, and rod end bearings. Along with that subject one has to investigate the issue of lubrication.

The purpose of this section is to familiarize you with how to tackle the selection process of each topic mentioned.

10.2.1 Sprocket / Chain Drive:

In general your basic chain drive is not used when the small sprocket RPM exceeds 1150 (ANSI 60 chain) and for the ANSI 240 chain that small sprocket maximum RPM is 250. Chain drive should only be considered for the low rpm power transfer such as the final main rotor drive.

Photo 10.2.1-1

Engineering:
  Sprocket Selection:
  Step 1. "Determine the Service Classification" whether it will be Uniform Load (generators, etc), Moderate Shock Load (conveyors, etc.) or a Heavy Shock Load (Presses, etc.).
  Step 2. "Find the Service Factor" for each of the above based upon the type of input power being either internal combustion engine with hydraulic drive (lowest service factor; range from 1 to 1.4), electric motor or turbine (range from 1 to 1.5), or internal combustion engine with mechanical drive (range from 1.2 to 1.7). Additionally an additional 0.2 must be added to the above

for unfavorable conditions such as multiple shafts, excessive speed ratios, heavy starting loads, high temperatures, poor lubrication conditions or operation in abrasive circumstances.

Graph 10.2.1-1

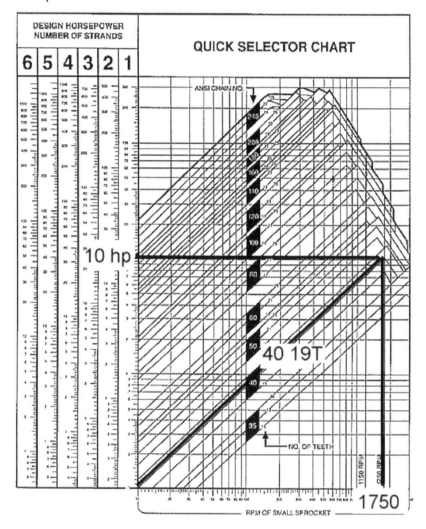

Note: As you can see from the chart the larger the chain pitch the lower the allowable RPM of the small sprocket. If the design horsepower we need at the RPM desired is greater than that stated for the largest pitch chain horsepower rating at its maximum allowable speed then a multiple chain drive should be considered. Horsepower values for multiple strand drives are shown in the left hand column of the chart. Any drive selection, which operates below the maximum rating stated here, will be noisier and have reduced operating life. (Shown 40 chain, 19T, 1750 rpm, transfers 10 hp.)

Step 3. "Determine the Design Horsepower" by multiplying the normal operating horsepower by the Service Factor found in Step 2.

Step 4. "Drive Selection" is developed by first working around a best guess approach by comparing the design horsepower calculated in step 3 with the generalized "Selection Chart" shown previously. At the present time we have not determined any rotational shaft speeds but if we had you would use the chart to determine the smallest pitch chain that has the required horsepower rating for a pinion sprocket rotating at the specified RPM.

Step 5. The "Driving Sprocket" should have a minimum of 17 teeth, however sprockets with a few as 7 teeth can be obtained. In any case all sprockets with 25 teeth or less should be hardened.

Step 6. The "Driven Sprocket" size is set by the driving sprocket selection for any fixed speed ratio needed to obtain the desired driven shaft RPM. As the driving sprocket number of teeth increases so does its pitch diameter and correspondingly increases the driven sprocket size, resulting in increased packaging space and overall drive weight. Shown below is a Speed Ratio Table to aid in selection of the driven sprocket size given the driver sprocket and the required speed ratio.

Some additional recommendations: the maximum speed ratio for a single chain setup is 7. If more reduction is required then a double reduction drive should be considered.

Step 7. The "Shaft Centers" calculation is next where you can determine the sprocket center distance given a specific chain length and the sprocket diameters, or you can calculate the "Chain Length" if you already know the shaft center distance and the sprocket dimensions. The actual formulas for this work are shown in the drive calculations found in Chapter 14.

Chart 10.2.1-1

Speed Ratios For Sprocket Combinations
Driver Sprocket Teeth

Driven Sprocket Teeth	9	10	11	12	13	14	15	16	17	18
9	1.00									
10	1.11	1.00								
11	1.22	1.10	1.00							
12	1.30	1.20	1.09	1.00						
13	1.44	1.30	1.18	1.08	1.00					
14	1.56	1.40	1.27	1.17	1.08	1.00				
15	1.67	1.50	1.36	1.25	1.15	1.07	1.00			
16	1.78	1.60	1.45	1.33	1.23	1.14	1.07	1.00		
17	1.89	1.70	1.55	1.42	1.31	1.21	1.13	1.06	1.00	
18	2.00	1.80	1.64	1.50	1.38	1.29	1.20	1.13	1.06	1.00
19	2.11	1.90	1.73	1.58	1.46	1.36	1.27	1.19	1.12	1.06
20	2.22	2.00	1.82	1.67	1.54	1.43	1.33	1.25	1.18	1.11
21	2.33	2.10	1.91	1.75	1.61	1.50	1.40	1.31	1.23	1.17
22	2.44	2.20	2.00	1.83	1.69	1.57	1.47	1.38	1.29	1.22
23	2.56	2.30	2.09	1.92	1.77	1.64	1.53	1.44	1.35	1.28
24	2.67	2.40	2.18	2.00	1.85	1.71	1.60	1.50	1.41	1.33

Data Sheets:

To get the complete picture of the amount of data to be explored during the sprocket selection process would require looking at over 100 pages in the vendor's product catalog. That amount of detail will not be included here. Nevertheless I will highlight the types of topics covered but before I list those I just have to mention one area that was particularly embarrassing to me as a young engineer.

At the time I had already completed all the engineering work and was ready to specify the sprocket. Basically, there are two ways you can order a sprocket either bored to size or one which accepts an interchangeable bushing bored to size. Here are the possible types of bushings: taper, QD type B, QD type C, weld-on hub type WA, weld-on hub type S, weld-on hub QD type 1, and weld-on hub QD type 2. To make a short story even shorter when the hub and sprocket I ordered came in they were the wrong match. Now, I have to say the assembler who called me into the shop just happened to be an okay guy and spared me the flack normally associated with this type of "screw-up". In fact, he even took some time to educate me a little on the subject. The point here is making sure you order the correct hub to mate with type sprocket you want. Check and double-check your selections before ordering.

10.2.2 Silent and HV Chain Drive:

Photo 10.2.2–1

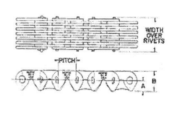

Pound for pound the HV Chain Drive was my preferred method for driving the main rotor shaft. Look at the comparison chart on the next page for confirmation of that fact.

Briefly, the HV link design, the compensating pin, rocker joint and involute hobbed sprockets enabled high chain speeds, 99.7% efficiency, higher transmitted horsepower per inch width with less vibration and virtually noiseless operation on high speed applications. What was not to like? Actually, I went thru all the engineering and ordered the drive for the main rotor shaft. Since this HV drive was so impressive I will just skip over "Silent Drives" in general and just limit my discussion to this particular drive.

Graph 10.2.2-1

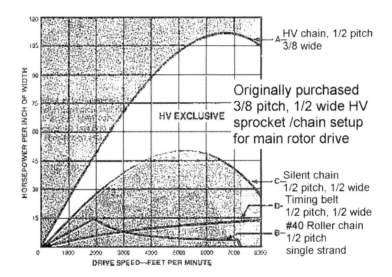

Engineering:
"HV Drive Selection" selection consists of the following steps:

Step 1. Determine the diameter of the high-speed shaft and its RPM.

Step 2. Determine the "Total Transmitted Horsepower" from your power source.

Step 3. Look up the appropriate "Service Factor" from the table given in the catalog.

Step 4. Calculate the "Design Horsepower" by multiplying your "Total Transmitted Horsepower" by this "Service Factor".

Step 5. Select the Small Sprocket by looking at the "Horsepower Rating Tables" to determine its pitch, number of teeth, width and maximum allowable bore. Note that you must make sure this maximum allowable sprocket bore will fit the intended shaft.

Step 6. Determine the drive ratio required, where the ratio is equal to the (RPM of the High Speed Shaft) divided by the (RPM of the Slow Speed Shaft)

Step 7. Calculate the Large Sprocket number of teeth by multiplying the "Ratio" by the "Small Sprocket" tooth count.

Step 8. Calculate Chain Length from the book formula given that HV drives use modified center distances to compensate for sprocket and chain tolerances.

Data Sheets
Basically there are six different pitches (3/8, 1/2, 3/4, 1, 1 ½, and 2) which range in width depending on the pitch. The 3/8 pitch has widths ranging from 3/4 to 3 inches whereas the 2-inch pitch starts off at a 3-inch width and goes up to 6 inches in one-inch increments. Horsepower rating tables,

service factor tables, chain length calculation sheets, lubrication requirements, hub types, sprocket physical dimension sheets and chain tensile strength and weight per foot charts cover all the specifics for this drive in the one vendor's catalog. Just for the record the smallest drive (3/8 pitch by 3/4 wide) chain has an average ultimate tensile strength of 5,625 pounds and has an average weight per foot of 0.65 pounds whereas the largest HV chain weighs 27.6 pounds per foot has an average ultimate tensile strength of 240,000 pounds and can transmit over 300 horsepower per inch in certain drive applications.

Conclusion! I never used the drive. I intended to but here's what happened. For my application the drive required bath type lubrication and to complicate things further my main rotor drive is horizontal so I didn't know exactly how to keep the chain from dropping off the sprockets without some elaborate design work.

10.2.3 V-Belt Drive
Photo 10.2.3-1

V-Belt drives are the ones of choice when transmitting power from shafts with high rotational speeds, in our case transferring power from the engine when operating at 6000 RPM or to the tail rotor. Needless to say their use is wide spread in the commercial world of machines and process lines. When sized properly they have a long operational life (when is the last time you replaced a car fan belt or a washing machine pump belt?), can be used on shafts that are at right angles to one another, can be used to clutch a drive into action, require no lubrication, are relatively inexpensive and can operate with variable speed sheaves.

Currently the Browning Company, which has been in business since 1886 is probably still the world's largest manufacturer of V-Belt, drives and their B5V product line serves 90% of all 10 to 125 horsepower applications. The classical belt sizes were A, A-B, C and D; the newer "Gripbelt" line belts sizes are 3V, 5V and 8V.

Engineering:

"Gripbelt Drive Selection" selection consists of the following steps:

Step 1. Determine the Belt Section from the follow chart:

Chart 10.2.3-1

Belt Section Selection Chart

Table No. 1

HP	Belt Section			
½	A	AX		
¾	A	AX		
1	A	AX		
1½	A	AX		
2	A	AX		
3	AX	A	BX	
5	BX	AX	B	A
7½	BX	AX	B	3VX
10	BX	B	AX	3VX
15	BX	3VX	AX	B
20	BX	3VX	B	
25	5VX, 5V	3VX	B	
30	5VX, 5V	3VX	B	
40	5VX, 5V	B	3VX	
50	5VX, 5V			
60	5VX, 5V			
75	5VX, 5V			
100	5VX, 5V			
125	5VX, 5V			
150	5VX, 5V			
200	5VX, 5V			
250	5VX, 5V	CX		

Horsepowers stated here are maximums and must be derated for pulley diameter and rpm. This size used for engine pulley

Step 2. Determine the Design Horsepower by multiplying the rated Horsepower by the Service Factor for the application (chart in catalog)

Step 3. Get "Drive Ratio" by looking it up in the Drive Selection Table (in catalog) or by calculation, where Drive Ratio equals Driver Speed divided by Driven Speed.

Step 4. Determine the sheave combination from the appropriate "Drive Selection Table" by looking at the "Horsepower per Belt" column to find the smallest sheave diameters with the least number of belts that satisfies the Design Horsepower requirements.

Step 5. Verify in the "Minimum and Maximum Number of Grooves" column of that same "Drive Selection Table" if the sheave combination you selected in Step 4 is available with the number of grooves you need.

Step 6. Determine the "Belt Part Number" for the approximate Center Distance and the "F" factor (horsepower correction factor for Browning belts)

Step 7. Calculate the Corrected Horsepower of the Drive by multiplying the Horsepower per belt by the "F" factor and by the number of belts.

Step 8. Get part numbers for the components selected.

Data Sheets

The one company has over 2100 stock sizes of single and multiple groove sheaves so we are not going to go there, rather I suggest obtaining one of the electronic catalogs available. These type catalogs will not only give you product information but also CAD (Computer Aided Drafting) template drawings.

10.2.4 Synchronous Drive:

Photo 10.2.4-1

The first time I "really" noticed a timing belt (synchronous) drive I was around twelve. It was in the late 1950's in the true peak of the "Hot Rod" era when chopped and channeled (lowered and customized) 1940 Mercs (Mercury cars) ruled the road (think ZZ Top's car or Stallone's car in the movie Cobra). This was a time when "Bondo" or fiberglass kits that you can now buy at stores as common as Wal-Mart, did not exist. Then the metal body work customization was performed by true artists working their craft using molten lead to create the necessary sculptured beauty of say bullet tail lights. In addition to the bodywork, it was the practice to bolt on centrifugal blowers to the big block V-8's (think of the hopped up Pontiac GTO in one of the "Fast and Furious" movies). The drive used to power these "Blowers" was via a 3 or 4" wide timing belt. From a kid's perspective it was an awesome sight to see and hear.

Over the years timing belt technology has progressed so far that a timing belt instead of chain now performs the engine crankshaft drive to the camshaft or shafts. Goodyear is now making the Eagle Pd Synchronous Belt drive system that virtually obsoletes metric and standard timing belt drives by transmitting more horsepower per face width. They are superior to chain drives in that they are more efficient, quieter, and do not require re-tensioning or lubrication. In comparison to V-belts they are more efficient, transmit more horsepower per unit space and stretch less.

After my review of all the advantages this drive seemed to offer I decided to use it. This is the drive discussed below. Bear in mind this drive is expensive especially the belt so it probably would not be first choice if packaging space and component weight was not a concern. It is my recommendation you also consider Gearbelt and HPT (metric) synchronous belt drives found in the referenced catalogs listed in the Chapter 12.4.

Engineering
Photo 10.2.4-2 The PD Synchronous Belt Sprocket

Right off the top you can see why this PD drive can transmit more power. The belt engaging surfaces are longer than your typical timing belt drive where grooves are just straight and perpendicular across the face. Now for the selection process bear in mind you will need the vendor's PD Engineering Manual to do this. You can obtain the manual from one of the vendors listed in Chapter 12.4.

Preliminary: Establish the horsepower for the drive, DriveR and DriveN RPM, the acceptable center to center distance range, hours of operation each day, type of DriveN equipment and load characteristics, space limitations for the drive (maximum diameter and width) and shaft sizes.

Step 1. Calculate the Speed Ratio

Step 2. Determine the Service Factor

Step 3. Calculate the Design Horsepower by multiplying the "Rated Horsepower" by the "Service Factor"

Step 4. Determine the Sprocket Combination by using the Drive Selection Table and your Speed Ratio as a starting point. Check selected possible combinations for NEMA minimum sprocket size restrictions if applicable. Verify that the center distance criteria is met, and insure that sufficient packaging space is available noting that a half inch should be added to each sprocket diameter for belt clearance. Then note the color combinations that are available for the selected sprockets.

Step 5. The Color Choice is determined by looking up the rated horsepower for all available colors. Then multiply the rated horsepower by the length correction factor to obtain the belt corrected horsepower for each color drive under consideration. If this value is greater than the horsepower determined in step 3. it can be used. Typically, the color drive horsepower value closest to the step 3 horsepower is the one that should be used since it probably will be the least expensive.

Step 6. Generate a list of Part Numbers for the items selected.

Step 7. Installation and Tensioning instructions should be investigated to make sure the tensioning required is possible for your application.

Data Sheets

Types of information contained in the Engineering manual is as follows: Horsepower ratings for the various color codes pitch diameters versus rpm; Belt Correction Factors; Service Factors; Drive Selection Tables; Center Distance Allowances; Belt Tension Deflection Forces; Tensioning Ratio Factor, Belt Dimensions and Sprocket Dimensions.

10.3 Bearings

10.3.1 Plain Bearings

The plain bearing is an inexpensive way to support a shaft whether rotating or sliding. There are many types brass, composites, etc, and some are self-lubricating. In some applications they are quite effective but have a higher coefficient of rolling friction than roller or needle bearings. Helicopter Application: Control box "Cam" inserts and Foot pedal inserts.

Photo 10.3.1-1

10.3.2 Needle Bearings:

Provide excellent radial load carrying capability in a small space. Helicopter Application: Main rotor hub Flap and Feathering bearings.

Photo 10.3.2 –1

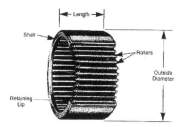

10.3.3 Ball Bearings

Good radial with some thrust load capability. Can be purchased with end seals for permanent lubrication. Helicopter Application: Main rotor shaft support bearings, Tail rotor shaft bearings, Internal Swashplate bearing, Idler roll bearings, gearbox output shaft support bearing.

Photo 10.3.3-1

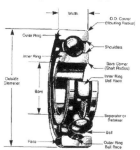

10.3.4 Ball Thrust Bearings

Far superior to ball bearings in handling a thrust load, but not used on our helicopter because the needle thrust bearings listed next packaged in a smaller space.

Photo 10.3.4-1

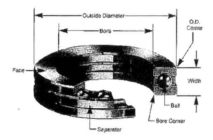

10.3.5 Needle Thrust Bearings

Excellent thrust load capability in a small package. Helicopter Application: Main rotor Feather bearings.

Photo 10.3.5-1

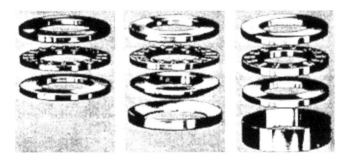

10.3.6 Straight Roller Bearings

Excellent radial load capability, not used on this helicopter.

Photo 10.3.6-1

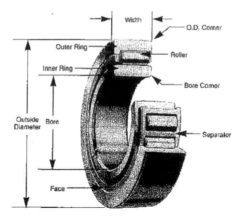

10.3.7 Tapered Roller Bearing

Photo 10.3.7 –1

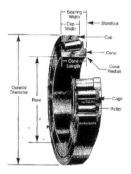

Taper roller bearings have superior radial and thrust loading capability. Timken is the major player in producing these bearings. This type bearing is used on all cars and trucks to keep the front wheel hubs on, but not used on this helicopter.

10.3.8 Spherical Plain Bearings

Neat looking bearing because the inner part can rotate off its centerline. The parts are both hardened and have lubrication groves to enable a grease layer to be maintained between mating surfaces. Helicopter Application: Swashplate swivel bearing.

Photo 10.3.8-1

 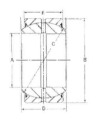

10.3.9 Rod Eye Bearings

The diagrams below are more or less self-explanatory. Helicopter Application: They are mounted at the ends of all the control rods. Can't imagine how hard it would be to duplicate their functionality if you couldn't just buy them.

Photo 10.3.9-1

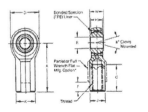 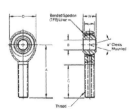

Chapter 11- Fabrication Machinery

   In this section the equipment used to fabricate the helicopter parts is discussed. The basic operating principle of all of them is to use cutting tools that are harder than the material to be cut. For example attempting to cut thru a drill bit or a bearing surface having a Rockwell hardness of 62 on a band saw with a blade having the same hardness just will not work. Fortunately over 95% of the parts you will be making will be from 6063-T6, 2024-T3 or 7075-T6 aluminum and although these are the hardened forms of these alloys, they are relatively soft (48 Rockwell) in comparison to the tool steels used in the cutting operations.

   For the unconventional 5% of the materials you must fabricate you need to go beyond standard machining practice used for the above stated materials. Just for discussion say you wanted to make a rotary score cutter, which is basically a round disc with a sharp knife-edge around the circumference. First you would obtain a rod of wear resistant steel having a good carbon content having a diameter slightly larger than the disc. I'd use 52100-grade ball bearing steel purchased in the normal state (RC 20). Then after cutting the discs one by one off round bar stock I would drill the arbor mounting hole, sharpen the edge, and heat treat the disc to produce a Rockwell hardness of 62 to a depth of 0.062". This hardness depth would be deep enough to allow for re-sharpening and yet minimize the effects of the heat treatment distortion since this process usually grows the part as much as 5%.

   Altering hardened parts can be accomplished by grinding but dimensional control is limited. Otherwise, that hardened part would have to be softened by re-heating and slow cooled to remove the heat treatment. Then after the required machining is performed, the part is re-heat treated to the original hardness.

   Fortunately, of all the parts manufactured for this helicopter only one hardened part required modification. That part was a spherical bearing used on the swash-plate assembly which required a slot. How this was accomplished will be discussed later in this section. The machinery and hand tools discussed were all that was required for this project.

   There is one other important point in part fabrication. It is after all a destructive process in that material is being removed from the work piece forcefully. This generates heat and if not controlled causes the cutting tool to dull, vibrate, and chatter and the modified material to expand such that the finished part when it reaches ambient room temperature is no longer within the allowable machined tolerance. This fact makes water flow cooling mandatory for lathe and milling operations.

11.1 The Band Saw

Photo 11.1-1

a. Function: The band saw's main purpose is to make straight cuts through aluminum and soft steels (both bars and square stock) at a cutoff length determined by the operator. The cuts can be made at an angle to the clamping direction varying up to 30 degrees from the 90-degree home position.

b. Machine Description: The band saw purchased for this project has the capability to cut raw stock up to 4" wide and 4" high. The machine consists of two subassemblies, the base and the pivoting saw arm. The base incorporates the material clamping vice, an adjustable spring tensioning device used to regulate the saw blade downward force on the material, and the on/off automatic end of cut shut-off switch.

Photo 11.1-2

The pivoting saw arm houses the electric motor, which drives a 62"
continuous loop 20 teeth per inch hacksaw blade, a saw blade tensioning
device, and a set of blade guide rollers to control blade to material
alignment. The machine weighs 120 pounds and is powered by a 120-volt
single-phase 1/2 horsepower motor.

   c. Setup: This machine has the capability to cut on an angle, which is great
if you have to make one. Over 95% of the time the angle desired is a direct
square cut, one that is perpendicular to clamped direction. Since there is no
guarantee the last cut was square or possibly the clamp shifted, the first part
of the set-up procedure is to verify the guides of the vise are where they
should be. A standard "tee-square" aligned with the back clamp guide and
the inside edge of the saw blade should enable a decent cut if adjusted
properly such that ambient light is uniformly visible between the "tee-
square" and the band saw blade surface.

   The next check is to make sure there is sufficient tension on the band saw
blade so that it doesn't stop (slip on the driving pulley) during the cutting
cycle. Care should be taken not to over tension the blade since this will
cause premature blade failure.

   Finally, the last setup is to adjust the forward blade guide roller assembly
as close to the work piece as possible. This is important for two reasons.
First it aligns the blade to the work piece better and secondly it provides the
band saw blade the most distance available to twist 90 degrees into cutting
position after leaving the driving drum. This also improves blade life.

   d. Operation: This is a manually operated saw so the length of the piece
you want cut has to be established thru manual measurement. Typically I
scribe a mark on the work piece indicating the desired length. Then I align
the work piece under the band saw blade so the cut to be made is just slightly
more than the length desired. Typically, the industry allowance for material
cutoff is equal to the length plus one-eight of an inch. Once the work piece is
in position it is clamped. Then the saw is switched on and the saw arm
lowered. If the arm blade downward force is adjusted properly you can
simply walk away and the band saw will cut the piece, then automatically
shut off. Just as a point of interest, the numerically controlled band saws
used in large manufacturing facilities allow the operator to simply enter the
cut-off length on the control panel interface device and once the material is
loaded into the queue the machine will continue to cut to length all of the
material. Last time I checked this type of machine is in the $100,000 plus
range.

   e. Consumables: band saw blade for aluminum 20 teeth per inch, band saw
blade for steel 24 teeth per inch, The useful life of any one particular band
saw blade depends upon luck, regardless of the brand name purchased. Some
blades just seem to last longer than others. The normal failure mode is at the
point where the two ends of the blade are welded together. I have saved
those and at some point may try re-welding them.

   f. Recommended Spare Parts: one v-drive belt and extra band saw blades

11.2 The Lathe

a. Function: The main purpose of the lathe is to transform mill delivered round aluminum or soft steel bar stock into dimensionally controlled round parts in conformance to some detail print. It is also used to square off a raw material edge resulting from the band saw operation. In many cases this is just one step a part takes before all operations are performed to get to the physical shape stated on the detail drawing.

Photo 11.2-1

Shown below is a part being formed for the tail rotor assembly. On the extreme right hand side the small diameter is for eventual installation of a radial bearing. As for the drilled holes on the left side nearest the chuck they were made using the drill press during a previous machining operation.

Photo 11.2-2

b. Machine Description: This is a light duty lathe capable of turning a 3" diameter bar of aluminum or a 1" diameter bar of soft steel. It can generate metric and SAE threads. The base machine houses a bearing mounted thru bore variable speed spindle. This spindle is threaded to accept various sizes and type chucks. The chuck, when mounted to the spindle, has its faceplate positioned at a fixed right angle with respect to hardened machine bed ways.

Mounted to the machine bed is a gibed slide, which provides the surface for the cutting tool holder. The gibed slide is free to move along the bed length via manual positioning or thru direct connection to a threaded shaft running the length of the bed. When "automatic operation" is desired the side lever on the slide is rotated 90 degrees to engage the screw then the speed selection lever on the main frame chuck housing is moved to the desired speed slot. During this "automatic" mode when the spindle is selected to rotate "clockwise" the slide moves "toward" the chuck, and away from the chuck in the reverse direction.

The slide tool holder is capable of being clamped in any position over a 90-degree range with respect to chuck centerline. This allows for proper adjustment of the cutting tool for optimized chip removal. In "automatic" mode once the tool is positioned no further attention is required. In manual mode the slide is positioned at the work piece and although the cutting tool can remain, as it was setup for automatic operation that assembly rests on another slide that moves the cutting tool forward.

There are various types of chucks that mate to the spindle screw thread, some have three holding jaws and some four, some are manually positioned and others mount on a scroll plate such that rotation of the scroll plate brings all jaws together concentrically whether moving in or out. This of course is a fantastic device. It's so easy to get the work piece on center every time. The other feature of the scroll plate mounted jaws is they are made to grip the outside work piece diameter when positioned on the scroll plate one way; and in the reverse position grip the inside diameter of a tubular work piece.

c. Operation: As stated before this machine makes round parts, has a scroll chuck which centers round parts on center, and has an automatic cutting tool indexing function. What more is there to know? All in life is not that simple and neither is the raw material you want to chuck. None of it will actually be round! Part eccentricity (out-of-round) varies greatly between bar stock and typically depends on the forming processes that made it. Chucking a piece of bar stock requires a certain amount of effort to minimize the irregularities. For this setup you need a dial indicator mounted on a magnetic base with an adjustable arm. Then after the bar stock is chucked, place the travel end of the dial indicator on the circumference and manually rotate the chuck. Once you obtain a reading around the entire circumference if the measured variance is 0.005 or less you are good to go. If not, you must repeat the process by re-chucking the material in incremental angles relative to the original chucked position to find the smallest measured variance. Note: once you start machining the opposite end "DO NOT REMOVE THE PIECE FROM THE CHUCK UNTIL YOU ARE FINISHED WITH THE INTENDED OPERATION" because it is very difficult to rechuck the piece exactly where you had it in the chuck clamps to maintain concentricity.

In the previous picture once the intended machining operation has been completed on the piece, the part "can" be reversed in the chuck. This enables the machining of the opposite raw material surface to be worked. When the part is reversed we are now clamping a theoretically round surface. Although I don't recommend taking the part out until the machining of this section is complete, it is possible to rechuck the part easily at any relative clamp angular position without losing previous obtained part concentricity.

The machining process for long parts is as follows: if the bar stock is longer than three or four inches and the diameter less than 1-1/2 inches, then on this lathe the common practice is to pass the piece thru the chuck to expose only a short section and repeat the centering procedure defined above. Once you have that end-centered use the drill attachment to drill a centering hole. Repeat the process on the opposite end. Now you have a piece that can be chucked at one end and supported at the opposite end at its center, and if the piece requires it can be supported in the middle by another lathe attachment.

Long shafts larger than 1-1/2 inch diameter cannot pass thru the chuck and require significantly longer setup time to add the centering holes to each end. In addition to having the chuck end centered, it must also have the mid supporting end centered.

Bear in mind without formal training machining a piece of raw material into a part will be a trial and error operation. Some key points: keeping the part "water cooled" is critical, making sure the cutting tools are sharp, essential; and insuring all clamps tight, fundamental. As for cutting speeds, feeds, and tool cutting angles, this you will start to get a feel for after you start making several of your round parts. In the construction industry they say, "If the foundation is right the building will be right." The same goes for this project. Every part you make must conform exactly to the print. If a part is out of specification just make it over and chalk it up to a learning experience! Do not try to use it because in the end it will cost you way more time and effort to make the wrong part work than if you just "made it over." Note: I ruined several parts on the last cutting pass because the cutting tool just seemed to cut differently when removing only 0.001" of material during a final pass than during the initial material removal when 0.010" of material was being removed. Maybe it was an inherent quirk of this particular lathe but I must say it was an extremely frustrating one. To some extent I was able to work around this peculiarity by decreasing the cut depth substantially as I approached the desired diameter.

Photo 11.2-3

Shown previously is a lathe operation drilling a centering hole in one end of a shaft to be reworked in a following operation. Note: If you look closely at the end of the shaft you will notice that it was already machined to remove any cutoff saw marks. This is done so the drill will not walk the saw mark grooves but remain on center during the initial drilling operation. Shown below is a lathe modification adding a hand held die grinder to the tool holder bed. I did this to be able to cut-to-length hardened steel shafts and to add snap ring grooves to hardened shafts. The added grinder modification worked perfectly.

Photo 11.2-4

d. Consumables: cutting tools and drill bits, size depending on manufacturing requirements

e. Recommended Spare Parts: at least two motor drive belts

11.3 The Milling Machine

Photo 11.3-1

a. Function: To make flat parts (face cutting), cut slots (peripheral cutting) or drill holes. The bits used for cutting use multiple teeth flush with each other at the cutting surface whereas drill bits lead into their cutting flutes from a single point.

Shown below are the milling cutters: be advised just picking one out of the box by the flutes will leave several paper cuts on your finger tips, use gloves.
    Photo 11.3-2

Shown below is a milled section (peripheral cutting) being made on an engine support base per print. This allowed easier access to the bolts holes (lower surface) which are also slotted to enable longitudinal adjustment.
    Photo 11.3-3

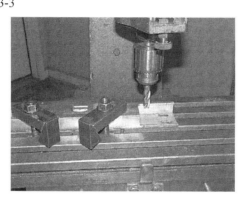

b. Machine Description: The milling machine shown and the lathe described previously were the workhorses of the project. Virtually every part made was fabricated by either of these two machines and very often both. The milling machine, in contrast to the lathe I purchased, was a heavy-duty commercial machine capable of handling steel or aluminum operations effortlessly. Where the lathe would choke-up (chuck would just stop under the load) if the cutting depth of the tool was too deep, the milling machine would do one of the following with ease: either cut the material, snap the tool bit or forcibly send the part flying from it's securing clamps. Of course all of these possibilities were my first hand experiences. The two horsepower motor definitely had the power to do all the above. In fact one of the first things you are told when operating a milling machine is not to wear loose clothing or rings on your fingers because their contact with the tool bit will either kill you or take body parts. Oh don't even think about wearing the tie.

Being in the engineering department I wore a tie and when called into the shop to view a part being machined I always had to be cognizant of the dangers of being in close proximity to this rotating stuff, tucking my tie into my shirt became automatic.

My milling machine is basically a glorified drill press. In fact it looks the same in that a vertically mounted motor driven spindle rotates a chuck. The main difference is the capability to control piece part movement in the X-Y-Z planes with micrometer control. In the Z-plane a gearbox mounted hand wheel located on the right side of the spindle housing lowers the chuck. Spindle travel is noted by a forward facing scale and by a micrometer mounted adjustable collar located on the hand wheel. When the hand wheel rotates so does the micrometer-scribed collar. At any time one can determine the cutter depth by looking at the depth indicated on the collar versus a scribed mark on the machine housing.

Due to its simplicity this Z plane hand wheel / micrometer adjustable collar arrangement is also used on the "X" and "Y" direction machine bed control. There you have it, the basic manually controlled X-Y-Z plane milling machine.

Now for the finer points of this machine, lets start with some of the features:

The entire spindle/motor housing can be raised and lowered in order to handle various size work pieces. First you loosen the locking nuts then use the spindle housing crank handle to raise and lower the unit to the desired position then retighten the locking nuts.

1)   The bed is "T" slotted along it's length to accept special clamp retaining nuts for various "Bridgeport" style clamp assemblies to screw into, this by the way makes anchoring parts to the bed an easy process for both securing them and for quick release when finished.

2)   In the X-Y travel direction an acme threaded shaft and mating bronze nut assembly control bed movement, and as in any threaded nut combination there is a certain amount of clearance between the mating threads (in general the control tolerances tighten as the grade of nut/bolt improves, for example a grade 8 is better than a grade 2 from both a clearance and tolerance stand point but to get the complete picture of this topic you should refer to Mark's Handbook). How much thread clearance (backlash) is acceptable? It is hard to imagine trying to machine a part to within 0.001", if there is that much possible bed travel just due to the slop in the threaded connection. This machine uses a double-nut combination to adjust out backlash that is great as the machine starts to wear.

3)   Each bed can be locked into position independently. This feature is neat if you are milling a slot say in the "X" plane and want to completely eliminate bed movement in the "Y" plane (bed movement due to backlash for example).

4)   For using the machine as a drill press the spindle mounted gear box driven hand wheel is quickly bypassed via a knurled knob on its side which transfers vertical chuck movement to a radial hub with three access handles mounted 120 degrees from one another. In this mode of operation it

functions exactly like a stand alone drill press even to the point of having an adjustable coil spring tensioning device to enable faster spindle return to home position after a drilled hole is completed.

5)   Depending on the type of operation being performed and the diameter of the bits being used the actual spindle can be quickly swapped out for a different size chuck or even an actual planer head.

6)   The spindle speed can be adjusted from 120 to 2500 rpm by varying two motor driven belt drives via 12 different pulley combinations.

7)   Typically stated for these milling machines is the parallelism of the bed surface as it travels to its end points in the X-Y plane relative to the chuck and the degrees of perpendicularity of the bed is to the chuck.

Shown below is a half-inch spindle/chuck assembly

Photo 11.3-4

c. Operation: Basically the work is moved past the cutter perpendicular to the cutter axis. For example if I wanted to mill a slot into a piece 0.10 deep, after securing the piece and making sure it was parallel to the bed surface, I would rotate this hand wheel until the milling cutter just contacted the part surface. Then the hand wheel adjustable collar would be zeroed out. Now I would start the machine so that the chuck was rotating in the forward position (clockwise) and begin to rotate the hand wheel so that the milling cutter started to penetrate the piece. I would continue to rotate the hand wheel until the depth indicated reached 0.10. Now the slot could be made two ways (for the sake of discussion lets make this slot in the "X" plane direction and the "Y" plane is locked out) either you make skim cuts the length of the slot until the desired depth is obtained or you cut to the depth you want right off the bat then proceed to obtain the slot length. In my experience no matter how hard I tried using the latter approach, the result was the cutter walked slightly off center. Unless you fare better, I recommend skim cuts for slotting.

The drill press function is rather simple to perform. Do all the same clamping as previously stated and drill the first hole. Then for the next hole use the X-Y hand wheels to get to the next position as indicated on the print repeating the process until all holes are drilled.

As you can tell by the above operational description this milling machine is manually operated and as a result of all the hand wheel action, you will, after a while, develop a very strong grip. In contrast NC (numerically controlled) milling machines enable you to just set the X-Y-Z travel distance on a data entry screen and the corresponding stepping motors do the rest. These type machines can get so sophisticated that we could download all the hole co-ordinates for an entire machine side frame and the NC milling machine would not only drill all the holes required but also automatically change drill or mill bit size to match the specified hole diameter.

Shown below is a milling machine modification I made to put weight saving "lightening" holes radially around the main drive gear. Another modification to the machine (not shown) was to mount the die grinder to the spindle housing so I could slot the hardened spherical bearing used on the swashplate.

Photo 11.3-5

Shown below is the gear described in the discussion to which lightening holes were added.

Print 11.3-1

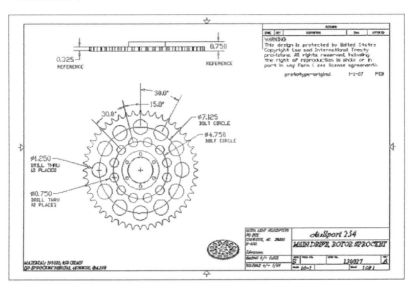

d. Consumables: SAE and metric drills and SAE milling cutters

e. Recommended Spare Parts: upper and lower drive belts

11.4 The Drill Press

Photo 11.4-1

a. Function: To manually drill perpendicular holes in material quickly. Originally I purchased this machine to do more than just drill holes but to also do some milling. To that end I purchased a compound slide and mounted it to the bed. Unfortunately, the whole setup was much too light to handle milling type operations. Just to highlight this point further where this machine weighed around 125 pounds the milling machine I purchased subsequently was in the neighborhood of over a 1000 pounds.

b. Machine Description: This machine stands about five feet tall and has a motor/spindle housing mounted at the top of a circular steel column which is anchored into a cast iron base. On the back of the vertical column is a moveable vertical gear rack that connects to the bed via a hand operated gear set. Upon losing the bed securing locks the hand wheel can be rotated to either raise or lower the working surface of the bed relative to the chuck. The ½ hp motor drives the spindle thru a single belt drive pulley arrangement. Final chuck speed is determined depending on which pulley diameter is selected on the motor shaft and the one on the spindle shaft.

Vertical chuck movement is by a right hand mounted radial hub with three access handles mounted 120 degrees from one another directly linked to the drive spindle. An adjustable coil spring tensioning device is also included to enable faster spindle return to home position after a drilled hole is completed.

c. Operation: In its most basic mode a piece to be drilled is placed on the bed and with the machine "off" the hand lever lowered so that the material hole to be drilled is centered with the drill bit. Once that alignment is made, the part is clamped, the machine turned on and the hole drilled. On occasion, you may think that since the material you want to drill a hole thru is so thin that you don't need to clamp it and may decide to keep in place with your fingers. If you do that plan on getting cut because somehow or another just before the drill cuts thru the last shred of material the drill bit binds with the piece and spins it out of your hand. The thinner the piece is the finer the cut to your hand.

Where the drill press gains a distinct advantage over an ordinary hand drill is if you have to locate the same hole in a lot of identical pieces. In this case you setup a fixture to locate the part against some hard stops and utilize a quick release clamp to keep it there during the drilling operation. When that hole is finished the clamp is released and another piece inserted into the fixture and the process continued until the job finished.

d. Consumables: SAE and metric drills

e. Recommended Spare Parts: one motor drive belt

11.5 The Bench Grinder / Polisher

Photo 11.5-1

a. Function: Polishing, grinding, de-burring, beveling shaft ends

b. Machine Description: The bench grinder is basically a 1/4 HP 3600 rpm motor, with the rotor shaft extending out of the stator housing at each end. Connected to the right output threaded-shaft a 6" diameter 1" wide grinding wheel is mounted and on the opposite shaft the polishing wheel. Both sides have adjustable sight guards, which always seem to be in the wrong position when using it, but they are included nonetheless. The unit weighs around 20 pounds and can be purchased with a stand-alone base. The stand-alone base offers the flexibility to move it around when working on long pieces of tubing.

c. Operation: Turn on and either polish or grind. The polishing operation is good for all the aluminum tubing used on this helicopter but removal of the sight guard is required for just about all of it. Best results are obtained by polishing a small area until the metal gets hot. Then continue the polish operation in small incremental steps over the preheated sections. Aluminum parts that require drilled holes should be polished first then drilled since the polish tends to thicken on one side of the hole and is very difficult to remove with the polishing wheel. When this occurs only strong hand polishing with a clean cloth will remove the build up around the hole. Typically, if the part is polished before drilling the part is easily hand polished to a bright finish after the holes are drilled. Gloves, safety glasses and a dust mask should be used since the material polished gets very hot and the polishing process generates grime and fumes.

Use of the grinder is rather straightforward. It is good at de-burring steel shaft ends, and to aid in that function a grooved guide is positioned 45 degrees from the grinding wheel centerline to support the shaft while beveling.

d. Consumables: polishing wheel, red & white polish, grinding wheel

e. Recommended Spare Parts: none

11.6 The Hydraulic Press

Photo 11.6-1

a. Function: To exert a concentrated force perpendicular to the horizontal support beams. This force can be used to press bearings into a housing or the opposite and is quite efficient in bending things.

b. Machine Description: As seen in the above photograph the press is a rather simple looking. The "C" channel structure supports a spring-loaded traveling base plate to which a 20-ton hydraulic bottle jack is installed. Just below the jack assembly is a moveable base plate adjustable by relocating steel dowel pins into one of several predrilled holes located parallel to one another on the vertical "C" channel uprights. Resting on top of this base plate are two cast steel part supports.

Shown below is a part bent using the hydraulic press.

Print 11.6-1

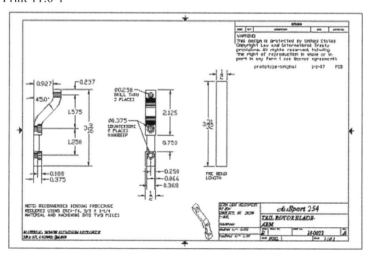

c. Operation: On this project this machine was rarely used but it did come in handy for certain processes. Anyone who has ever tried to install a bearing by lightly tapping the outer race into a bearing housing knows rather quickly that there has to be a better way. Just think about the hammer approach for a second. What you are attempting to do is drive one side of the bearing diagonally into a housing where the recommended fit tolerance can be as low as 0.0005 or 5 ten thousands of an inch. After the first hammer tap your bearing is wedged into the housing requiring a heavier tap on the opposite side to make it straight. Right around now you realize you must tap the bearing back out and start the process over again. The hydraulic press alleviates the above problem by providing a uniform insertion force, and yes the initial setup and piece part alignment to the press-driving shaft requires a certain amount of finesse. Be that as it may, once all the pieces are staged, the bottle jack activation screw closed, and the hand operated handle cycled to increase internal pressure on the bottom of the piston, the piston end then pushes against the top of the framework. The result of this causes the entire jack support structure to move downward guided by the side frames against the resisting bearing to housing interface. Once the bearing is installed to the proper depth the jack activation screw rotation is reversed and the entire jack assembly returns to its original starting position as a result of the developed tension in the support springs.

If you want to bend things setup is critical. You have to be aware of the elastic limit of the material you are bending. If you exceed it the part will snap. In the section on aluminum, bend radii are given for the various sheet thickness, which should be followed. Also be aware steel has an elastic limit about three times that of aluminum so don't expect to duplicate what you did in steel on a piece of aluminum.

Additionally aluminum parts that I attempted to make but cracked I found some success in preheating the material first. The exact changes in yield strength as a result of that maneuver I do not know so I wouldn't recommend it on high stressed or critical-use parts. One last point; don't expect to duplicate anything you make identically. The only way I was successful to that end was to use material a little more that twice what I needed. Then after the bending operation split the piece in half with a band saw and milling both together got the exact width I wanted. Tedious indeed! Oh, by the way, don't pre-drill holes before bending because they will either elongate or the part will snap at the reduced cross-section.

d. Consumables: none

e. Recommended Spare Parts: none

11.7 The Welder

a. Function: Lets start with the definition of what it means to weld something. A weld is a localized coalescence (fusion) of metal produced by heating to a specific temperature ($800^{O}F$ or higher) under conditions with or without pressure and or use of filler materials. Apparently the $800^{O}F$ temperature separates a welded joint from a mechanical or adhesively bonded joint and within this definition there are 37 different processes grouped into six general categories. Those are gas welding, resistance welding, brazing, solid-state welding and "other" processes. In industry it's the first four processes that are most widely used. As for its function, lets say you want to add a steel angle iron to a steel plate. One way would be to drill holes thru both pieces and bolt it together which by the way is time consuming or you could simply clamp the bracket in position and weld it together in a matter of minutes.

My welder is a MIG (Metal Inert Gas) welder that uses a consumable wire for the electrode and the process is called GMAW (gas metal arc welding). This unit can generate weld temperatures up to 10,000^{O}F and has the capability to inject an inert gas cloud at the electrode to allow welding to take place without atmospheric oxygen to interfere.

This welder was a fantastic adder to the project. Anybody can weld with this unit especially if you use the auto darkening eye protection helmet, the details of which I'll discuss in the operation section. You may ask why was this a great adder? Even though I never welded any of the aluminum parts used on the helicopter it enabled me to make many of the fixtures necessary to make the parts, specifically the 60 ton hydraulic press built to fabricate the rotor blades.

Photo 11.7-1

b. Machine Description: This unit requires a 230-volt single-phase circuit (think household electric dryer circuit) to operate. It weighs about 50 pounds and has a front panel containing four operator controls: two for the electrode heat setting, one for wire feed speed, and an on/off switch. Opening the wire feed access panel allows one to install the wire spool, adjust wire feed drive tension, and set the welder up for either inert gas or non-inert gas operation.

Photo 11.7-2

At the rear of the unit a space is provided for the inert gas cylinder which when installed has a pressure regulator mounted to the tank on/off valve body. Opening the tank valve sends gas thru a tube which runs parallel to the wire feed tube contained within the five foot section of hose leading to a second valve in the hand held welding electrode "gun". Shown previously is the "gun" with my hand wrapped around the black trigger mechanism. Upon depression of the trigger the motor feed is activated forcing the welding wire out the center of the nozzle (barely visible in picture) and the gas valve at the "gun" opens sending a cloud of inert gas thru the nozzle and around the center electrode feed tube.

To complete the circuit for the feed wire electrode to arc the work piece must be grounded back to the machine. This is accomplished by a separate five-foot long copper cored cable connected to a spring loaded "battery jumper cable type" clamp.

C. Operation: Obviously to melt something you have to bring "The Heat". Once you are ready to weld I typically depress the trigger to get a small section of wire to extend past the nozzle. Then using a pair of wire cutters I leave about a half inch exposed and cut off the rest. Then I lower my auto-darkening helmet and get myself into a comfortable welding position such that the exposed weld wire is just touching the start point of the weld. Getting into a comfortable position is important in that once you start the arc any jerky hand movement while trying to readjust your position translates into poor weld bead formation.

At this point depressing the trigger brings "THE HEAT". The electric arc that forms when the feed wire electrode contacts the grounded work piece can generate, as said before temperatures as high as $10,000^O$ F with accompanying light as bright as the sun, and it's not like there is any time lag involved. For example the auto-darkening feature of this welding helmet switches into eye protecting darkening mode in $1/30,000^{th}$ of second. Needless to say that's fast.

Now for the best part of welding with this machine, as the molten electrode wire is transformed into the weld bead the motorized feed function keeps replacing it at your set bead-forming rate. This means your trigger hand always stays approximately the same distance with respect to the forming arc, and as a result, there is less eye hand coordination required. In contrast with "stick" welding where your hand holds a clamped one foot long rosin coated fluxed core welding rod "called the stick" you must not only do everything described above but also adjust for the welding rod length loss as this metal is being lost to the bead formation.

At this point it is good to remember that welding is not only a certified skill, where examining boards qualify you based upon your abilities (like not every welder passes the criteria to have nuclear certification) but also an engineering field. Just think about it for a moment. There must be over a thousand different types of steels out there and each alloy type requiring different methods in order to achieve some standardized strength.

Case in point, I remember working on a Navy project where the alloy steel could not be welded unless both parts were elevated to some preheat temperature and then the reference parts soaked at that temperature for some time period prior to welding. After welding the rate of the cool down temperature had to be regulated.

The subject of welding is so broad a topic that the AWS (American Welding Society) publishes a technical set of books on the subject where you can not only look up the material you want to weld but it will also tell you the weld temperatures to use, the weld process, the filler material, whether or not a gas is required, and if so what type. Also included in that series of books is a fifty-page section devoted to just "Symbols for Welding and Nondestructive Testing".

My welder can weld the aluminum alloys used on this helicopter if argon gas is used but I never produced a weld that I felt I could trust so I defaulted to basic mechanical connections using nuts and bolts. I even purchased a spot welder that according to the material specification sheet could be used to fuse parts together. Unfortunately I just vaporized the places where I attempted to join them so I returned the device.

d. Consumables: spooled rosin cored weld wire, argon gas cylinder

e. Recommended Spare Parts: electrode wire feed brass guide tips for several wire diameters, for example 0.020 and 0.032 inch and welding wire.

11.8 The Rotary Table

Photo 11.8-1

a. Function: To precisely rotate an object mounted on its faceplate 360 degrees in minute or second intervals. The unit is perfect for use with the milling machine to accurately position drilled holes on the circumference of round parts, for example in adding radial holes to the swashplate housing or the tail rotor bearing housing not to mention drilling lightening holes on the main drive sprocket.

b. Machine Description: The rotary table is a substantial mechanism weighing about 25 pounds. It can be setup to provide either a vertical or a horizontal rotating surface. In either case each mounting surface is machined to close tolerances for parallelism and perpendicularity. In addition to rotary position indication there is one lever to reduce rotational backlash (unwanted rotary bed movement) and another lever to lock the bed in position while work is performed on the surface mounted part.

c. Operation: Lets set the rotary table up to drill holes radially around the main drive sprocket in accordance with the detail print. This operation requires the rotary table to be mounted with the bed in the horizontal position, which is easily accomplished with the milling machine clamps. Next the sprocket is centered on the bed and anchored to it. Then the entire milling machine bed is maneuvered such that the center of the sprocket is aligned with chuck center. At this point either the "X" or "Y" bed axis is clamped and the free bed axis moved the radial distance required for the first group of holes to be drilled. After each hole is drilled the rotary table hand wheel is rotated to the accumulative angular position for the next hole. This procedure continues until the last hole is drilled. Then as a cross check rotate the hand wheel as if to drill one more hole. If everything was performed properly the drill will line up exactly with the first hole you made.

d. Consumables: none

e. Recommended Spare Parts: none

11.9 Arbor Press

Photo 11.9-1

a. Function: To press things together, or to generate stampings when used with a tool and die set.

b. Machine Description: a vertical moving gear rack is connected to a hand operated gear all housed in a support structure where the driving force of the gear rack is directed to the integrated horizontal surface below it.

c. Operation: I used this device to make the main rotor blade aluminum stiffener stampings. In that mode of operation I would place a predetermined size of flat aluminum sheet into the die base, then I would insert the tool atop the aluminum sheet guided by the fixture framework. This setup was then transferred to the arbor base where I would rotate the arm levers until the tool and die bottomed out. At that point the formed aluminum part was removed from the tooling and the process repeated until the lot was finished.

d. Consumables: none

e. Recommended Spare Parts: none

11.10 Hand Tools:

11.10.1 Hand Drill:

Photo 11.10.1-1

Two types of electric hand drills are required, a 3/8-drive 3600-rpm variable speed drill for small holes and some polishing and a 1/2- drive 900-rpm variable speed drill for the larger size holes. Both drills should have mechanical locking chucks since they secure the drill bit substantially better than the quick locking twist type and should operate on 120 volt single phase 60 cycle power.

1.10.2 Metric & SAE Wrench Set:

Metric sizes 5,6,7,8,9,10,11,12,13,14,15, and SAE sizes 1/4, 5/16, 3/8, 7/16, 1/2, 9/16, 5/8, 11/16, 3/4, 13/16, 7/8, 15/16 and 1". Need I explain why?

Photo 11.10.2-1

11.10.3 Metric & SAE Allen Wrench Set

Metric sizes from 0.7 mm to 10 mm and SAE sizes from 0.035 to 0.375. These wrenches are used mainly for internal setscrews (swashplate bearing retention) and socket head cap screws (collective lever clamps).

Photo 11.10.3-1

11.10.4 Metric & SAE Socket Set

Metric sizes 5,6,7,8,9,10,11,12,13,14,15, and SAE sizes 1/4, 5/16, 3/8, 7/16, 1/2, 9/16, 5/8, 11/16, 3/4, 13/16, 7/8, 15/16 and 1". No comment.

Photo 11.10.4-1

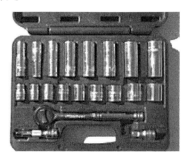

11.10.5 Metric & SAE Tap & Die Set:

Photo 11.10.5-1

Metric sizes 5,6,7,8,9,10,11,12,13,14,15, SAE sizes 1/4, 5/16, 3/8, 7/16, 1/2, 9/16, 5/8, 11/16, 3/4, 13/16, 7/8, 15/16 and 1". This set can either put a bolt thread on a plain steel rod or tap a hole for a standard size bolt to be threaded into. Be advised "spend the money" for a quality set because the cheap sets dull after the first use.

11.10.6 Die Grinder:

Photo 11.10.6-1

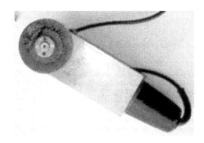

The 4" 3600 rpm die grinder is one of the most useful hand tools money can buy for this overall helicopter build project. In its' purchased out of the box mode it has two general uses. With a 1/16" wide abrasive disc it works well to cut thru sheet steel up to 1/8" thick or 1/2" round bar stock, and with the 5/16" wide disc is a great de-burring tool. The die grinder comes complete with a steel guard, which completely covers the abrasive disc with the exception of a small section where the disc cuts. This is also the most dangerous hand tool to use. Any tool that can cut thru hardened steel as if it was cutting thru butter has to be respected. Never attempt to use this tool without leather gloves and safety goggles. In use sparks in the form of hot molten steel, stream off the disc sometimes greater than five feet from the contact point. Rags, flammables or anything you don't want coming in contact with molten metal should be moved before ever starting to grind. Never turn this device off by simply pulling the plug out of the socket because the next time you plug it in it will do damage to you or the work piece guaranteed!!! This grinder was also used in a modified form in the rotor blade aluminum sheet-cutting fixture. Those modifications are discussed in detail in that section covering the rotor blade fabrication.

b. Consumables: cutoff and die grinding discs

11.10.7 Hand Rivet Tool
Photo 11.10.7-1

This is another must have tool. Its function is to rivet parts together where you can only access one side. For example attaching the Lexan windshield to the cockpit aluminum-tubing framework. In this case I didn't want to weaken the tubing any further then I had to by drilling a thru hole and then bolting the Lexan on. Using the rivets gave a better appearance, were less expensive, and substantially lighter than a bolted connection.

The rivets used are called blind rivets. After being inserted thru a predrilled hole of a specified diameter thru the part being attached, and the supporting piece. The hand rivet tool is positioned such that it bottoms out on the head of the rivet while griping the center mandrel. Squeezing the handles extracts the mandrel from the center of the rivet expanding its external diameter in the process until it reaches the internal surface of the supporting piece. At that point the force exerted by the tool is sufficient to snap the mandrel so that it is removed when the tool is removed.

Once again as with anything else stated in this book there is more than one choice of what to use and with each possibility comes the instruction guidelines. I'm not going to go into all the details of all the rivet options other than to say don't buy the $5.00 box of 100 rivets from Harbor Freight and think they will survive all the vibrations associated with helicopter operation. Stick with a good quality rivet (Cherry), which leaves a steel core when the mandrel snaps off since they are the strongest.

11.10.8 Brake Rivet Tool

Photo 11.10.8-1

This is the tool to use when installing a standard rivet. What is a standard rivet? One end has predefined shape and the opposite end, which is inserted thru the material to be fastened, must be formed. This is the device used exclusively to fasten the rotor blade trailing edges together.

Although tedious to get a perfectly formed rivet, it is quite simple to use. A perfect rivet is one in which while turning the screw the mandrel stays centered on the rivet end being formed. The final shape produced makes a perfectly concentric circle relative to the hole centerline.

To begin I setup a fixture on an old drill press I had where I could clamp the two rotor blade trailing edges together, drill the exact diameter required for the rivet size, then rivet the two pieces together. This process was repeated until the entire 10.5 foot long blade was assembled.

11.10.9 Hand Files

Photo 11.10.9-1

Everything you make will have a sharp edge and as drills get dull every hole will have some too. That's where the round and flat files come in to debur everything so you can handle the parts without getting cut. In fact, it's not a bad idea to have a note on the detail prints stating "break all sharp edges".

11.10.10 Snap Ring Tool

Photo 11.10.10-1

Snap rings are one of the best ways to keep circular parts together. Installing or removing them without destroying them is virtually impossible without the tool and it's associated end inserts. Once again it's a must have tool for your inventory.

11.10.11 Vise & Clamps

Photo 11.10.11-1

Shown above is one of two machinists' vises used extensively on this project. The one shown simply clamps. The other vise not only clamps but also can rotate to change the clamp angle. As for the two gage blocks on the lower right they are indispensable for aligning parts mounted on the milling machine bed to be parallel with bed travel in either "X" or "Y" direction.

Next is the old standby. I can't tell you how many times this old fashioned, heavy, antiquated looking vise was just what was needed to made a part, hold a part, or bend one. I highly recommend this thing being included in your shop.

Photo 11.10.11-2

Shown below is the Bridgeport knockoff milling machine clamp set. Photo 11.10.11-3

Somehow or another this set wasn't supplied with the milling machine but it was soon apparent it was as vital to the milling operation as was the electrical plug which turned the machine on. Unfortunately, I struggled for several weeks trying to secure parts to the machine bed before I came to the conclusion there had to be a better way. My answer came while looking thru a Wholesale Tool catalog when I saw the clamp set and ordered it. It was the best-spent money for an accessory on the entire project.

11.10.12 The Shop Vac

After every part you make on either the milling machine or the lathe you will generate a substantial amount of extremely wet aluminum scrap. The shop Vac makes cleanup so much faster and safer than manually attempting to remove this scrap from your machines or workshop floor.

Once again here is one more case where you should spend the extra money for a top end unit. The better motors use ball bearings to support the internal stator shaft, which enables the Vac to last much longer than those with plain bearings. You should also buy the top of the line replacement filters because they can handle the cooling water on the scrap better and again last longer.

11.11 Air Compressor

Photo 11.11-1

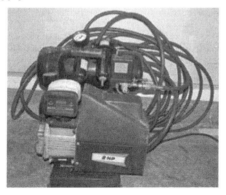

Actually, I didn't really need this thing. Having it around enabled me to repaint our outdoor furniture set and tend to a persistent slow air leak on the rear tire of my wife's SUV. Once and a while I'd use it to blow off some residual material from a part just machined and it could be used with a device I purchased which converted air pressure into a vacuum source (by the way that device never saw much use either). Don't get me wrong if we had an assembly line making helicopters the first thing I would do is slap an air wrench to it to speed up the fastening process.

11.12 Electronic Scale

Photo 11.12-1

I used a structural epoxy / Kevlar based composite to anchor the steel rods in the nose section of the main rotor blade. This scale enabled me to get exact resin to hardener proportions correctly. Just for the record, there are two ways to obtain these proportions either by weight or volume and the manufacturer gives that information. Automatic dispensing machines use the volume approach whereas I utilized the weight method since it was substantially easier to use and less expensive.

11.13 Consumables:

I've mentioned milling bits, welding wire and drills before in their appropriate section so this is just a repeat. These items get expensive and somehow they must be included in the total project cost. One thing I recommend you purchase which in the long run would have saved me money is a commercial drill bit sharpener. As for the milling cutters they seem to last a long time provided you keep them water cooled during the milling process. Unfortunately, in this learning process wear was never an issue since I managed to snap mine off at the shank due to one of the following reasons: the first rash of failures came because the part wasn't secured well to the machine bed, this problem cured itself when I purchased the Bridgeport "knockoff" clamp set. Next type of cutter failure was due to excessive machine bed movement, in this case the part was secure enough to the bed but the play in the travel screw was enough to cause the cutter to dive into the work piece and self-destruct. The reason this excessive bed play came to be was I was adding too much bed clamping force to the bed ways. For clarification the machine bed rides on a beveled base whereby both contacting surfaces are machined to tolerances of less than 0.001. A set of adjustable side clamps prevents any movement. The problem is you should loosen them before attempting to rotate the hand wheel to change bed position, as a result I wore down the brass nut connected to the drive screw to the point the mating thread clearance became too much. Later, I realized I could adjust out the backlash and that problem went away.

Finally, during the learning experience I found out about the detrimental effects of heat buildup and the milling operation. Here's the pitch, what you are going to do is take a strong piece of a hardened alloy aluminum and commence to drive a sharp wedge along its surface at high speed physically shredding any material in its path. Yes, one could readily come to the conclusion heat will be generated. Initially I didn't use a water-cooling system and the result was the heat generated between the part and the milling cutter reached temperatures so high aluminum welded itself to the cutter. Now on occasion I would realize something was wrong, stop the machine and remove the aluminum mass. If I didn't this welded aluminum filled the milling cutter flutes. It would grow the cutters diameter until the side load on the shank at the chuck exceeded its dynamic strength and it would snap off.

As for the welding wire it just doesn't last as long as you think it will and it will always run out in the middle of something you are welding. The good part is welding anything is always a satisfying experience even if you have to grind a lot of it down to get something that looks professional.

One more side note, shown below is a sheet of electrical wire labels. This not only adds a neat finish to your helicopter wiring but if followed thru on the wiring diagram it makes trouble shooting easier as well as any future modifications, such as adding more instrumentation.

Photo 11.13-1

11.14 Safety Items

The most important thing you must have is a good fitting OSHA approved set of safety glasses. "DON'T START ANYTHING WITHOUT THEM"! Sure the other obvious things are gloves, both the leather faced type for manual work and the soft cotton type for handling the highly polished rotor blades but the glasses are mandatory. A hard hat is a good adder during the testing phase but one thing I overlooked is the respirator mask. You'd be quite surprised how filthy your face gets polishing aluminum. Even if you think you can handle polishing without it, which I did on numerous occasions I didn't escape getting chemical pneumonia. Chemical pneumonia is like everyday pneumonia in that it physically affects your lungs and respiratory system but is the result of toxic fumes versus bacterial exposure. I had this happen to me while running the engine in the shop with the garage doors open. Unfortunately, the amount of two-stoke exhaust gas not exiting the garage was sufficient to impact me and kept me bed ridden for several days. After that experience I never ran the engine inside the open garage and used the respirator when polishing, hand grinding structural composites, and welding anything giving off toxic fumes such as the welding of the galvanized parts on the helicopter transport trailer.

Photo 11.14-1 The Respirator

Chapter 12- Suppliers

12.1 Tools

Wholesale Tool Co., Inc
Harbor Freight

12.2 Machinery

Wholesale Tool
Harbor Freight

12.3 Raw Materials

Aircraft Spruce & Specialty Co.
Wicks
Online Metals .com

12.4 Commercial Parts

Aircraft Spruce & Specialty Co. (Nuts, bolts, instruments etc.)
Wicks (Nuts, bolts, instruments etc.)
Applied Industrial Technology (power transmission)
McMaster-Carr (snap rings & power transmission)
American Ring & Tool Co. (snap rings)
Martin (power transmission)

TEK (power transmission)
EPT (power transmission)

12.5 Engines & Transmissions

Rotax, CPS (California Power Systems, Inc.)
Wankel-Rotary, Atkins Aviation Co.
Kawasaki
Hirth
Compact Radial Engines

12.5.1 Engine Rebuilders

SBT

12.6 Electrical

Atlantic Jet Sports

12.7 Software

AutoCAD
Softworks
TurboCad

12.8 Built-to-Print Part Fabricators

None Used

PART 4- DESIGN DETAIL

At this point our efforts are now concentrated on designing the parts needed to make a functional helicopter. Before starting there is some groundwork that must be attended to in order that the project stays manageable. Mind you I didn't come up with any of these procedures its just proven methodology I lifted from industry.

Chapter 13- Visualizing the Design

13.1 General Comments:

Back in the day when computer inventory systems were just starting to be implemented in businesses across the country the saying got to be "garbage in equals garbage out." Actually the saying used stronger language than I care to repeat but what it really meant was your actual inventory was only as reliable as the people inputting the data. To support that statement I can remember going into a Jeep dealer's parts department with my son to buy some engine mounts. After giving the year, model, engine type etc., the counter man not only came out with the wrong parts but also insisted I gave him bad information about the vehicle. Next day I brought the vehicle in question with me and the fellow realized that this Jeep had mid-model-year updates that were not recorded on his micro-fiche. Not a favorable example for Jeep's inventory control and record maintenance! All this applies to Computer-Aided Drafting or CAD as well in that the reliability of the full scale three-dimensional model you create will only be as good as your tenacity in keeping your updates accurate.

Lets discuss the concept of that full-scale 3-D model you create a little further. If you build it where every hole lines up accurately then when you are finished every 3-D part you pull off the model can be detailed with the satisfying knowledge that the part made will fit with the others as intended.

13.2 The Three Dimensional Model

CAD Dwg 13.2-1 The 3-D CAD Model

Every project has to start someplace. In our case we developed the overall parameters for our helicopter several chapters ago, defining gross weight, main and tail rotor size, rotor hub configuration and engine horsepower. Now we get to package all those components in some viable layout where the design helicopter weight and accurate positioning of the aircraft's center-of-gravity must be attained. Basically I started by getting the profile of a 95 percentile male (actually, this is easier to obtain than I thought) to establish hand reach, back angle, and foot pedal locations. In the end to get the real feel for all this I just sat on the floor and documented what I personally felt comfortable doing. Once I got control of my seating position I started the layout. What followed next was developing the various subsystems to the point that they were mechanically sound and their individual 3-D shape could be incorporated into the Main Model that formally in the old drafting paper days was referred to as the General Layout.

13.3 Details

In the paper drafting days the designer would have to redraw the part referenced on the General Layout then determine the lengths so it could be dimensioned and then hand draw all the letters. It was this redraw phase that typically introduced errors. In CAD drafting this redraw phase is eliminated since that part is simply lifted from the model and the dimensions accurately measured by the program. Any notes can be just typed in.

For some additional depth to the subject of CAD, I can remember when our company attempted to make the transition into using it. First order of business was for us to decide what system to use. To determine that the owner, the head of finance and myself traveled to various vendors to view the systems available. Finally it came down to either Hewitt Packer (HP) or AutoCAD. HP's system utilized a central mainframe host computer supporting satellite workstations. The advantage of such a system was the ease of backing up all the engineering work done in a day. Designers worked with large 25-inch displays and workstation-processing speed was good as long as the number of connected workstations didn't exceed some predefined number. The disadvantage was the initial setup cost was exorbitant and if the host computer went down all the workstations were inoperative. The AutoCAD system used stand-alone independent workstations driven by computers with 286 processors handling small 20-inch monitors. Needless to say these stations were slow but the cost of a workstation relatively cheap for the time. I say this because what you can buy today for $500 dollars cost over $2500 then (1987) and the one 25-inch monitor we purchased cost over $5000. Our collective decision was to go with AutoCAD for the following reasons: most of our key layout designers were older fellows who would never embrace this technology and we weren't sure of the long-term benefits. After all, the cost of this equipment had to be paid by realizing increased productivity so to limit our financial exposure we went with the least costly approach.

Initially we started with three workstations for our 45 man-engineering department. The number three was derived from the number of designers who desperately wanted to learn CAD to the point they signed up for AutoCAD classes to become proficient even before the first station arrived. When the stations were setup they worked overtime to make sure they completed what they were assigned.

The CAD results were that initially on any particular machine design project the generation of the main layout took longer but detailing time took less. All in all it was a wash but now I was able to use less experienced designers to do critical layout work. Furthermore as our CAD database of motors, transmissions, bearings and other hardware expanded, machine development time shortened. It was a good decision to go CAD.

Before we look at some general details you should know there are drafting standards to be complied with. For the military it is Metric Standard MIL-STD-100, covering engineering drawing practice. I used the GE standard for the basic English detailing instruction set and for ISO drawings a geometric standard is used for indicating parallelism, concentricity and perpendicularity. Note that each standard defines its own set of detail drawing sizes. I used the standard A, B, C and D size with the B size predominantly used.

13.4 Bill of Materials

Photo 13.4-1 The Actual Swashplate

My first 3-D model completed was the swashplate assembly. This was because it was simple and contained few parts. To keep track of the parts required, cost and weight I used a standard intended "Bill of Materials" (BOM) format. This conforms to ISO 9000 standards.

The basic thought behind an indented BOM is that all items follow a hierarchy of importance based upon indented location. "No indent" then that's the complete part. "One indent" an individual part or subassembly is necessary to make the part. "Additional indents" continue until you get to basic raw materials or basic commercial parts.

There is a BOM for every subsystem along with its own drawing series two-digit preface. For example 11,xxx Main Rotor, 12,xxx Swashplate, 13, xxx Tail Rotor, 14,xxx Control, 15,xxx Main Drive, 16,xxx Tail Drive and 18,xxx Frame. All 19x,xxx series numbers are dedicated to commercial parts and those are also further segregated by type. For example 191,xxx are bearings, 192,xxx are bolts, 194,xxx are eyebolts etc.

I also separated the way the indents were presented in the BOM such that all fabricated parts were presented first with the commercial parts listed last. This enabled me to quickly get to a part number I needed for a detail print of a part I wanted to fabricate next or for use to note any changes on. As far as the two-digit preface, that just made generating part numbers easier and organizing the manual faster.

Shown below is the BOM spreadsheet for the Swashplate generated from Microsoft Works that not only does a great job in listing the items, but also totals all the weights of the individual components to come up with a final weight for the assembly. It performs the same type calculations for total cost.

Chart 13.4-1

Microsoft Works- swashplatePROTOTYPE.wks

File Edit View Insert Format Tools Window Help

Times New Roman C 12

F42 "Swashplate Assembly Control Rod

	A	B	C	D	E	F	G	H	I	J	K	L	M	N
1				Part #	#Pcs	Part Description	Wt	Total Wt		$/FT	FT/PC	$	Total Cost	
2								4.86					$199.71	
3	1	120001				Assembly Swashplate		4.10					$170.41	
4														
5			1	120010		Swashplate Housing, Outer Bearing								
6	O				1	6061-T6 alum plate 5 x 1/2, 5 lg	0.51			14.04	5/12	$5.85	$5.85	
7			1	120011		Swashplate housing inner bearing								
8	O				1	6061-T6 alum rd 4 dia 2-5/16 lg	0.62			59.96	25/16/12	$11.56	$11.56	
9			1	120012		Swashplate bearing linear								
10	S				1	2024-T3 alum tubing 1-3/4 od x 0.25 wall, 3-5/8 lg	0.21			17.99	3.625/12	$5.43	$5.43	
11			1	120013		Swashplate brg, modified spherical								
12	A				1	191003 17SF28 Torrington spherical bearing	1.42					$32.51	$32.51	
13			1	120016		Pin, modified shoulder screw								
14					1	192034 Holo-krome shoulder cap screw 0.248 dia x 1-1/8 lg thread 10-24 UNRC-3A	0.01					$1.00	$1.00	
15			1	191004		Swashplate bearing								
16	A				1	KP-49B roller bearing bore od	0.78						$93.50	
17			1	193301		Swashplate snap ring								
18	R				1	RS306 spirolox external snap ring 3.062 shaft	0.03					$0.50	$0.50	

AIRSPORT 254 - SWASHPLATE Issue date: 04/05/05, rev0

			Part#	#Pcs	Part Description	Wt	Total Wt		$/FT	FT/PC	$	Total Cost
							4.86					$199.71
1	120001				Assembly Swashplate		4.10					$170.41
		1	120010		Swashplate Housing, Outer Bearing							
O				1	6061-T6 alum plate 5 x 1/2, 5 lg	0.51			14.04	5/12	$5.85	$5.85
		1	120011		Swashplate housing inner bearing							
O				1	6061-T6 alum rd 4 dia 2-5/16 lg	0.62			59.96	25/16/12	$11.56	$11.56
		1	120012		Swashplate bearing linear							
S				1	2024-T3 alum tubing 1-3/4 od x 0.25 wall, 3-5/8 lg	0.21			17.99	3.625/12	$5.43	$5.43
		1	120013		Swashplate brg, modified spherical							
A				1	191003 17SF28 Torrington spherical bearing	1.42					$32.51	$32.51
		1	120016		Pin, modified shoulder screw							
				1	192034 Holo-krome shoulder cap screw 0.248 dia x 1-1/8 lg thread 10-24 UNRC-3A	0.01					$1.00	$1.00
		1	191004		Swashplate bearing							
A				1	KP-49B roller bearing bore od	0.78						$93.50
		1	193301		Swashplate snap ring							
R				1	RS306 spirolox external snap ring 3.062 shaft	0.03					$0.50	$0.50
		2	193302		Swashplate snap ring		0.02					$1.00
R				1	RS175 spirolox external snap ring 1.75 shaft	0.01					$0.50	
		1	193303		Swashplate snap ring							

13.5 Revision Documentation:

No matter how good you think you are at some point in the project you will make changes. Those changes may be a result of commercial parts becoming obsolete, parts couldn't be installed as intended, parts failed, or for a myriad of other reasons. You must have a methodical way of handling changes. The upper right hand corner of every print has a place to record the change made, the date and the reason why. Keeping a general log of the changes is helpful especially if you want to see if the modifications impacted other areas of the design.

Chapter 14- The Subsystems Analyzed

In this phase of the project the parts designed must survive the real world stresses generated during the operational limits placed on this helicopter. Here are the overall helicopter specifics as developed in Chapter 8 that define the helicopter and our starting point:

- gross wt., 525#
- engine, 50 hp
- main rotor diameter, 21 ft
- main rotor rpm, 438
- tail rotor diameter, 3 feet
- tail rotor rpm, 2600
- maximum allowable vertical landing speed; 15 mph

We will begin developing assembly drawings one subsystem at a time until all can be packaged in the resulting frame. Once the frame is sized the main and tail drive trains will be added. I will give you additional design specification information subassembly by subassembly as we progress. At this point lets put some target weight numbers on the subsystems in order to arrive at the ultralight empty weight limit of 254 pounds. Those values are:

- Swashplate_____5#
- Main Rotor Assembly____33#
- Controls_____3#
- Tail Rotor Assembly_____7#
- Engine_____70#
- Main Frame_____90#
- Drive Train_____44#
- Electrical_____2#
- Total = _____254#

14.1 Swashplate

I started doing this assembly layout first because I had a good basic understanding of its function and there were only a few parts involved.
- Basic Function Description: to transfer Control Rod (joystick) pilot directional motion from the frame based reference plane to the rotor shaft rotational operating plane where this pilot input is transmitted to the

individual rotor blades. This device is to be able to perform its function without any singularities or disturbances in the directed motion while the Collective Lever is being raised or lowered.

CAD Dwg 14.1-1

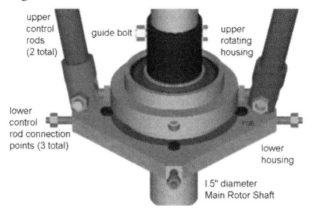

We have already discussed what the Swashplate does in section 3.2.4.

The functional design description is as follows: the Upper Rotating Housing packages the spherical bearing on its inside diameter and a radial bearing on its outside diameter. Press fit into this spherical bearing is an aluminum sleeve with a slot on the exposed end where the guide bolt passes thru it and locates it to the main rotor shaft. Want the visual? Check out the assembly drawing following this verbal description, see Print 14.2-2. In addition to the sleeve being press fit into the spherical bearing, snap rings are added to keep it there. The upper housing contains two fastening points 180 degrees from one another for the upper control rods to connect to, and on its lower surface a radial bearing is mounted.

The Lower Housing bore fits over the outside diameter of this radial bearing. Three shoulder bolts protrude from the housing at 90-degree intervals for the lower control rods to mount.

Once assembled the entire Swashplate can move vertically the length of the slot in the sleeve (for Collective function) and pivot about the spherical bearing centerline (for Cyclic function). The key point with designing this assembly is the control rod dynamic forces on it are small, probably less than 5 pounds if your rotor blades are perfectly balanced about the feathering axis. Since both the spherical and radial bearing must be sized to fit over the 1.5 diameter main shaft any bearing you select will be overkill for the application.

Design Issues: 1) Given the main rotor shaft diameter of 1.5 inches select and package the spherical bearing and internal rotor bearing such that minimal vertical distance exists between the spherical bearing centerline and the radial bearing centerline. This is to minimize non-linear swivel movement to the upper control rods. 2) Establish control diameter rods sizes. Note these control rod calculations are given in the section dealing with shafting.

3) Conclude how to allow the inner and outer spherical bearing parts to swivel yet not rotate with respect to each other. I solved this problem by slotting the spherical bearing and using a dowel thru the outer race to keep them aligned together and yet still support the required swivel action.

The following are assembly drawings of what my Swashplate looked like:

Print 14.1-1

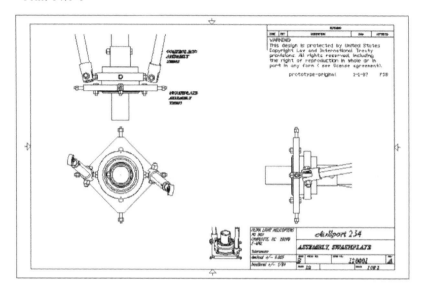

Print 14.1-2

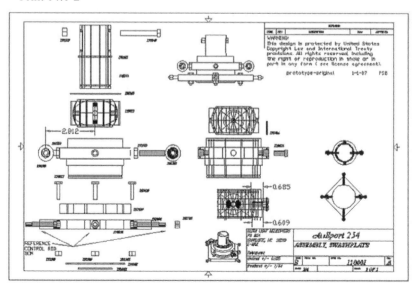

14.2 The Rotor Blade

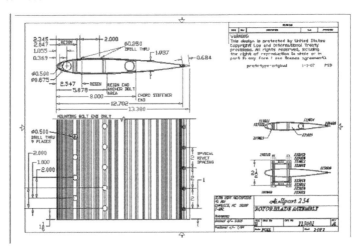

   It has to be apparent by now the rotor blade is the key to a helicopter's
capability to provide Lift. You could make the point that the main focus of
this entire book is centered on how the rotor blade performs in various flight
conditions. It seems now is the time to discuss how to design and fabricate
one.

   Needless to say I spent many hours to come up with the rotor blade design
as shown in Print 14.2-1. I would describe its critical features as follows:
The Center of Gravity of the blade and the Center of Pressure coincide at the
25% chord distance leading back from the blade's leading edge. This is also
the location of the blade's feathering axis, which as a result theoretically
eliminates any moments to develop with Angle of Attack change. The
exterior skin is made from 7075-T6 aluminum bent into the shape of an
NACA 0012 planform. At the trailing edge of the material where the two
ends meet the top sheet surface has a preformed recess in it so when the ends
are riveted together the required shape of the NACA profile is maintained.
This top trailing edge recess is then filled with epoxy to smooth out the
surface.

   One design issue that consumed a significant amount of time to solve was
getting the blade CG correct. What had to happen was the internal blade
supportive structure CG had to offset the outer skin CG in order to get the
entire blade assembly to balance out at the 25% chord location. As you can
easily realize the center of gravity of this aluminum skin was approximately
at the mid point of the cord or approximately 3.35 inches from the leading
edge. The 25% chord position is 1.67 inches from the leading edge. A lot of
weight had to be added near the leading edge for this entire blade to balance
correctly.

   Eventually the solution was to add two steel rods (a 1/4 inch & a 7/16
inch) to the inside leading edge. To ensure the rotor blade kept its shape
under load internal aluminum stiffeners were added uniformly along the

entire length of the blade held in place by an aluminum tube in the spanwise direction. Both the steel rods and the stiffeners were anchored to the skin with a precise amount of structural Kevlar/epoxy composite.

Sounds simple enough now lets investigate the reality of it all.

Lets start with calculating the weight and CG of this Rotor Blade. In the design I have steel rods, a Kevlar/epoxy composite, aluminum cord stiffeners, an aluminum tube to keep the cord stiffeners positioned together and of course the aluminum skin. I'll start by showing you the mass property data extracted from the Rotor Blade 3-D CAD model for the steel rods. In an effort to minimize redundancy for the remainder of the materials used I have simply summarized the important information such as mass and CG data below in Calculation 14.2-1 section A2. The result of these calculations give the Rotor Blade weight at 13.857 lbs. and the CG at 1.7 inches (note that to be theoretically perfect this number should be 1.67" however since the variance is only 1.6% off I decided to keep with this design).

Calculation 14.2-1

170MM NACA0012 ROTOR BLADE 7075-T6; 0.032 tk.

A1. ROTORBLADE MASS PROPERTY DATA

Mass: 23.132   Volume: 23.132
Bounding box: X: 0.059 -- 0.746, Y: -0.219 -- 0.219, Z: 0.000 -- 116.000
Centroid: X: 0.443, Y: 0.000, Z: 58.000
Moments of inertia: X: 103756.891, Y: 103761.940, Z: 5.511
Products of inertia: XY: 0.001, YZ: 0.164, ZX: 594.594
Radii of gyration: X: 66.973, Y: 66.974, Z: 0.488
Principal moments and X-Y-Z directions about centroid:
       I: 25939.396 along [1.000 0.000 0.000]
       J: 25939.902 along [0.000 1.000 0.000]
       K: 0.967 along [0.000 0.000 1.000]

A2 SUMMARY OF MASS PROPERTY DATA:

Definition of Terms:
p_4130 = steel density = 0.28 lbs/ in^2
m_4130 = mass of rotor blade steel (1/4 and 7/16 rods) = 23.132 in^3
cg_4130 = center of gravity of the steel on chord axis = 0.443 in.
p_comp = composite density = 0.032 lbs/in^2
m_comp = mass of rotor blade composite = 51.238 in^3
cg_comp = center of gavity of the composite on chord axis = 1.134in.
p_chord = 2024-T3 alum. density of cord stiffener = 0.098 lbs/in^3
m_chord = mass of all chord stiffeners = 1.355 in^3
cg_chord = center of gravity of all the chord stiffners = 2.402 in.
p_stiff = 6061-T6 aluminum density of rear stiffener = 0.098 lbs/in^3
m_stiff = mass of rear stiffener = 4.807 in^3
cg_stiff = center of gravity of rear stiffener = 3.664 in.
$p\_airfoil$ = 7075-T6 aluminum density of airfoil skin = 0.100 lbs/in^3
$m\_airfoil$ = mass of rotor blade skin = 50.034 in^3
$cg\_airfoil$ = center of gravity of airfoil skin = 3.290 in.

Variables Defined:

$\rho\_4130 := 0.280$	$m\_4130 := 23.132$	$cg\_4130 := 0.443$
$\rho\_comp := 0.032$	$m\_comp := 51.238$	$cg\_comp := 1.134$
$\rho\_chord := 0.098$	$m\_chord := 1.355$	$cg\_chord := 2.402$
$\rho\_stiff := 0.098$	$m\_stiff := 4.807$	$cg\_stiff := 3.664$
$\rho\_airfoil := 0.100$	$m\_airfoil := 50.034$	$cg\_airfoil := 3.290$

B..Total Rotor Blade Weight Calculation:

$WT\_4130 := \rho\_4130 \cdot m\_4130$ $\qquad$ $WT\_4130 = 6.477$

$WT\_comp := \rho\_comp \cdot m\_comp$ $\qquad$ $WT\_comp = 1.64$

$WT\_chord := \rho\_chord \cdot m\_chord \cdot 2$ $\qquad$ $WT\_chord = 0.266$

$WT\_stiff := \rho\_stiff \cdot m\_stiff$ $\qquad$ $WT\_stiff = 0.471$

$WT\_airfoil := \rho\_airfoil \cdot m\_airfoil$ $\qquad$ $WT\_airfoil = 5.003$

$WT1 := WT\_4130 + WT\_comp + WT\_chord$

$WT2 := WT\_stiff + WT\_airfoil$

$WT\_rblade := WT1 + WT2$

$WT\_rblade = 13.857 \quad$ lbs

C. Rotor Blade "Center of Gravity Calculation: noteTheoretical CG equals 25% of the chord

$$CG\_170mm := 0.25 \cdot \frac{170}{25.4}$$

$CG\_170mm = 1.673 \qquad$ This is the desired CG location

$$CGa := \frac{cg\_stiff \cdot WT\_stiff + cg\_comp \cdot WT\_comp}{WT\_rblade}$$

$$CGb := \frac{cg\_airfoil \cdot WT\_airfoil + cg\_4130 \cdot WT\_4130}{WT\_rblade}$$

$$CGc := \frac{(cg\_chord \cdot WT\_chord)}{WT\_rblade}$$

$CG := CGa + CGb + CGc$

$CG = 1.7 \quad$ within 1.6% of theoretical

Next up would be to determine if this design will handle the radial stress involved. In the section covering the Main Rotor design I used the Rotor Blade weight calculated here to generate the axial force (F_axial) this rotor blade will be subjected to. If you look at Calculation 14.3-1, a3 you will see how F_axial was calculated but for now I have extracted that information and F_axial = 5746 lbs.

How can we determine the stress on the fabricated Rotor Blade when we have three different materials involved including three different types of aluminum? My solution draws on Castigliano's method of determining the displacements of a linear-elastic system based upon the energy used to cause the displacement. I will use a simple formula to calculate elongation of each material type making up the 116-inch long rotor blade by incrementally increasing the applied force. This of course takes some trial and error to find that combination where all the deformations equalize and the sum of the applied force on the steel rods, the composite and the aluminum skin totals approximately 5746 Lbs. At this point you must realize in the blade's axial direction only the steel rods, the composite and the aluminum skin resist this force.

Here's the result: Under "F_axial" the rotor blade elongates 0.037 inches with a load distribution as follows:

1900# (steel) +1430# (skin) +2420# (composite) = 5750#.
The actual stresses are stated at the end of the calculations.

Calculation 14.2-2

Calculation of Rotor Blade Stress

Definition of Terms::
   E_steel   = 30x10^6 psi; Steel modulus of elasticity
   E_alum   = 11x10^6 psi; Aluminum modulus of elasticity
   E_resin   = 0.52x10^6 psi
   E_kevlar = 19x10^6 psi; Kevlar modulus of elasticity

   A_steel   = 0.199 in^2 = 0.049 in^2 (1/4" rod)+ 0.150 in^2 (7/16" rod)
   A_alum   = 0.4316 in^2; rotor blade skin area
   A_kevlar = 0.398 in^2 ; crossectional area of kevlar composite

P_steel   = load carried by the steel rods
P_alum   = load carried by the alum skin
P_ kevlar = load carried by the resin-kevlar composite
Pt   = total load radial load on rotor blade
Fr   = total resultant load (axial / radial direction)
δ_steel   = deformation of the steel
δ_alum   = deformation of the aluminum
δ_kevlar = deformation of the kevlar
Rr   = actual rotor blade length = 10.5 - 0.75 = 9.75ft (117in)
σ_steel   = rotor blade steel stress psi
σ_4130   = 4130steel, yield stress 100,000 psi
σ_ alum   = rotor blade aluminum stress
σ_7075   = 7075-T6alum, yield stress 58,000 psi
σ_kevlar = rotor blade kevlar stress
σ_kevlar49 = kevlar49, yield stress 43,000 psi

Variables defined:

$Rr := 116$ $A\_steel := 0.199$ $\sigma\_4130 := 100000$

$A\_alum := 0.432$ $A\_kevlar := 0.398$ $\sigma\_kevlar49 := 43000$

$E\_steel := 30 \cdot 10^6$ $E\_kevlar := 19 \cdot 10^6$ $E\_alum := 10.4 \cdot 10^6$

$\sigma\_7075 := 58000$

Variable Ranges defined: $P\_steel := 1900, 1910 .. 2000$

$P\_alum := 1400, 1410 .. 1500$ $P\_kevlar := 2340, 2350 .. 2440$

Formulas for the Deformation of the Steel, Aluminum and Kevlar:

$$\delta\_steel := \frac{P\_steel \cdot Rr}{A\_steel \cdot E\_steel}$$

$$\delta\_alum := \frac{P\_alum \cdot Rr}{A\_alum \cdot E\_alum}$$

$$\delta\_kevar := \frac{P\_kevlar \cdot Rr}{A\_kevlar \cdot E\_kevlar}$$

The Calculations:

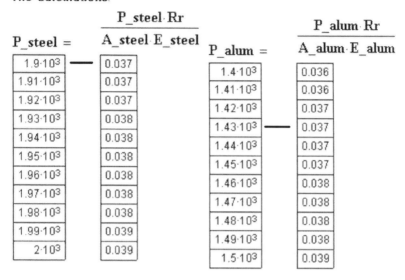

$$P\_steel = \frac{P\_steel \cdot Rr}{A\_steel \cdot E\_steel}$$

$1.9 \cdot 10^3$	0.037
$1.91 \cdot 10^3$	0.037
$1.92 \cdot 10^3$	0.037
$1.93 \cdot 10^3$	0.038
$1.94 \cdot 10^3$	0.038
$1.95 \cdot 10^3$	0.038
$1.96 \cdot 10^3$	0.038
$1.97 \cdot 10^3$	0.038
$1.98 \cdot 10^3$	0.038
$1.99 \cdot 10^3$	0.039
$2 \cdot 10^3$	0.039

$$P\_alum = \frac{P\_alum \cdot Rr}{A\_alum \cdot E\_alum}$$

$1.4 \cdot 10^3$	0.036
$1.41 \cdot 10^3$	0.036
$1.42 \cdot 10^3$	0.037
$1.43 \cdot 10^3$	0.037
$1.44 \cdot 10^3$	0.037
$1.45 \cdot 10^3$	0.037
$1.46 \cdot 10^3$	0.038
$1.47 \cdot 10^3$	0.038
$1.48 \cdot 10^3$	0.038
$1.49 \cdot 10^3$	0.038
$1.5 \cdot 10^3$	0.039

$$P\_kevlar = \frac{P\_kevlar \cdot Rr}{A\_kevlar \cdot E\_kevlar}$$

$2.34 \cdot 10^3$	0.036
$2.35 \cdot 10^3$	0.036
$2.36 \cdot 10^3$	0.036
$2.37 \cdot 10^3$	0.036
$2.38 \cdot 10^3$	0.037
$2.39 \cdot 10^3$	0.037
$2.4 \cdot 10^3$	0.037
$2.41 \cdot 10^3$	0.037
$2.42 \cdot 10^3$	0.037
$2.43 \cdot 10^3$	0.037
$2.44 \cdot 10^3$	0.037

$$Fr := P\_steel + P\_alum + P\_kevlar$$

$$1900 + 1430 + 2420 = 5.75 \times 10^3$$

Summary

Deformation of steel $= \dfrac{1900 \cdot Rr}{A\_steel \cdot E\_steel} = 0.037$ in

Deformation of aluminum $= \dfrac{1430 \cdot Rr}{A\_alum \cdot E\_alum} = 0.037$ in

Deformation of kevlar $= \dfrac{2420 \cdot Rr}{A\_kevlar \cdot E\_kevlar} = 0.037$ in

Stress on steel $=$

$\sigma\_steel := \dfrac{1900}{A\_steel}$ $\qquad \sigma\_steel = 9.548 \times 10^3$ psi

Stress on aluminum $=$

$\sigma\_alum := \dfrac{1430}{A\_alum}$ $\qquad \sigma\_alum = 3.31 \times 10^3$ psi

Stress on kevlar $=$

$\sigma\_kevlar := \dfrac{2420}{A\_kevlar}$ $\qquad \sigma\_kevlar = 6.08 \times 10^3$ psi

Safety factor of steel $=$ $SFsteel := \dfrac{\sigma\_4130}{\sigma\_steel}$

$$SFsteel = 10.474$$

Safety factor of aluminum $=$ $SFalum := \dfrac{\sigma\_7075}{\sigma\_alum}$

$$SFalum = 17.522$$

Safety factor of kevlar $=$ $SFkevlar := \dfrac{\sigma\_kevlar49}{\sigma\_kevlar}$

$$SFkevlar = 7.072$$

As you can determine from the above summary if the Rotor Blade Kevlar/ epoxy composite can keep the internal structure together, then all the stresses developed on the individual components have a relatively large working safety margin. Initial testing proved that it does.

The next calculations determine the Rotor Blade's flexibility or in the vernacular it's Modulus of Elasticity.
Calculation 14.2-3

E1. Calculation of Actual Rotor Blade Modulus of Elasticity from Measured Data.

Definition of Terms:

$\Delta 1$ = Measured deflection with 0.996# load = 0.066

$\Delta 2$ = Measured deflection with 2.008# load = 0.131

I_c   = Combined rotor blade moment of inertia from dwg AERO1ROTORBLADEmoment= 0.077

P1   = Load 1 at center = 0.996#

P2   = Load 2 at center = 2.008#

L_rbs   = Length between rotor blade supports 116 - 2(2"offsets) = 112

E_rb   = rotor blade elasticity

Variables Defined:

$$P1 := 0.996 \qquad P2 := 2.008 \qquad L\_rbs := 112$$

$$\Delta 1 := 0.066 \qquad \Delta 2 := 0.131 \qquad I\_c := 0.077$$

Formula for a Centrally Loaded Simply Supported Beam:

$$\Delta := \frac{P \cdot L^3}{48 E \cdot I}$$

Rearranging terms in this formula to solve for rotor blade modulus of elasticity $E\_rb := \dfrac{P1 \cdot L\_rbs^3}{48 \cdot \Delta 1 \cdot I\_c}$

Calculations:

Solving for E $E\_rb = 5.736 \times 10^6$

Using this value of E_rb calculate the deflection and see how it compares with the measured value of $\Delta 2$

$$\frac{P2 \cdot L\_rbs^3}{48 \cdot E\_rb \cdot I\_c} = 0.133$$

NOTE: For a simply supported beam this calculated value of 0.133 compares well with the measured value of 0.131.

Now that we know the Rotor Blade's Modulus of Elasticity we can calculate how much the blade will deflect when cantilevering from the main rotor hub and the resultant stresses generated.

Calculation 14.2-4

Using our calculated blade modulus of elasticity lets determine the blade deflection and stress in the cantilevered mode.

Calculation of Cantilevered Rotor Blade in a No Load Condition

Definition of terms::

BL   = Blade length = 116" (from section B this document)
L_arb = Unsupported blade lenght = BL - BktL = 116 - 5 = 111
BktL   = Bracket length = 5"
WT_rotorblade = Weight of the rotor blade without the mounting
              bracket 13.886
WT_rbu = Rotorblade weight of unsupported section
I_c   = Rotorblade moment of inertia = 0.077
E_rb   = Average Modulus of elasticity calculated for the rotor
         blade cross-section 5.736x10^6
Δ_rb   = Maximum deflection of the rotorblade due to its own
         weight
w_rb   = Rotorblade weight per unit foot (lbs/ft)
c_rb   = Distance from rotorblade neutral axis to outer fiber =
         0.8blade thickness / 2 = 0.4
M_rbn   = Bending moment due to rotorblade weight (no load)
σ_7075 = Minimum yield for 7075-t6 aluminum = 58000psi

Variables Defined:

$$BL := 116 \qquad WT\_rotorblade := 13.886$$

$$w\_rb := \frac{WT\_rotorblade}{BL} \qquad\qquad BktL := 5$$

$$L\_arb := BL - BktL \qquad c\_rb := 0.4 \qquad E\_rb = 5.736 \times 10^{6}$$

$$I\_c = 0.077 \qquad I\_skin := 0.034 \qquad w\_rb = 0.12$$

$$WT\_rbu := (BL - BktL) \cdot w\_rb$$

$$M\_rbn := WT\_rbu \cdot \frac{L\_arb}{2}$$

Formulas for Bending Stress on a Cantilevered Beam with Uniform Load:

This is the maximum deflection which occurs at the free end:

$$\Delta\_rb = \frac{-w\_rb \cdot L\_arb^4}{8 \cdot E\_rb \cdot I\_c}$$

Total Bending Stress on Rotor Blade: $\sigma\_rbu := M\_rbn \cdot \dfrac{c\_rb}{I\_c}$

Bending Stress on Aluminum Skin: $\sigma\_rbul := M\_rbn \cdot \dfrac{c\_rb}{I\_skin}$

Results:

Deflection, in.	Rotor Blade Stress, psi.	Skin Sress, psi
$\Delta\_rb = -5.143$	$\sigma\_rbu = 3.831 \times 10^3$	$\sigma\_rbul = 8.676 \times 10^3$

This completes the calculations, so now lets see how to build the blades.

Even to the casual observer making anything over ten feet long requires special tools. Additionally two such rotor blades are required and both have to be identical to one another in order for the rotor system to perform in a predictable way. As we are well aware, in the real world, versus the theoretical one, there are some absolutes such as "nothing is perfect". For example normal engineering fabrication drawings typically allow for errors (called a tolerance) in the part manufacturing process to be within plus or minus 0.005 of an inch. On extremely specialized parts some critical dimensions may be held to four decimal places with a temperature stated for measuring the part during quality control inspection. Before we even attempt to build a rotor blade we must have some idea of what the weight variation between blades should be and how close both rotor blades should balance both statically and dynamically.

The tolerance for the weight variation between blades I have set at 5 grams, static imbalance (the equalization of mass distribution) less than 5 grams, and dynamic imbalance at less than 1/16 inch differential (ideally blade track should remain constant as collective is increased).

Our first concern is obtaining quality-grade rotor blade materials. In the engineering section the stresses the rotor blade would encounter were determined for the steel rod, aluminum stiffeners, aluminum airfoil skin, and the Kevlar/epoxy composite filler. Various materials were reviewed during this process and those providing optimal results selected. Even though the stated book rating for particular steel may be given and used in the engineering calculations all steels of a particular designation (say 4130) are not the same. Steel, aluminum, Kevlar, epoxy are all made by raw materials and a process, it is these two items that vary from batch to batch. The FAA in an effort to reduce catastrophic part failure and to better understand why it occurs requires a formal paper trail on all materials used in aircraft parts and structure. This paper trail not only traces the raw material origination but also the process batch characteristics in which the material was formed in its delivered form. In sheet form even the grain direction is specified. Before ordering the materials, we need a vendor list that will supply the necessary paperwork to support the FAA requirement. In this regard the cost of this material certification is expensive and every new order of the same material results in another certification requirement therefore it behooves you to order all the same material at one time to reduce cost and traceability issues.

Traceability in our limited production scenario of building only one set of rotor blades the material needed for the aluminum airfoil skin comes as one 4 ft x 10 ft sheet. The size of this sheet allows for three individual blades to be made which is great if during your first attempt at producing your first rotor blade something goes wrong during the learning curve. If you were successful on your first try and later on you decide to build two more sets of rotor blades and another aluminum sheet must be ordered more then likely it's certification document will be unique. One set of blades would be made from one material certification and the third set having each blade with different material certifications. Keeping track of which blade was produced from which material is the issue called traceability,

If we have a vendor that can supply certification sheets for all the materials specified resulting from the engineering process, the big question is can the rotor blade be fabricated to meet the exacting dimensions stated on the detail drawings. The two steel rods are used "straight" (within a MIL specification manufacturing tolerance) as purchased and the Kevlar/epoxy composite mix achieves its rated strength in the first twenty minutes of the resin curing state, so these two rotor blade elements required no additional concern.

My main focus was on making the "internal stiffeners", bending the aluminum sheet to produce the "airfoil leading edge" and forming the "trailing edge contour". All the above required some experimentation to see if any of this was possible. I started by making fixtures that I felt could accomplish my objectives. Preliminary testing of the possible forming techniques dictated use of a 20-ton manual hydraulic press and a 5-ton arbor press so both were purchased.

Lets start with the forming operations on the airfoil skin. Bending the leading and trailing edge of 7075-0 aluminum in the un-heat-treated form where the yield tensile strength is only 21,000 psi would be easily accomplished. The follow up heat treatment of the 10 ft blade to obtain the required T6 state is not so easily managed. Noting that the mechanical yield tensile properties increase to 58-64,000 psi in this condition and it was this value used in the engineering calculations. In addition to the high probability that the rotor blade may distort during the heat treatment process it would be very expensive to crate, ship and freight the rotor blades back from one of the very few places that exist today having furnace capability to handle parts of this size.

In an effort to control cost and minimize the risk of part distortion the approach taken was to cold form the part. This means the raw material would be purchased already in the T6 condition and the fixture we would design would be sized accordingly to generate sufficient controllable brute force to form the part at an ambient temperature of say 75 degrees F. Based upon the mill information given for 7075-T6 aluminum the stated allowable bend radius is 0.125 for a material thickness of 0.032. This data was used in making the male and female parts of the forming equipment. The next step was to obtain enough material to conduct the testing in the chord wise direction of the rotor blade without having to purchase an expensive full size 4 ft x 10 ft sheet. After a experimenting on 4 inch wide pieces with various combinations of roller diameters and mandrel travel distances both the leading and trailing edge forming operations were possible and repeatable without any indication of stress cracks in the formed regions.

Photo 14.2-1 A Successful Leading and Trailing Edge Forming Operation

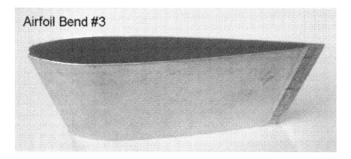

Airfoil Bend #3

Photo 14.2-2 Airfoil Bend #3 Top View

These forming experimental operations were complete now all I had to do was scale up the testing equipment in size to handle the ten-foot long sheet.

Now we're ready to tackle the development of the fabrication technique to produce the longitudinal "Rotor Blade Stiffeners". Before we begin note that producing this part involves the full extent of engineering, cad/cam and machinist experience. The stiffener, as the name implies maintains the desired shape of the rotor blade aft of the steel rod-Kevlar/epoxy composite leading edge portion of the airfoil. The engineering work establishes the proper spacing, the dimensions and the desired material thickness of the stiffeners such that it can withstand the anticipated stress without crushing under load. The CAD portion generates the detail drawing of the "stiffener" whereby the desired width of the part "y" is stated as a function of chord length "x".

Once all this detail part information is established a "die and punch" to form the part can be designed. Allowances are made so that as the punch forces the material into the die portion cracks in the formed part are not generated in the tight radiuses. CAM (Computer Aided Machining) comes into play if a numerically controlled milling machine makes the punch and die we designed. This specialized machine has the capability to act on the co-ordinates entered by the operator and the machine then completes the entire machining operation without additional operator interaction. Unfortunately a fully integrated CAM controlled milling machine can cost over a quarter million dollars. That of course was not an option so the fabrication of this punch and die setup filtered down to the capabilities of a machinist to produce these parts using a milling machine.

This was the first formal punch and die tool I ever made and the fact that my experience in forming was limited, I expected several design changes before getting the draft angles correct so after the part was made it could be easily removed from the die. Eventually, it all came together but the original 7075-T6 material never worked due to its inherent stiffness, nor did the other aluminum alloy 6061-T6. What did work well was 2024-T3 x 0.020 thick. This of course resulted in generating a new engineering analysis. The end result was this material could be used if additional stiffeners were added, so they were.

Photo 14.2-3

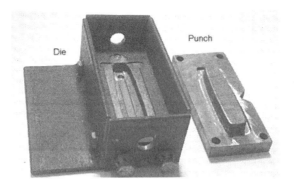

Here's some informational tid-bits on the subject. The structural stiffeners are used to maintain the proper airfoil shape aft of the theoretical rotor blade CG and to prevent deviations to that shape when subjected to loads on the cantilevered rotor blade in both static and dynamic states. When deviations to the surface between stiffeners occur the term is commonly referred to as "oil-canning" in reference to the old days when oil purchased in tin cans was poured thru only one opening. The resulting flow would pulse as air rushed past the flow to fill the partial vacuum created. Typically, as this occurred the thin tin wall of the can would experience a cyclical flex, which in turn generated a unique sound. To avoid such problems as airfoil skin-flex the stiffeners must be placed correctly along the supportive 1/2-inch diameter 0.032-wall tube as stated on the assembly drawing.

Now that the stiffener fabrication process was under control the next step was to scale-up the preliminary blade forming test fixtures to produce the full size 10-foot blade. The total list of fixtures required to complete the job are listed in order of manufacturing use, a brief fabrication description of each fixture follows:

1.   "Precision Cutting Table", this is used to cut the 4 x 10 ft long 7075-T6 aluminum sheet into three 14.5 inch wide x 120 inch long strips

2.   "Trailing Edge Forming Press", this tooling modification to the 11-foot long 60-ton hydraulic press I designed adds the trailing edge contour to one edge of the 10-foot long 7075-T6 aluminum strip.

3.   "Leading Edge Forming Press", this tooling modification to the 11-foot long 60 ton hydraulic press centers the 13.5 inch width of the 10-foot long aluminum strip then bends the strip in half such that the appropriate leading edge shape is formed.

4.   "The Structural Stiffener Spacing Fixture", after using our structural stiffener "punch and die" tooling to fabricate all the required stiffeners we need to space them correctly in the rotor blade. Problem was you only have a half hour to do this because the Kevlar/epoxy composite these stiffeners were anchored with must still be soft and pliable. It was intuitively obvious a half hour for this work was totally unrealistic. I needed a simple way to insert all the stiffeners at once, and then add the composite. This was accomplished by adding a hole to the preformed stiffeners so an aluminum tube could slip thru. Now all the stiffeners along the entire longitudinal length of the blade could be pre-positioned on this small diameter aluminum tube then installed just after the steel rods were placed. The "Structural Stiffener Spacing Fixture" was made to insure "both" rotor blades had the same stiffener spacing.

5.   "Kevlar/Epoxy Curing Fixture", after the leading edge is formed the piece is removed from the hydraulic press and secured into this fixture. Here the steel rods and the structural stiffener assembly are epoxy bonded to the internal leading edge aluminum skin. Then within the half-hour time frame after the entire composite is added the trailing edges are clamped together so the 24-hour curing process can begin.

6.   "Drill Jig and Riveting" fixture, when blade is removed from the curing fixture clamps. If you did everything correctly, the upper and lower trailing edges will still be in contact. All that is necessary for this operation is to position the trailing edge into this fixture then sequentially drill and rivet every inch along the 120-inch length.

7.   "Rotor Blade Final Machining Fixture", now we go back to the milling machine, setup some guide blocks and mill the rotor blade to an exact 170 mm overall chord length dimension. This will give us a rotor blade with a

chord variation with no more than a 0.005 width variation.

8.   "Rotor Blade End Cap" forming fixture, we use this fixture to epoxy seal the ends of the rotor blade so no foreign material (insects, water, dirt, etc) can alter the blade's weight and CG.

The final procedure is to add a filler material to the trailing edge contour section just above the rivet heads to smooth out the transition.

Fixture Fabrication Overview:

1.   The Precision Cutting Table:

On a limited budget nothing cuts thru 0.032 inch thick aluminum faster or easier than a 4 inch die grinder (cost $10) with a 1/16 wide cutoff wheel, so that tool was purchased first. This tool is typically hand held so it was modified to incorporate a flat surface similar to a standard Skill saw where the cutting blade is perpendicular to it's flat supporting surface. In addition having excellent cutting ability the die grinder is small and very controllable as long as the direction you are cutting the aluminum has the grinding wheel rotating down into the material. Working the grinder in this way enables the cut material's molten edge to be supported after the cut. Otherwise the grinder will want to "walk-up" the material and quickly tear across the surface thereby destroying it if proper downward force is not applied. I highly recommend cutting some scrap material first until you get the hang of the cutting operation. This grinder "walk-up" could be controlled by using a more elaborate cutting table, which held the grinder more securely, but I couldn't justify the added complexity and cost when with a certain amount of patience and proper setup my parts were slit easily and as anticipated.

As for the cutting table itself a steel frame was made and mounted on a piece of 3/4" marine plywood and that structure secured to a 10' section of a commercial aluminum extension ladder, this piece added extreme stiffness to the assembly without adding too much weigh and at minimal cost. The steel framework was setup on three sides with a 1 x 1 angle iron providing side guide supports and a forward stop for the 10 foot long 7075-t6 aluminum sheet to butt against. Located 0.040 above the plywood at a distance of precisely 13.438 from the rear stop the forward 1x1 angle iron was located. Precision spacers were added uniformly along the length of the rear and forward angle iron to maintain accurate width control. After sliding the 10-foot long aluminum piece under the forward angle iron and butted up lengthwise against the rear angle iron stop, the sheet was now in position to be cut. At this time the grinder was turned on and the forward angle iron was used as the guiding edge for the traversing die grinder to travel adjacent to. The result was a 10-foot long aluminum strip with less than a 1/32-width variation.

For 2. & 3. "The 60-ton Hydraulic Press:"

Fortunately designing and building this fixture was relatively easy when compared to the effort involved with producing the "stiffener punch and die". This fixture is massive with respect to that one. To begin, the approach was to scale up the equipment used to produce the successful bends in the 4-inch wide sample. The envisioned device to duplicate the results on the 10-

foot rotor blade outer surface sheet is described as follows. Several hydraulic presses would be positioned in a straight 10-foot line, all equidistant from one another. Then each press would be setup such that the moveable intermediate framework was adjusted to the same respective height. Finally each press would be floor-screw adjusted so the top surface of the intermediate support frame of each press was level with respect to the others. On this level surface of each press and on center, a 12-foot long I-beam would be added perpendicular to each press. Next, the flange side of a 10-foot long T-beam is under-slung from each press's driving mandrel such that the leg portion is directly above the top I-beam surface. This was the basic concept of the hydraulic press.

Now for the details. Calculations were generated to determine the force required to form the 7075-T6 aluminum sheet. This dictated the purchase of two additional 20-ton hydraulic presses, bringing the total to three. Next was to size the I-beam linking the three presses together such that total deflection at any one point under load was limited to less than ten thousands of an inch. The mandrel I-beam selected was a W16 x 36, which was further modified to include wood stiffeners spaced every 6 inches between upper and lower flanges on both sides of the I-beam to provide additional stiffness. The final part required was the ten-foot long ram. The specifications for this piece were: it could not buckle under load, project at least 7 inches from its base, have a smooth curved surface at the forming edge, and its cross-sectional width not to exceed five-sixteenths of an inch. The above requirements suggested the use of a rather straightforward mill T-beam. Upon closer investigation of the standard sizes none satisfied all the requirements so an I-beam was selected so it could be modified to accommodate these design constraints.

Modification to this ram I-beam required slitting the web (section between the flanges) along its entire 10-foot length. To accomplish that task a 3-hp cutoff saw was purchased. After rearranging the saw position relative to its base, the saw was capable of making a straight sustained cut on the horizontal web as the saw base frame was guided by the inside flange of the I-beam. On paper this seems like an easy thing to do but in reality using a 3-hp cutoff saw having a 16-inch diameter abrasive wheel and grinding thru a quarter inch of steel ten feet long was far from easy. In fact it required the use of four abrasive wheels and over ten hours to accomplish. Within seconds into this manual operation the deafening sound generated by grinding metal into molten steel along with all the residual grit and smoke made using ear plugs and a respirator type mask a necessity.

After drilling the required holes through the flange for mounting to the presses the remaining task was to round off the saw cut edge of the web. This was necessary to avoid destroying the inside surface of the 13.5 inch wide aluminum strip as it was being formed into the desired shape. After various methods were experimented with it was determined the easiest way to obtain a uniformly round surface was to simply purchase a 10 foot long quarter-inch diameter alloy steel rod and weld it to the web leading edge. Welding such a thin rod anywhere but at the ends would instantly distort it, so to make this work a one-eighth-inch groove was machined into the center of the web along its length and the rod secured in place with epoxy and welded to the web at the extreme ends. Although this would not be considered a very strong mechanical bond the rod held in place in use. This completed the basic press.

2. The Trailing Edge Tooling: now that we have the press defined we can discuss the topic of the "Trailing Edge Tooling". In this press setup mode both the I-beam and the Ram are shifted to the right-hand side of the 60-ton press assembly. This is required so the edge of the aluminum strip will align properly with the ram and yet still fit within the hydraulic press framework. Next an angle iron is added longitudinally to the I-beam. One-quarter inch from this angle iron a steel plate (1/4" thick) is mounted parallel to it.

This setup provides the forming space for the ram to travel. The edge contouring process consists of placing the 10-foot long aluminum strip up against the angle-iron stop and hydraulically lowering the ram. Note: the contact edge of the longitudinally mounted steel plate must be rounded off and any repositioning of the ram and I-beam necessitates the use of a 2-ton engine hoist.

3. Leading Edge Forming Press Tooling:

The setup begins by moving the ram and I-beam back to the 60-ton press centerline. Once verified the press is setup correctly the focus is on preparing the aluminum sheet. In this process initial sheet alignment is critical as 10 feet of extremely expensive 7075-T6 material is ruined if the resultant bend is off center.

Before the bending process can begin the desired centerline must be established at each end of the sheet, as indicated on the engineering drawing, and a scribe mark placed. Then exactly at the scribe mark, at each end, a one-eighth-inch diameter slot is milled into the material one-sixteenth inch deep. This modification is performed by the milling machine where the sheet is accurately positioned parallel to the bed using edge guides and clamps to insure exact placement of the slot. As with all milling operations the material must be firmly secured in place before attempting to make the slot or the material will shift. It is recommended a trial run be made on a scrap piece of material to determine how much clamping force is required before permanent scuffmarks are left on the material surface, which should be avoided.

The overview of this forming process is to center the 10-foot strip of material on top of a space existing between two, 2" diameter rollers, which are aligned parallel to the I-beam of the press. At each end of the I-beam, a locator bracket is mounted which anchors a dowel pin at the exact theoretical airfoil bend centerline. It is on these extreme end dowel pins the milled notches at each end of the 10-foot strip are positioned. After placement of the sheet and all pre-bend checklist data entries are recorded, the forming process can begin.

Utilizing all three hydraulic jacks the ram is lowered onto the centerline of the material. If the proper alignment is achieved the ram is lowered further driving the center of the material between the two rollers. The ram travel terminates when it reaches the end stops. The ram is then retracted and the formed ten-foot open "U" shaped airfoil carefully removed from the press.

4. The Structural Stiffener Spacing Fixture:

Review of the I-beam base on the hydraulic press showed that along its entire 10-foot length the 16-inch web was clear of obstruction. The upper web surface closest to the top flange was determined as the ideal location for the Stiffener Spacing Fixture. On that surface a tape measure was mounted, which extended the entire length and at each inch line position where a stiffener should be located a scribed mark was made on the web. Brackets were then added to support the stiffener tube at a height such that the top of the inside edge of the wide portion of the stiffener would rest next to the tape measure. This easily allowed the stiffeners to be shifted on the support tube until they aligned with all the appropriate scribe marks. Once all stiffeners were in position a small tab of contact glue was added to prevent any movement when this final stiffener assembly was moved to the Kevlar/ Epoxy curing fixture.

5. The Kevlar/Epoxy Curing Fixture:

Utilizing the same logic previously used for the "Structural Stiffener Spacing Fixture" the free space on the hydraulic press I-beam was used. The top section of the I-beam web was dedicated to the "stiffener" fixture but the lowered part near the lower flange provided more than sufficient space for mounting the Kevlar/Epoxy Curing Fixture.

The purpose of this fixture is to provide a place to retain the 10-foot "U" shaped airfoil skin produced in the hydraulic press while the internal blade structure is added. In this position the 3/16 and 7/16 inch diameter steel rods are inserted into the forward portion of the "U". The diameters of these rods as well as the way they stack above each other conform exactly to the desired internal dimensions of an NACA0012 leading edge shape. Securing these rods in this position the Kevlar/Epoxy composite is added. After the entire Kevlar/ Epoxy process is completed, and while still in the semi soft curing state (which is about fifteen minutes into the process) the Structural Stiffener Assembly is accurately positioned into the composite. Gage blocks and clamps are spaced along the edge of the I-beam flange to verify the installed overall stiffener assembly height with respect to the leading edge. Once this final check is completed and while the composite is still pliable the trailing edges are clamped together for final curing.

Note: The production of the Kevlar/Epoxy composite is not a random association of materials and liquids. The procedure for adding the hardener to the resin is a very precise process as well as its addition to the Kevlar material. Its use and workability is time dependent. The end product must produce a material with a specific density of 0.032 lb/in^3 so the location of the theoretical rotor blade CG will be maintained.

The details of this fixture are as follows. The rear support of the fixture is provided by the lower I-beam web, the front support by a modified 2 x 2 x 1/8 inch thick angle iron. The intent of these two vertical support surfaces is to ensure the overall maximum height of the NACA 0012 profile is maintained at the quarter chord point, which is 0.803 inches. The separating distance between the web and the inside leg of the angle iron is equal to 0.0803 plus 0.060 inches. The 0.060-inch distance allows for an inner plastic film liner to be added protecting the leading edge surface of the airfoil as it is being pushed into the fixture.

6. The Drill Jig and Riveting Fixture:

If you have been following the progression of the rotor blade fixture development, the next process on the rotor blade, as it is removed from the Kevlar\Epoxy Curing Fixture, is to remove the curing clamps from the airfoil trailing edge and rivet these two edges together. This riveting process involves adding a scribed line the entire airfoil length 1/4-inch inboard of the outside edge of the trailing edge. To facilitate seeing this scribed line it is suggested marking that localized surface with a blue ink marker first. Then at each rivet location, as specified on the detail drawing, another scribed mark is made. At each of these rivet locations a 0.125 inch diameter pilot hole is drilled, then countersunk on both sides of the blade, a rivet added into the hole and the free end compressed flush with the top edge of the countersunk surface by use of a brake rivet installation tool.

Practically, this is a rather straightforward hand tool operation but in light of the care taken thus far, another fixture was designed to eliminate any aberrant drilling alignment errors. This fixture starts with the purchase of a small quarter-inch drill press, which is mounted to a workbench. Then the drill press base is modified to accept an adjustable support to hold the ~ 6.7-inch wide rotor blade in the exact riveting position. At the right and left of the workbench an adjustable height single roller conveyor support stand is positioned to support the free ends of the rotor blade.

On the press base, at exactly two inches from the left side of the drill bit centerline, the brake-riveting tool is attached. In this setup, after the first sets of holes are drilled and countersunk, rivets are sequentially added one at a time. Although it would seem to be more efficient to drill and countersink all holes in one operation then rivet, the problem with this approach is the localized airfoil distortion during each rivet operation has a tendency to misalign the upper and lower drilled surface of the next rivet location. Although a fabrication issue, the alignment errors do not totalize if the drilling, countersink and riveting process is completed in small steps.

7. The Rotor Blade Final Machining Fixture:

This is a Quality Control operation to bring the blade chord length into specification. Final milling of the trailing edge eliminates any chord length variations that were produced by the bending and riveting process. The fixture is not much more than a straight (leading edge) guide surface mounted to the milling machine where the offset distance from guide to the mill bit cutting edge equals the chord length plus the allowable tolerance. Outboard of each end of the milling machine adjustable height single roller conveyor support stands are positioned to support the free ends of the rotor blade as it is being milled.

8. The Rotor Blade Epoxy Fixture:

Finally the last steps in completing the rotor blade are accomplished with this fixture. Those are as follows, Step 1, the structural epoxy is applied to cover the exposed rivets on the formed side of the trailing edge and match the shape of the non-formed trailing edge and Step 2, is to make the rotor blade epoxy end cap.

Both of the above processes utilize the Hydraulic press I beam as a base. In Step 1 the hydraulic ram is removed, and the I-beam moved closer to the hydraulic press vertical frame support. In this position brackets are added along the length of the I-beam such that the trailing edge side to be epoxied is horizontal with respect to the I-beam surface. Epoxy is added and allowed to cure in this position.

Step 2 requires the purchase of a shallow plastic pan 2x 8 inches long and the addition of two holding clamps mounted to the hydraulic press vertical support frame to secure the rotor blade vertically. Manually lowering the 13.8-pound blade into the pan and allowing epoxy to seep inside the rotor blade end to a depth of 3/8 inch form the end cap. The blade is held in position by the press-mounted clamps. Keeping this epoxy from bonding with the external blade surface, a special plastic coated "release tape" is wrapped around it locally prior to placing the blade in the pan.

There is one more subject to discuss before I conclude this Rotor Blade section and it has to with blade dynamics. The calculations already performed indicate this rotor blade as designed will handle the anticipated forces within an acceptable stress range. There are four more terms used in discussing rotor blade function and they are as follows:

Aerodynamic stability: The stability of a body with respect to aerodynamic forces.

Aeroelasticity: The study of both the static and dynamic effects of aerodynamic forces on elastic bodies. The flutter of flags, aircraft wings, and sails are examples of the interplay of aerodynamic forces, inertia forces, and the elastic properties of the structures.

Aero-isoclinic: The ability of an airfoil to maintain, or tends to maintain, the same angle of attack at all airfoil sections, especially when flexed or bent.

Elastic center: A point within a section of a structure or member, such as an airfoil section, at which the application of a load will cause lengthwise or span wise deflection but not torsional deflection, hence, a point in a section about which torsional deflection occurs.

The fact that this rotor blade deflects in the cantilevered position downward under a no load condition and upward when in flight indicates we have a flexible structure. What is unknown is if this degree of blade flexibility is detrimental? Will these rotor blades be stable while in flight?

14.3 Main Rotor

We have already discussed the Main Rotor in section 8.3 Defining the Helicopter's Rotor Hub Type. The functional design description is as follows: While driven from the Main Rotor Shaft provide an Anchor point for the two Rotor Blades to Feather, Flap and change Angle of Attack without self-destruction of bearings and or structure while subjected to the dynamic loads of flight.

CAD Dwg 14.3-1 Rotor Design Nomenclatures

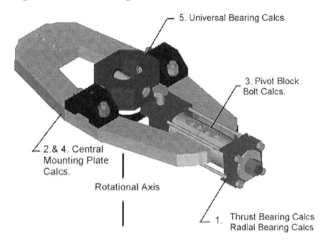

Design Issues:

1)   The Bearings: Resolve how to keep the rotor blades attached to the hub assembly while feathering and flapping. To start, we need to calculate the centrifugal force generated by the weight of the individual rotor blades at the design rpm since that force will be a substantial contributor to the stresses involved for adequately sizing the bearings. Then we need to size the pitch and thrust bearings.

2)   The Plate Angle: Earlier in this book it was decided to utilize a Semi-Rigid Rotor with an under-slung hub (teetering rotor). To optimize this hub type it would be a good idea to angle the plate ends such that it equals the tangent angle (approximate cone angle) generated by the vertical blade lift force and the horizontal centrifugal force. By doing this, the structural part of the hub handles most of the moments developed from steady state coning loads instead of the rotor blade support bearings. These calculations determine the cone angle.

3)   The Pivot Block: The description of the hub type states, "under-slung". This means our design must establish the offset distance this hub has to be such that blade feathering takes place along an axis where the blade Center of Gravity intersects the Universal Bearing Axis in order to reduce the Coriolis forces. Also we need to determine the bolts required to retain the pivot block to the Central Mounting Plate.

4)   The Central Mounting Plate: The design issue here is to determine the cross-sectional area of the central mounting plate to handle the design loads with a safety factor of at least 5.

5)   The Universal Bearings: These must be selected and sized based upon the lift forces generated.

Additionally provisions must be included to limit blade flap such that the ends of the rotor blades do not contact the helicopter tail structure. It shouldn't be a hard stop whereby strong contact would induce destructive impact stresses to blade and structure.

1. Pitch Bearing Thrust and Radial Load Calculations:
   • Definition of Terms:
   A. Thrust Bearing Load Calculation:

Definition of Terms:

RPM = 455 rpm (Note: 437.4 rpm is the approximate minimum design rpm)

m1 = 13.9 lbs blade (13.89 + 0.01 end cap)

m2 = 3.64 lbs rotor parts;
   (1.14 lbs straps + 0.61 lbs levers + 0.90 lbs fasteners + 0.40 lbs anchor plates + 0.59 lbs pivot blocks)

R1 = 10.5 ft, rotor blade radius

R2 = 0.75 ft, distance from hub centerline to cg of rotor parts

BL = rotor blade length = 116

TBL = thrust bearing load on each pivot block (inner and outer) thrust bearing

TBA = Torrington TRC 1427 thrust bearing dynamic rating 3420#, static 6869#

Vtcg = tangential speed at cg of rotor blade (1/2 blade length + blade mounting edge to centerline)

Vtr = tangential speed at cg of rotor parts

Fcc = centripetal force at center of rotor blade, use 1/2 rotor diameter

Fcr = centripetal force at cg of rotor parts (straps, pivot blocks, lever arms, etc)

   • Procedure: We know the weight (m1 = blade wt. & m2 = rotor part wt.) and Center of Gravity (1/2 R1 "blade" & R2 "parts") of both the rotor blade and its pivot block parts, and how fast they are rotating (RPM). Using this information we can calculate a1, the tangential speed (Vtcg & Vtr) and a2, centripetal force (Fcc & Fcr) of each. Those sum to obtain a3, the total axial force acting on the two thrust bearings (F_axial). After investigating possible bearing choices to package in the space available, a Torrington thrust bearing was selected that has a 3420-pound dynamic rating. Dividing the bearing rating by the axial force on that bearing gives us the Safety Factor for this thrust bearing application. The thrust bearing selection is based upon two bearing sets handling the resultant axial load TBL = 2873 lbs which is directed radially outward at an approximate 3 degree angle. This θ is pre-bent into the central mounting plate thereby eliminating any stress by having the radial force act along the line of action. The thrust bearing selected is a Torrington NTC1427, static rating 6860 lbs and a dynamic rating of 3420 lbs.

   Initiating the radial load calculations we will make one simplifying assumption. Instead of using the lift equation where F(lift) = coefficient of lift (at that pitch angle) x air density x blade area x tangential speed at the "center of lift" / 2 g_c ; we will just say the blades lift 600 pounds, 525 pounds as per design and 75 pounds for pure vertical acceleration. Using the first pivot block radial bearing as a reference point the moment the rotor blade generates with respect to that point is calculated. Dividing this moment by the distance to the second bearing gives us the radial force on the bearing. The Torrington radial bearing selected has a dynamic rating of 1750 pounds. Resulting from this information the Safety Factor can be calculated. I start with the pitch bearing calculations.

Calculation 14.3-1

1. Pitch Bearing Load Calculations

A Thrust Bearing Load Calculation:

Variables Assigned

$$BL := \frac{116}{12} \qquad L\_bo := \frac{10}{12} \qquad m2 := 3.64 \qquad R1 := 10.5$$

$$g := 32.2 \qquad RPM := 455$$

$$ml := 13.9 \qquad R2 := 0.75 \qquad R\_cg := \frac{BL}{2} + L\_bo$$

a1. Tangential Speed Calculations: Formula

$$Vtr := \pi \cdot 2 \cdot R2 \cdot \frac{RPM}{60} \qquad\qquad Vtr = 35.736 \text{ ft /sec}$$

$$Vtc := \pi \cdot 2 \cdot R1 \cdot \frac{2}{3} \cdot \frac{RPM}{60} \qquad Vtc = 333.532 \text{ ft /sec}$$

$$Vtcg := \pi \cdot 2 \cdot R\_cg \cdot \frac{RPM}{60} \qquad Vtcg = 270.002 \text{ ft/sec}$$

a2. Blade Force Calculations:
  a2a Centripetal Force on Blade

$$Fcc := \frac{ml}{g} \cdot \frac{Vtcg^2}{R\_cg} \qquad Fcc = 5.553 \times 10^3 \text{ lbs}$$

a2b. Centripetal Force on rotor parts

$$Fcr := \frac{m2}{g} \cdot \frac{Vtr^2}{R2} \qquad Fcr = 192.481 \text{ lbs}$$

a3. Resultant Force on Thrust Bearings
(thrust bearing load equal on inner and outer pivot bearing blocks)

$$F\_axial := Fcc + Fcr \qquad F\_axial = 5.746 \times 10^3 \text{ lbs}$$

$$TBL := \frac{F\_axial}{2} \qquad\qquad TBL = 2.873 \times 10^3 \text{ lbs}$$

$$SF\_tb := \frac{3420}{TBL} \qquad\qquad SF\_tb = 1.19 \quad \text{Safety Factor}$$

Summary: As you can readily see the safety factor of 1.19 is not over whelming for the selected Torrington NTC1427 bearing (static rating 6860 lbs, dynamic rating of 3420 lbs). I decided to use the bearing because it never rotates in the feathering mode more than several degrees so its rating is probably closer to the static rating. In addition this bearing packaged in a very small space which was critical in keeping the rotor weight low.

B. Radial Bearing Load Calculation:

- Procedure: We stated above each blade will Lift 300 pounds at a 2/3 radial distance from the hub. To counteract that moment we have one bearing at a specific distance (Dis2) and the second at (Dis3). The assumption will be both bearings have equal force applied to them.
- Definition of Terms:

Frb = radial bearing load on each pivot block (inner and outer) radial bearing = Lx Dis1 / (Dis2 +Dis3)

RBL = the radial bearing selected is a Torrington B1412, static rating of 5400 lbs and a dynamic rating of 3420 lbs.

Dis1 = distance from rotor hub to rotor blade center-of-lift = 84 inches = 10.5 (12 x 2/3)

Dis2 = distance from hub to first radial bearing = 3.125 inches

Dis3 = distance from hub to second radial bearing = 8.263 inches

Calculation 14.3-2

B Radial Bearing Load Calculation:
Variables Assigned

$$Dis1 := 84 \qquad Dis2 := 3.125 \qquad Dis3 := 8.263$$

$$L := 300 \qquad RBL := 3420$$

a1.Radial load on Bearings

$$Frb := \frac{0.5 L \cdot Dis1}{(Dis2 + Dis3)} \qquad Frb = 1.106 \times 10^{3} \text{ lbs}$$

$$SF\_rb := \frac{3420}{Frb} \qquad SF\_rb = 3.091 \text{ Safety Factor}$$

Summary: The selected Torrington B1412 (static rating of 5400 lb, dynamic rating of 3420 lbs) has a reasonable safety factor so it was utilized. This bearing was also selected because it kept the rotor weight down by packaging in a small space.

2. Calculation of Blade Vector Angles in Vertical and Tangential Directions
- Procedure: The best way to eliminate stress on the main rotor bearings is to pre-bend the central mounting plate to the anticipated vector angle generated from the Lift and Centripetal forces acting on it. In part A that angle is determined from the actual equations for Lift and Centripetal Force. In part B the magnitude of the Drag angle is calculated. NOTE: The resultant angle calculation does include the rotor part centripetal force. This assumption was made because these rotor blade reactions will be used in subsequent calculations.
- Definition of Common Terms

Vtc = tangential speed at center of lift (usually calculated at 2/3 radius)

roe = 0.0752 lbm/ft3 , mass density of air @ 68degrees F

S = blade area = (10.5ft-9/12ft =9.75ft)x 6.69/12ft = 5.44ft^2; Note: the 9/12 fraction represents the rotor hub dimension which is not part of the blade area.

g = 32.2 (lbm x ft) / (lbf x sec^2)

M = 0.56 figure of merit (design)

L = 300 lbs = 525 lbs (design) + 75 lbs (maneuvering load). Note for comparison L1 is calculated from formula

A. Resultant Angle of "Lift versus Centripetal Force":
- Definition of Terms:

F_axial = centripetal force value calculated above as = 5746 lbs.

L1 = 1/2 (Air x V2 x S x CL). Note center of lift is at 2/3-rotor diameter.

Angle = arc tangent (Lift) / Centripetal force (F_axial)

Angle_xy = arc tangent (Lift (L1) / Centripetal force (F_axial)

Cl = lift coefficient = 0.8 (for a section angle of attack of 8 degrees utilizing a NACA 0012 airfoil. See Graph 4.3-1

Calculation 14.3-3

Calculation of Blade Vector Angles in the Vertical and Tangential Direction

Variables Assigned: $L = 300$ $F\_axial = 5.746 \times 10^3$

$M := 0.56$ $\rho := 0.0752$ $Cd := 0.008$ $S := 5.44$ $Cl := 0.8$

A. Calculation of Blade Vector Angle as a function of Lift and Centripetal Force

a1 Blade Lift Angle Calculated

$$L1 := .5 \cdot M \cdot \rho \cdot Vtc^2 \cdot S \cdot \frac{Cl}{g} \qquad\qquad L1 = 316.581 \text{ lbs}$$

$$Angle\_xy := atan\left(\frac{L1}{F\_axial}\right) \qquad\qquad Angle\_xy = 3.154 \text{ deg}$$

$$Angle := atan\left(\frac{L}{F\_axial}\right) \qquad\qquad Angle = 2.989 \text{ deg}$$

Summary: Both angles were calculated for L and L1. I decided to round off the angle to 3 degrees for the vertical bend in the main rotor support plate at each rotor blade anchor point.

B. Resultant Blade Angle as a function of "Drag and Centripetal Force":
• Definition of Terms:
Cd   = drag coefficient = 0.008 (for a section angle of attack of 8 degrees utilizing a NACA 0012 airfoil. See Graph 4.3-2
D   = blade drag = 1/2 M (Air density x V^2 x S x Cd)
Angle_zx = arc tangent (Drag (D1) / Centripetal Force (F_axial)

Calculation 14.3-4

B. Calculation of Blade Angle as a function of Drag and Centripetal Force
a1 Blade Drag Angle Calculated

$$D1 := .5 \cdot M \cdot \rho \cdot Vtc^2 \cdot S \cdot \frac{Cd}{g}$$

$$D1 = 3.166 \quad \text{lbs}$$
(rotor blade drag force)

$$Frxy := \sqrt{L1^2 + F\_axial^2}$$

$$Fr := \sqrt{Frxy^2 + D1^2}$$

$$\left(\frac{Fr}{2}\right) = 2.877 \times 10^3$$

$$Angle\_zx := atan\left(\frac{D1}{F\_axial}\right)$$

$$Angle\_zx = 0.032 \, deg$$

a2. Magnitude of Lift- Drag Vector:

$$RBL := \sqrt{L1^2 + D1^2}$$

$$RBL = 316.597 \quad \text{lbs}$$

Summary: The resulting drag angular deflection of only 3 hundredths of a degree implies a very small disturbing angle which is desirable. What hasn't been discussed is how stable the flexible rotor blade will be when reacting to these Lift / Drag forces. Also these calculations are based upon a constant 8 degree angle of attack while in "hover". How will the rotor blade react when the helicopter is in forward motion and cyclic feathering is involved? In that case the drag forces will fluctuated with every rotor rotation and those dynamics have yet to be determined. This type of data will need to be gathered and analyzed during field testing.

3. Main Rotor Outer Pivot Block Bolt Stress Calculations:
• Procedure: In these calculations we want to optimize the bolted connection of the pivot blocks to the central mounting plate such that the rotor blades are held in place during all dynamic loadings. As you can see in the previous sketch the "inner pivot block" is not as much a concern as the pivot block mounted at the extreme end of the central mounting plate since its possible travel is restricted because it "butts against" the plate. Our analysis is focused on the outer most pivot block. The goal here is to maximize the pivot block retention force and minimize the diameter of the drilled holes through the central mounting plate.

In (A), the calculation is to determine if the clamping force generated by two 1/4 –28 bolts is sufficient to handle half the axial force. The next calculation (B), checks to see if those two bolts can resist the axial force based upon bolt shear strength alone.

Note: The axial load "Fr" is distributed equally between the inner and outer pivot blocks, each of these blocks has two contact surfaces for the bolt-normal force to act on. Bear in mind only 75% of the bolt yield is utilized and the formulas reflect this. Also the endurance limit for this aluminum material 6061-T6 is 14,000 psi. By designing this part to be subjected to stress less than the endurance limit of 14000 psi the part will withstand an almost infinite number of loadings without experiencing failure.

A. Using 2, 1/4-28 bolts to clamp, the Safety Factor is:
• Definition of Terms:
μ = 1.05 coefficient of friction of aluminum on aluminum (dry-static) from Marks handbook section 3-34
Fr   = Total rotor blade resultant load (calculated as 5746#)
RSy25   = 3130#, yield at root diameter
F_axial   = resultant force on pivot block connection
Fpby   = axial force developed with clamping with design bolting (force normal to each pivot block mounting surface) where 1/2 Fr = μ x Fclamp.
SFpby   = safety factor of clamped pivot block connection

B. Using 2; 1/4-28 bolts in double shear, the Safety Factor is:
• Definition of Terms:
Fpbs   = axial force developed with design bolting in shear (force normal to each pivot block mounting surface) where 1/2 Fr = μ x Fclamp
RSs25   = 3680#, shear at diameter

SFpbs   = safety factor of bolted pivot block connection in shear

C. Plate Tear-out Stress Calculations:
• Definition of Terms:
A_fpbto = cross sectional tear out area of forward pivot block with 1/4-28 bolt = 0.588in**2

S_fpbto = tear out stress on forward pivot block
S_6061 = yield stress for 6061-t6 aluminum =35000 psi
SF_fpbto = tearout stress safety factor

Calculation 14.3-5

3. Main Rotor Outer Pivot Block Stress Calculations
   Variables Assigned

$$\mu := 1.05 \qquad RSy25 := 3130 \qquad RSs25 := 3680 \qquad d25 := 0.25$$

A. Using 2, 1/4-28 bolts to clamp, the Safety Factor is:

$$Fpby := 2(2RSy25 \cdot 0.75)\mu \qquad Fpby = 9.86 \times 10^3 \text{ bolt clamping}$$
force capability

$$0.5 \cdot F\_axial = 2.873 \times 10^3 \qquad \text{axial force}$$

$$SFpby := \frac{Fpby}{0.5F\_axial} \qquad SFpby = 3.432 \quad \text{Safety Factor}$$

B. Using 2; 1/4-28 bolts in double shear, the Safety Factor is:

$$Fpbs := 2(2RSs25) \qquad Fpbs = 1.472 \times 10^4 \quad \begin{array}{l}\text{strength of 2 bolts}\\\text{in double shear}\end{array}$$

$$SFpbs := \frac{Fpbs}{0.5F\_axial} \qquad SFpbs = 5.124 \quad \text{Safety Factor}$$

Summary: Use of (2) 1/4-28 bolts in double shear while clamping the outer pivot block in place is acceptable.

4. Calculation of Main Rotor Central Mounting Plate Stresses:
• Procedure: The first calculation (A) determines if the smallest cross-sectional area of the Central Mounting Plate can withstand the rotor blade force FR trying to cause it to fly apart. In section (B) the calculation focuses on the compressive and tensile stresses at the extreme edges of the Central Mounting Plate at the "bend". The upper and lower extreme edge in either case is the same distance (c_cmp) from the neutral axis. In this case the Moment of Inertia of the material at the cross-section is evaluated.
   It should be noted that from a metallurgical point of view when fabricating the Central mounting Plate it is critical that the molecular grain direction is aligned with the intended applied force direction. This fact was taken into consideration when the plate was machined. Finally, in (C) we note the end deflection of the Central Mounting Plate due to the bending moment (assuming both rotor blades equally loaded as in Hover), to determine if this will impact the dynamics of the rotor disc.

A. Main Rotor Central Mounting Plate Stresses at Cross-Section
- Definition of terms:

Fr = Total rotor blade resultant load (calculated)

Aplate = 2(1.539 x .75) = 2.625 in^2 ; total cross-sectional area that radial force acts

S_6061T6 = 35000 psi = yield for 6061-T6 aluminum

S_cmpt = central mounting plate (cmp) stress due to axial load on cross-sectional area

Sf_cmpt = safety factor for radial stress on central mounting plate

B. Main Rotor Central Mounting Plate Stress at Bend
- Definition of Terms:

m_armcmp = moment arm which Fr acts on central mounting plate = 0.294 inches (3 degree bend)

M_cmp = bending moment on central mounting plate, Fr x m_armcmp

b_cmp = width of each bent cross-section of the central mounting plate (cmp) 1.5"

h_cmp = height of cmp 0.75"

c_cmp = distance from neutral axis to outer fiber

I_cmp = moment of inertia of the central mounting plate at the smallest cross-sectional area

S_cmpb = central mounting plate stress due to bending

C. End Deflection of Central Mounting Plate

E_alum = elasticity of aluminum 11 x 10^6 psi

L_cmp = overall length of the central mounting plate

D_cmp = deflection at end of cmp = moment (M) x force (L) / 2 x elasticity (E) x moment of inertia (I)

Calculation 14.3-6

4. Calculation of Main Rotor Central Mounting Plate Stresses
Variables Assigned

$Fr = 5.754 \times 10^3$ $Aplate := 2 \cdot (1.5 \cdot 0.75)$ $E\_alum := 11 \cdot 10^6$

$L\_cmp := 14$ $m\_armcmp := 0.294$ $c\_cmp := .375$

$b\_cmp := 2 \cdot (1.5)$ $h\_cmp := 0.75$

$S\_6061T6 := 35000$ $M\_cmp := Fr \cdot m\_armcmp$

A Main Rotor Central Mounting Plate Stresses at Cross-Section

$$S\_cmpt := \frac{Fr}{Aplate} \qquad S\_cmpt = 2.557 \times 10^3 \text{ psi}$$

$$SF\_cmpt := \frac{S\_6061T6}{S\_cmpt} \qquad SF\_cmpt = 13.687 \text{ Safety Factor}$$

B. Main Rotor Central Mounting Plate Stress at Bend

$$I\_cmp = \frac{b\_cmp \cdot h\_cmp^3}{12}$$

$$S\_cmpb = M\_cmp \cdot \frac{c\_cmp}{I\_cmp} \qquad S\_cmpb = 6.014 \times 10^3 \qquad \begin{array}{l}\text{Stress at}\\\text{outer fiber}\\\text{of plate}\end{array}$$

$$SF\_cmpb = \frac{S\_6061T6}{S\_cmpb} \qquad SF\_cmpb = 5.819 \qquad \text{Safety Factor}$$

C. End Deflection of Central Mounting Plate

$$D\_cmp = \frac{M\_cmp \cdot (0.5L\_cmp)}{2E\_alum \cdot I\_cmp} \quad \text{deflection}$$

$$D\_cmp = 5.103 \times 10^{-3} \quad \text{inches}$$

Summary: The design of the Central Mounting Plate utilizing a three quarter inch thick plate one and a half inches wide at each side is more than adequate to handle the centripetal forces involved.

  5. Main Rotor Universal Bearing Load Calculations:
• Procedure: This lift force is handled by two bearings, one at either end of the universal. Since this is a pinned joint and no moments must be accounted for the load each bearing must accommodate is L/2 or 300 lbs. The bearing selected is a Torrington B810 basic static rating 2590 lbs and a dynamic rating 1750 lbs.
  • Definition of Terms:
RBA   = Torrington radial bearing dynamic rating 1750#, static 2590#
L   = 600 lbs, actual lift (see formula stated above)
  SFu   = Safety Factor universal bearing

  Calculation 14.3-7

5. Main Rotor Universal Bearing Load Calculations

$$L = 600 \text{ lbs} \qquad RBA = 1750 \text{ lbs}$$

$$SF2 = \frac{RBA}{0.5L} \qquad SF2 = 5.833 \quad \text{Safety Factor}$$

Summary: The Torrington B810 bearings (basic static rating 2590 lbs, dynamic rating 1750 lbs) are more than adequate for the application.

Print 14.3-1

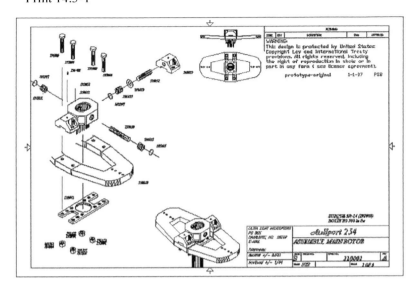

Print 14.3-2

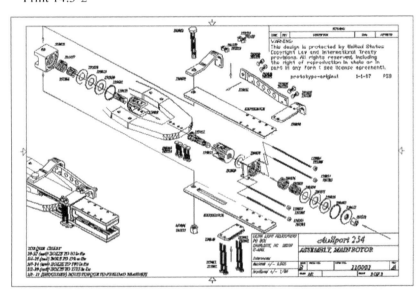

14.4 Control System

We have already discussed the Control System in section 3.2.3. The functional detail descriptions are as follows:

CAD Dwg 14.4-1 The Hand Controls

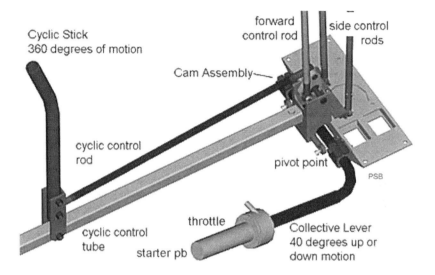

CAD Dwg 14.4-2 The Cam Assembly

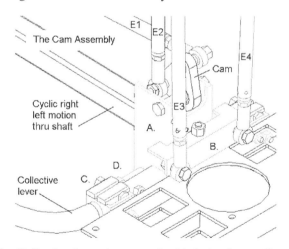

1) The Collective Lever increases the Pitch Angle equally on both rotor blades via mechanical linkage and control rods when manually raised by the pilot's left hand. As an aid to the pilot a Pitch Angle indicator is installed to provide the actual Pitch Angle as a function of Collective Lever position. To secure the Collective Lever in the desired spot a positive retention device is used. Engine speed is also controlled from this lever via a handgrip throttle

assembly mounted at the foremost end of the handle. The starter button is nested within the throttle assembly and protrudes at the extreme end of the lever.

2)   The Cyclic Control Stick (Joystick), this "360 degrees of motion" stick is used by pilot's right hand for directional flight by ultimately differentially feathering the individual rotor blades via the same mechanical linkage and control rod setup as used with the Collective Lever.

3)   The Cam Assembly "A", (Diagram 14.4-2) on the AIRSPORT 254 is the heart of the control system. It functions as follows: on the side of the Cam Assembly two rod eye bearings are mounted, one on each side "D". Correspondingly shoulder bolts slip through these rod eye bearings and bolt to the horizontal tubular section of the Collective control lever. As the pilot changes the Collective lever's angular position the entire Cam Assembly moves up or down in direct response to the lever's degree of motion via the shoulder bolt / rod eye relationship. The three externally mounted Swashplate control rods "E2, E3 and E4" track accordingly. A friction brake "C" keeps the Collective lever set once positioned.

As for the Cyclic lever operation, this mechanical conversion is more complex in that the lever (Joystick) has the capability of 360 degrees of motion that the Cam Assembly must translate up to the Swashplate. This is accomplished by breaking down the 360 degrees of motion into two basic component parts of either Joystick "forward or aft" "E1" (helicopter movement forward or aft) or Joystick "right or left" "B", (helicopter movement left or right). All other Joystick motion travel is a combination of these two and the same control rods mentioned above are used. When the Joystick moves "forward or aft" that motion is transferred to the top of an internal mechanical cam in the assembly which in turn connects to the Swashplate control rod "E2" that tilts the Swashplate either down (forward helicopter motion) or up (rearward helicopter motion). Right or left motion of the Joystick rotates its support shaft, which extends thru the assembly. At the end of this shaft is a centerline-mounted angle bracket "B" to which the right and left Swashplate control rods "E3 and E4" are attached at either end. As the right rod "E4" moves "up or down" the left rod "E3" travels in the opposite direction and this motion is directly reflected in the Swashplate right / left positioning.

The above description of the Cam Assembly details Swashplate movement resulting from pilot Cyclic & Collective input with a totally mechanical operating system. Hydraulic and electrically assisted Swashplate control systems have similar coordinated function.

Design Issues: 1. Ergonomics was the first issue to be resolved. It was important to determine the range of motion a pilot's right and left hand could move comfortably while anchored in the seat by a four point safety harness. Bear in mind that a 180-pound pilot's shifting movements in a helicopter that has a dry weight of only 254 pounds has major impact on the resultant CG of the helicopter's gross weight. If that CG doesn't align with the "rotor mast hub centerline" aircraft instability results. Once the range of motion is established for the pilot the next would be to equate rotor blade Angle of

Attack change with a control rod vertical travel distance. For example what is the control rod movement required to effect an Angle of Attack change from 6 degrees to 11.2 degrees (our design operational limit for the NACA 0012 profile)? Once those two areas of concern have been addressed the mechanics must be developed to translate this pilot input into a directed output to the Swashplate assembly. IMPORTANT: The Cyclic Control Stick is only functional if there is any Angle of Attack range "left over" once the Collective Lever is raised to initiate a Hover. If your "as build" helicopter requires an 11-degree of Angle of Attack to Hover don't expect to go anywhere. If you are hovering with an 8 degree Collective lever position then the Cyclic Control Stick can use that additional 3.2 degrees of Angle of Attack (11.2 degrees total) for directed flight.

Shown below in Figures 14.4-3 thru 4, are the ergonomic right and left hand range-of-motion I used, along with control rod travel distances associated with these hand movements.

CAD Dwg 14.4-3

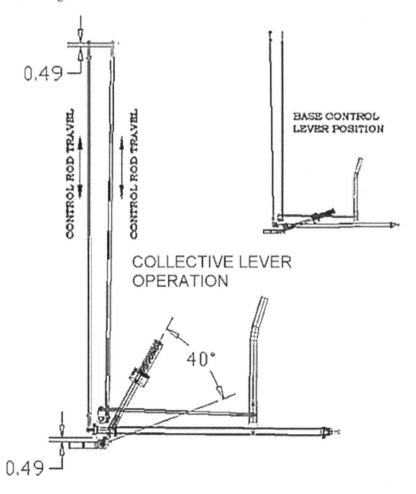

CAD Dwg 14.4-4

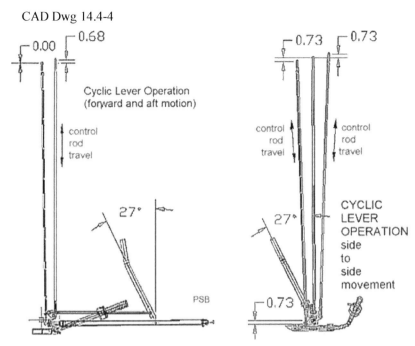

2)   There are four cockpit control rods involved with swashplate orientation and two on the upper rotating portion controlling rotor blade positioning. All of which must be sized to handle both the pilot's input and also the feedback dynamics of the rotating rotor blades. The lower control rods are setup as follows: one is connected to the Cyclic Control Stick / Cam assembly (E1) and three connected to the Swashplate / Cam assembly (E2, E3 and E4). The control forces at the rods themselves I found to be extremely small somewhere in the 1 to 5 pound range. The 6063-T6 tubing used had a 0.5 od x 0.035 wall. The upper swashplate control rods used a 0.75 od x 0.035 wall. The calculations of both rods are given.

3)   Design a mechanically operated Collective Lever Pitch Angle indicator, if that is not possible use encoders and electronic display.

4)   Size the Collective Lever position retention device.

Control Rod Buckling Analysis: Given: The Euler load is the theoretical maximum load that an initially straight column can support without buckling. For columns with frictionless or pinned ends, this load is given as $F = pi^2 (E \times A) / (L / K)^2$, (L/k = length of rod divided by its radius of gyration is called the slenderness ratio). Euler's equation to be valid the critical slenderness ratio range is between 80 and 120.

Conclusion: This yaw control rod will fail due to buckling and not due to reaching the yield limit. That occurs when the force on the control rod exceeds 56 pounds. Since our force is less than 5 lbs this size tube should work fine. The upper swashplate control rods can handle a load of over 1000 lbs. so they are sized appropriately also.

Definition of Terms:

1. Yaw Control rod (6061T6 aluminum yield 35000 psi = S_6061)

ro   = Outside radius = 0.25

ri   = Inside radius = 0.215

L_yaw   = 49-1/4 long = C x L where C=1 for the column pinned at both ends and L is the rod length = 49-1/4

I_yaw   = Material 1/2od x 0.035wall = pi(ro^4 - ri^4)/4

A_y   = Cross sectional area of yaw control rod

k_y   = Radius of gyration = (I / A)^1/2

SF_yaw   = Safety factor for the yaw control rod = 2

Fe   = Maximum force allowable based upon Euler buckling eq.

L_yaw/ k_y = slenderness ratio, must be > 80 for Euler's eq. to be valid

2. Swashplate Control Rod (6061T6 aluminum yield 35000 psi = S_6061)

ro_sw   = Outside radius = 0.375

ri_sw   = Inside radius = 0.34

L_sw   = 5" long = C x L where C=1 for the column pinned at both ends and L is the rod length = 5

I_sw   = Material = pi (ro_sw^4 - ri_sw^4)/4

A_sw   = Cross sectional area of swashplate control rod

k_sw   = Radius of gyration = (I / A)^1/2

SF_sw   = Safety factor for the yaw control rod = 2

Fe_a   = Force if Euler equation can be used; L_sw / k_sw < 80

F_sw   = Maximum force allowable on swashplate control rod

F_dsw   = Design load on Swashplate Control Rod = Fe / SF_y

Calculation 14.4-1

Control Rod Buckling Calculation
a Yaw Control Rod
  Variables Assigned:

L_yaw := 49.25   ro := 0.25   ri := 0.215   S_6061 := 35000

$$E\_6061t6 = 1 \times 10^7$$

Formulas:

$$A\_y := \pi \cdot \left(ro^2 - ri^2\right)$$

$$I\_yaw := \frac{\pi}{4} \cdot \left(ro^4 - ri^4\right)$$

$$k\_y := \sqrt{\frac{I\_yaw}{A\_y}}$$

$$Fe := \pi^2 \cdot E\_6061t6 \cdot \frac{A\_y}{\left(\frac{L\_yaw}{k\_y}\right)^2}$$

$$Syaw := \frac{Fe}{A\_y}$$

Results: Fe = 56.549

$$Syaw = 1.106 \times 10^3$$

$$\frac{L\_yaw}{k\_y} = 298.725$$

b. Swashplate Control Rod

Variables:

$$L\_sw := 5 \quad ro\_sw := 0.375 \quad ri\_sw := 0.34 \quad SF\_sw := 2$$

Formulas:

$$A\_sw := \pi \cdot \left(ro\_sw^2 - ri\_sw^2\right)$$

$$I\_sw := \frac{\pi}{4} \cdot \left(ro\_sw^4 - ri\_sw^4\right) \qquad k\_sw := \sqrt{\frac{I\_sw}{A\_sw}}$$

$$I\_sw = 5.036 \times 10^{-3} \qquad A\_sw = 0.079 \qquad k\_sw = 0.253$$

$$Fe\_a := \pi^2 \cdot E\_6061t6 \cdot \frac{A\_y}{\left(\dfrac{L\_sw}{k\_sw}\right)^2} \qquad \text{Euler load equation}$$

The slenderness ratio =

$$\frac{L\_sw}{k\_sw} = 19.756$$

Therefore since this value is less than 80 Eulers load equation can not be used

The maximum load on the swashplate control rod occurs when the material is stressed to yield therefore that force is:

$$F\_sw := S\_6061 \cdot A\_sw \qquad F\_dsw := \frac{F\_sw}{SF\_sw}$$

$$F\_sw = 2.752 \times 10^3$$

$$\boxed{F\_dsw = 1.376 \times 10^3} \quad \begin{array}{l}\text{SWASHPLATE CONTROL ROD} \\ \text{MAXIMUM DESIGN LOAD (lbs)}\end{array}$$

3. Design a mechanically operated Collective Lever Pitch Angle Indicator. In the photo on the top of the next page you can see the arrangement I used for Collective Angle indication. The basic parts consisted of a clear plastic sleeve with an inside diameter slightly larger than the supportive cockpit tube, a stainless steel clamp, a return spring, and a coated one-eighth inch diameter cable. The cable was connected to the Collective Level and to the stainless steel clamp on the slide. Slide return to home tension was provided by the spring. Marks on the support tube were calibrated to actual blade pitch position. When the forward edge of the slide aligned with the angle you wanted that would be the Collective Pitch Angle of the blades. This works quite well I might add.

Photo 14.4-1

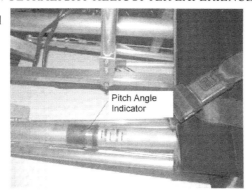

4. . Size the Collective Lever position retention device:
 In this design I simply over bored the Collective lever frame support
clamps to accept an asbestos liner. Tightening on the clamps increased the
lever retention force. This also worked quite well.
 Print 14.4-1

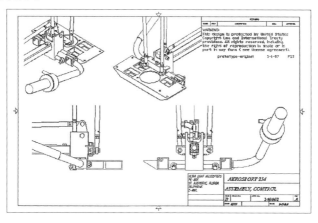

Print 14.4-2

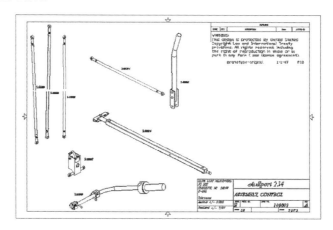

14.5 Tail Rotor

Basic Function Description: The Tail Rotor is used for Yaw control (cockpit turning right, left or centered) by means of the Foot Pedals. Basically Yaw control is a function of applied force produced by the Tail Rotor to negate the counter rotational torque developed by the Main Rotor.

Design Issues: 1. We know that increasing rotor blade Angle of Attack increases blade thrust and as a result increases the counteracting torque required to maintain a set helicopter orientation. The most counteracting torque is demanded at maximum power and maximum Angle of Attack. What must be determined is the range of counteracting torque that must be accommodated to keep the helicopter in Yaw control. Once that is established we can start to design the Tail Rotor so that the forces generated by it will produce the counteracting torque necessary to have a balanced, stable system.

Given the above information we must "set" the diameter and rpm of the Tail Rotor and its' variable range of Angles of Attack to enable Yaw control. Additionally, since Yaw is controlled by the pilot operated Foot Pedals this also becomes an ergonomic issue in that there is a limited range a leg can move comfortably to effect the Tail Rotor Blade angle change.

2. The mechanism must be designed to transfer Foot Pedal operation to Tail Rotor blade Angle of Attack change.

3. Design how to enable tail rotor blade Angle of Attack change. Additionally, the resultant design must accommodate any aberrant wind gusts from over stressing the subsystem.

4. All bearings must be sized and selected.

Calculations:

Tail Rotor Yaw Control Calculations:

Procedure: The thrust "T" required to hover a helicopter weighing 525 pounds was determined in chapter 8. Using that number as a basis, the horsepower to develop the main rotor shaft horsepower is calculated from:

Power $= (T/M \times g)(T/(2 \times Air \times pi \times r^{**}2)^{\wedge}0.5)$.

We also know there is a relationship between shaft horsepower, torque and rpm as (hp = torque (ft-lbs) x rpm / 5250). In this equation the shaft torque is obtained since we already have the horsepower (hp) and rotor rpm values. Shaft torque is the reason a helicopter wants to rotate in the opposite direction of rotor rotation. Yaw control prevents that from happening by applying a tail rotor force to oppose this shaft torque. The tail rotor force multiplied by the moment arm (the distance between it and the centerline of the main rotor shaft) produces the counter rotational torque. The first set of calculations walks us through the procedure to obtain this tail rotor force (thrust).

The second calculation determines if the tail rotor diameter, chord and the rpm selected will produce the necessary range of force (thrust) to not only keep the helicopter in a straight orientation but also turn right or left. To simplify everything I have jumped ahead and used the finalized numbers for rotor diameter, chord and rpm (that actually worked on this helicopter) in order to save you the tedious task of iteration of the formulas given below and just generated the flow path to follow.

The first set of calculations (A) tells us the thrust required from the tail rotor. The next set of calculations (B) tells us if we actually produce that thrust. For the setup of the equation we'll use the same NACA 0012 airfoil shape as the main rotor since we already know the drag data and the same basic angle of attack range of 8 to 10 degrees for yaw control. Plugging in the numbers we obtain the "tail rotor thrust" versus "angle of attack range".

Summary: Close inspection of that data indicates Yaw control at ~ 9 degree angle of attack, at 8-degrees the helicopter yaws to the left and at 10 degrees it yaws right. All the specific information on the tail rotor can be found reading the definition of variables defined below.

1a. Main Rotor Torque Calculation:
Definition of Variables:

Air   = air density = 0.0752 lbm/ft^3
M   = 0.55 figure of merit (tail rotor design)
g   = 32.174 ft / sec^2
hp/g   = 550 lb-ft / sec
Length   = distance from rotor hub to tail rotor blade = 11.47 ft (12) = 137.64 inches
T   = thrust {calculation 8.4-3 case c: T (45hz, 10.5ft) =16780 lb-ft / sec^2
RPM   = 455 rev per minute (rotor rpm used versus design rpm)
r1   = main rotor radius = 10.5 feet
Pt   = main rotor shaft power
Torque   = main rotor shaft torque
ForceTail = tail rotor force required to counteract main rotor torque

1b. Tail Rotor Available Thrust Calculation:
Definition of Variables:

k   = 1.17 induced power factor
cdo   = NACA 0012 drag coefficient values 0.014, 0.016, 0.018 corresponding to 8,9,10 degrees
N   = tail rotor number of blades = 2
c   = tail rotor chord length = 0.313 feet
r   = tail rotor radius = 1.503 feet
rpm_tr   = tail rotor rpm (tip speed equal to approximately one-half speed of sound) = 3120 rpm x 2 pie/60 = 326.7 Hz
Ttr(cdo) = tail rotor force (thrust) generated as a function of angle of attack (lb-ft /sec^2)
Thrust(cdo) = tail rotor force (thrust) generated as a function of angle of attack (lbs)
Pt (cdo) = tail rotor power required as a function of angle of attack

Calculation 14.5-1

1.HORIZONTAL TAIL ROTOR BLADE DESIGN CALCULATIONS:

A. Main Rotor Torque Calculation:

Variables Assigned:

$$M := .55 \qquad Air := 0.0752 \cdot \frac{lb}{ft^3} \qquad \frac{hp}{g} = 550.001 \frac{lb\,ft}{sec}$$

$$T := 16780 \cdot \frac{lb \cdot ft}{sec^2} \qquad Length := \frac{137.64\,ft}{12} \qquad g = 32.174 \frac{ft}{sec^2}$$

$$RPM := 455 \qquad rl := 10.5ft$$

Formulas:

$$Pt := \frac{T}{M \cdot g} \cdot \sqrt{\frac{T}{2 \cdot Air \cdot \pi \cdot rl^2}}$$

$$Torque := Pt \cdot \left(\frac{1}{\frac{455 \cdot 2\pi \cdot rad}{60 \cdot sec}} \right)$$

$$ForceTail := Torque \cdot \left(\frac{1}{Length} \right)$$

Results: Torque = 357.184 lb ft   ForceTail = 31.141 lb

I stated before Hp = rpm (torque in "ft lbs") /5250. In the above calculation for main rotor power "Pt" units are in ft lb /sec. This requires the RPM term to be converted to a radians per sec. The stated formula for "Torque" makes that conversion and the result is ft lbs.

B. Tail Rotor Available Thrust Calculation:

Variables Assigned:

$$RPM := 455 \qquad N := 2 \qquad k := 1.17 \qquad \pi = 3.142 \qquad c = .313 \cdot ft$$

$$r := 1.503 \cdot ft \qquad angle := 8,9 \, .. \, 10 \qquad cdo := 0.014 , 0.016 \, .. \, 0.018$$

Formulas:

$$rpm\_tr := 3120 \cdot \left(\frac{2\pi \cdot rad}{60 \cdot sec} \right) \qquad chord := c \cdot \frac{12}{ft} \qquad Dia := 2 \cdot r \cdot 12$$

$$Ttr(cdo) := 0.5 \cdot Air \cdot \left(\pi \cdot r^2 \right) \cdot (rpm\_tr \cdot r)^2 \cdot \left[\frac{\left(\frac{N \cdot c}{\pi \cdot r} \right) \cdot cdo}{k} \right]^{\frac{2}{3}}$$

$$Thrust(cdo) := \frac{Ttr(cdo)}{g}$$

$$Pt(cdo) := \frac{Ttr(cdo)}{M \cdot hp} \cdot \sqrt{\frac{Ttr(cdo)}{2 \cdot Air \cdot \pi \cdot r^2}}$$

Results: rpm_tr = 326.726 Hz

Dia = 36.072 in

chord = 3.756 in

Thrust(cdo)	Pt(cdo) =	cdo =	angle =
27.204 lb	2.575	0.014	8
29.737	2.943	0.016	9
32.166	3.311	0.018	10

2. Shown below is the finalized foot pedal cockpit setup where two tie rods connect to a pivoting tie bar. When one pedal travels forward the other moves back and vice versa. Not shown are the cables that connect to the foot pedals adjacent to the tie rod connection.
CAD Dwg 14.5-1

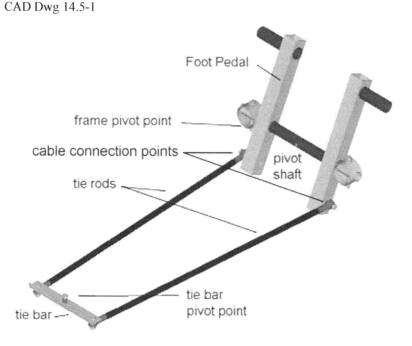

The basic idea was simply to make a foot pedal such that above the pivot point you have your foot action and below the pivot point your cable action. Originally I designed this to save as much weight as possible and didn't interconnect the right and left foot pedals. I figured if the cables to the "tail rotor blade pitch-changing device" were taunt enough I wouldn't need to add the extra complexity to the design.

Turned out I was totally wrong on that one. You definitely need to interconnect the two pedals, if for nothing else it makes setup a lot easier. Shown in CAD Dwg 14.5-1 is the finalized foot pedal cockpit setup where two tie rods connect to a pivoting tie bar. When one pedal travels forward the other moves back and vice versa. Not shown are the cables that connect to the foot pedals adjacent to the tie rod connection.

Shown next is the tail mechanism designed to transfer Foot Pedal cable motion to Tail Rotor blade Angle of Attack change.

Photo 14.5-1

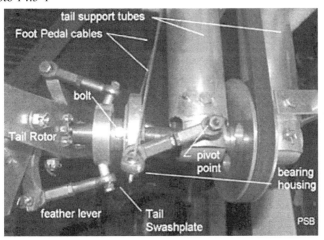

The first thing you notice is the "Swashplate" term is used again. As before the Swashplate is used to transfer motion from a fixed reference plane to a rotating one and accomplishes that task by use of an internal radial bearing and sliding along a rotating shaft. The main difference between this one and the main Swashplate is it doesn't swivel in response to control rod inputs.

Whoever came up with this idea was a genius. The whole Angle of Attack lever operation is based upon the hypotenuse distance change of a lever arm. When the foot cables move the swashplate bearing housing rotates either clockwise or counterclockwise. Since that housing is being held in place by a connecting rod tied to a frame mounted pivot point, as the housing rotates the entire swashplate shifts along the shaft in proportion to the rotation. A thru bolt in the center of the assembly keeps it in position relative to the shaft. Correspondingly the shaft is slotted to allow for the assembly to slide.

3. At this juncture we must figure out how to translate that sliding motion of the Tail Swashplate into an Angle of Attack change by the tail rotor blades. Once again I relied on a previous design to accomplish this (see Photo 14.5-2). Here's how it works. Each rotor blade housed two internal spherical bearings separated by about three inches. At the end of the blade assembly the actuation lever was mounted. This blade subassembly was then sandwiched between two side plates and kept centered by spacers. All this was kept together by the retaining thru bolts. The side plates then fastened to

a collar that was bolted to the tail shaft.

As you can see in the photo the tail rotor blade subassembly is angled with respect to the tail rotor shaft. Under normal operation the rotor blades would be at right angles to the shaft. In the position shown this simulates a wind gust and how the stress on the shaft is reduced by the ability of the subassembly to swivel. Upon inspection of the mounting collar you realize the bore is oversized relative to the shaft diameter and the hole for the shaft bolt over sized as well. When installing this subassembly to the shaft a bronze bushing is inserted into each hole before the shaft bolt is introduced. The length of these bushings is calculated such that when the shaft bolt is tightened the bronze bushings extend past the collar internal diameter and contact the tail rotor shaft outside diameter. This action centers the subassembly on the shaft and allows for swivel motion. Additional detail is shown in Print 14.5-2.

Photo 14.5-2

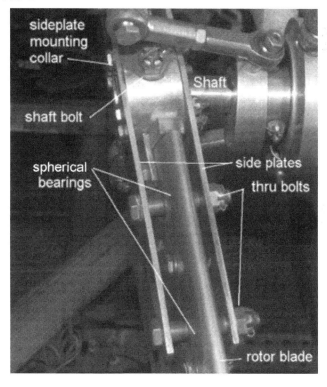

4. There are three bearings that must be sized, the radial bearings that support the tail rotor shaft on the main structural tail tubes, the radial bearing internal to the Tail Swashplate assembly and the bronze bushings in the Tail Rotor blade mounting housing that addresses aberrant wind gusts and keeps it centered.

a)   The bearings that support the tail rotor shaft are sized in accordance to the shaft diameter used but be aware the speeds involved are in the 2500-

rpm range and cheap bearings will not survive. Make sure you select a radial bearing that can handle the tail rotor thrust load (plus some safety factor) at the rpm you calculate the tail rotor to operate.

b)   Any radial bearing you use in the Tail Swashplate to enable transfer from the fixed reference plane to the tail rotor-rotating plane will more than likely be over kill from a thrust standpoint. This is because to fit over the housing diameter the bearing will be oversized with respect to the loads involved. Once again make sure the bearing selected can handle the rpm. See Print 14.5-3 for an exploded view of how I packaged this bearing.

c)   The bronze bushing internal diameter depends on the bolt diameter and the outside diameter selected based upon standard bushing sizes. The length of the two bushings plus the shaft diameter must equal the outside diameter of the side plate-mounting collar.

Print 14.5-1

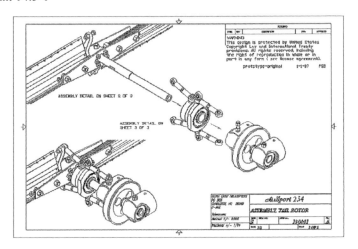

Print 14.5-2

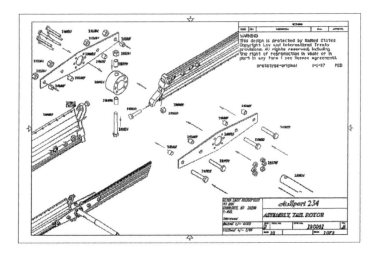

Print 14.5-3

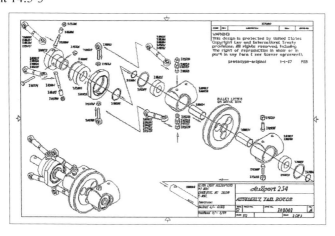

14.6 Engine

Basic Function Description: This is rather obvious as the engine powers the whole operation.

Design Issues: Virtually there are no design issues if you use either the helicopter version of the Hirth 625 or the Compact Radial MZ202 both of which are air-cooled, dual cylinder, two-stroke engines.

In contrast using the open loop water-cooled Kawasaki 650sx Jet Ski engine like I did proved to be a thermodynamic nightmare. This turned out to be so complicated I decided not to add all the conversion details in this manuscript because ultimately the end product added too much weight to the helicopter and as such would not be used in an ultralight application. Be that as it may it was a great engine for the prototype to debug the helicopter.

Shown below is the Kawasaki 650SX engine installed in the helicopter. Note that at this point in time only the expansion pipe section of the exhaust was being used.

Photo 14.6-1

For the record, here are some of the issues I had to overcome using this engine. The first was how to add a closed loop water circulation system to an engine designed to use an open loop system, where some of the water from the jet pump is forced thru the engine at one end, discharged out the other end and mixed in with the exhaust. Eventually what worked was adapting a motorcycle water pump to the front end of the engine, belt driven from the drive pulley and interconnected to a lightweight radiator cooled by a thermostatically controlled electric fan.

The next issue was the exhaust. The rated horsepower of this engine was based upon its carburetion and also its water-cooled packaged exhaust. When I kept the expansion pipe but eliminated the muffler and the cooling water injection there was a substantial horsepower loss. In fact it was so severe the engine would start to stall out just as the helicopter started to hover. I corrected the loss of power issue by adding a two-foot long extension pipe to the expansion output pipe. This eliminated the sonic shock choke point occurring at the expansion pipe exit restricting exhaust gas flow. As of this writing I'm still not sure of the actual output of the engine but I am hovering.

Just these two engine issues cost me a year of my spare time trying to resolve.

Calculations: none

Details:

A) Kawasaki 650SX

Liquid Cooled, Dual cylinder, Two-stroke Specifications: Bore = 76.0 mm, (2.992"), Stroke = 70.0 mm, (2.756"), Displacement = 635 cc, Compression 125 - 195 psi, Max torque = 47.7 ft# @ 5500 rpm; (64.7N-m; 6.6 kg-m @ 5500 rpm)

Max HP = 52 @ 6000 rpm (38.2 kW), Fuel consumption = 19 liters/hr @ full throttle; 19x0.2642= 5.02 gallons/hr, Cooling water = 2.4 L/min (0.625 gpm) @1800 rpm; 7.0 L/min (1.75 gpm) @ 6000 rpm

Weights:

Intake manifold complete (carb, reeds, manifold) = 1780g	
Exhaust head pipe	= 1891g
Exhaust expansion pipe & exhaust	= 2908g
Starter	= 1361g
Engine block, flywheel, magneto	= 21773g
Battery- ULTRA START RED battery 4.4#	= 1995g

Total engine less cooling system 31708g, (69.9#)

Cooling:

Small radiator	= 2051g
Large radiator	= 3282g
10" fan	= 954g
14" fan	= 1955g
Water pump/w bracket & hoses	= 4540g

Total engine weight with cooling = 31708 + 9777 = 41485g (91.4#)

B) Hirth 625 Air Cooled, Dual cylinder, Two-stroke Specifications:
  Bore   = 76.0 mm, (2.992")
  Stroke = 69.0 mm, (2.717")
  Displacement = 625 cc
  Max torque   = 52 ft# @ 4500 rpm
  Max HP   = 44.6 @ 4500 rpm
  Fuel consumption = 15.9 liters/hr @ full throttle, 4.2 gph
  Weight   79# w/ electric start and exhaust

C) Compact Radial Engine 626 Air Cooled, Dual cylinder, Two-stroke
Specifications:
  Bore   = 76.0 mm, (2.992")
  Stroke = 69.0 mm, (2.717")
  Displacement = 626 cc
  Max torque   = 62 ft# @ 5200 rpm,
  Max HP   = 60 @ 5800 rpm
  Fuel consumption =
  Weight   87# w/ dual carbs, electric start, gearbox, clutch and
exhaust Note: Gearbox included in this weight.

14.7 Main frame

Basic Function Description: To provide structural integrity to the
helicopter. There is one more point beyond structural integrity to be
addressed and that is dealing with helicopter landing dynamics. Looking
back in this book you will recall seeing the Hiller "Rotorcycle (built in
1958)" and upon inspection you will notice it has a three-point landing gear
setup. The ultralight "Mosquito" and some plans I have seen, also use the
same three-point system. In fact from a mechanical standpoint the three-
point support has the benefit of automatically self-aligning to any surface.
With your traditional four leg table setup unless the surface is perfectly flat
the table will inadvertently rock. In fact the end tables I purchased for my
slate covered pergola came with two adjustable legs to handle the rough
irregularities in the surface.

Gut feel told me that landing my helicopter with all the inertia developed
by the 21 foot rotor high above my head would require a really large landing
gear. Especially since I was expecting to use this as a trainer. From day one I
knew the landing gear setup would be the conventional type having two
skids just shy of eight feet wide (96 inches). Why the 96 inch maximum?
That's because wider than that you wouldn't be able to trailer it on the
highway without a permit.

Now that I have completed some tethered hovering I can assure you this
lightweight helicopter is a beast to control close to the ground. There is no
smooth calculated landing like possible in an airplane. Of course in the
commercial and military helicopters I have been in the landings seem more
controlled but in this helicopter motions are very jumpy, skittish. To
maximize control over the landing the three-point landing setup was not

considered since it can tip easily in at least three directions whereas with the same width a skid setup, only two.

   Design Issues: 1. Establish aluminum tube sizes for cockpit, structure and tail. 2. Working in tandem with drive train development, package the engine, main rotor, tail rotor, their drives in an aluminum framework that allows for easy maintenance, part replacement, belt & chain adjustment, and easy pilot ingress and egress. Finalize cockpit layout to include cyclic stick, collective lever, Yaw foot pedals, instruments, seat, & windshield. Accomplish all the above whereby the center-of-gravity summation of all parts, including the pilot aligns with the main rotor shaft centerline.

   Calculations:

   1. Structural Tubing Calculations:

A . Mast Tubing Stress Analysis:

   Given: From a tensile strength only perspective, determine the best aluminum round tube size to use for the helicopter mast based upon a design load of 525# supported by two tubes. Use a 6061-T6 alloy having yield strength of 35000 psi:

   The calculations follow but here's the summary of those results: All tubes selected (2.5 x 0.065 wall, 2.5 x 0.049 wall, 2.5 x 0.035 wall and 1.75 x 0.035 wall) satisfy the tensile strength requirement. The 2.5" diameter x .035 wall is the best from a weight standpoint but is pricey. I decided to use the 2.5" diameter x 0.065 wall for the prototype to address all the vibration induced stress issues which were undefined at the time.

   Definition of terms:

L = 600 lbs. Note that all hover calculations use 525 pounds.

S_6061 = Yield stress for 6061-T6 aluminum is 35000 psi

          Tube 2.5 od x 0.065 wall

OD_065 = 2.5 inch outside diameter tube with 0.065 wall thickness

ID_065 = 2.37 inch inside diameter of above tube

wt_65 = 0.5847 lbs weight of 0.065 wall, 2.5 tube

A_065 = cross-sectional area of 0.065 wall, 2.5 tube

S_065 = tensile stress on 0.065 wall, 2.5 tube

SF_065 = Safety Factor using this 0.065 wall, 2.5 tube

tw_65 = total weight of 2 (0.065 wall 2.5) tubes 4 feet long

          Tube 2.5 od x 0.049 wall

OD_049 = 2.5 inch outside diameter tube with 0.049 wall thickness

ID_049 = 2.402 inch inside diameter of above tube

wt_49 = 0.444 lbs weight of 2.5 tube with 0.049 wall

A_049 = cross-sectional area of 0.049 wall, 2.5 tube

S_049 = tensile stress on 0.049 wall, 2.5 tube

SF_049 = Safety Factor using this 0.049 wall, 2.5 tube

tw_49 = total weight of 2 (0.049 wall 2.5) tubes 4 feet long

          Tube 2.5 od x 0.035 wall

OD_035 = 2.5 inch outside diameter tube with 0.035 wall thickness

ID_035 = 2.43 inch inside diameter of above tube

wt_35 = 0.3254 lbs weight of 2.5 tube with 0.035 wall

A_035 = cross-sectional area of 0.035 wall, 2.5 tube
S_035 = tensile stress on 0.035 wall, 2.5 tube
SF_035 = Safety Factor using this 0.035 wall, 2.5 tube
tw_35 = total weight of 2 (0.035 wall 2.5) tubes 4 feet long
            Tube 1.75 od x 0.035 wall
OD2_035 = 1.75 inch outside diameter tube with 0.035 wall thickness
ID2_035 = 1.68 inch inside diameter of above tube
wt2_35 = 0.2264 lbs weight of 1.75 tube with 0.035 wall
A2_035 = cross-sectional area of 0.035 wall, 1.75 tube
S2_035 = tensile stress on 0.035 wall, 1.75 tube
SF2_035 = Safety Factor using this 0.035 wall, 2.5 tube
tw2_35 = total weight of 2 (0.035 wall 1.75) tubes 4 feet long

TUBE DATA
2.5 x 0.065 @ .5847# /foot; $4.68; 2.5 x 0.049 @ .444# /foot; $3.92
2.5 x 0.035 @ .3254#/foot; $5.61; 1.75 x 0.035@ .2264# /foot; $3.48

Calculation 14.7-1

1. MAST TUBE DESIGN CALCULATIONS:

Variables Assigned:

$S\_6061 := 35000$ $L := 600$

$OD\_065 := 2.5$ $OD\_035 := 2.5$

$ID\_065 := 2.37$ $ID\_035 := 2.43$

$wt\_65 := 0.5847$ $wt\_35 := 0.3254$

$OD\_049 := 2.5$ $OD2\_035 := 1.75$

$ID\_049 := 2.402$ $ID2\_035 := 1.68$

$wt\_49 := 0.444$ $wt2\_35 := 0.2264$

Formulas

$$A\_065 := \frac{\pi \cdot \left(OD\_065^2 - ID\_065^2\right)}{4}$$

$$twt\_65 := 4 \cdot 2 \cdot wt\_65$$

$$A\_049 := \frac{\pi \cdot \left(OD\_049^2 - ID\_049^2\right)}{4} \qquad twt\_49 := 4 \cdot 2 \cdot wt\_49$$

$$A\_035 := \frac{\pi \cdot \left(OD\_035^2 - ID\_035^2\right)}{4} \qquad twt\_35 := 4 \cdot 2 \cdot wt\_35$$

$$A2\_035 := \frac{\pi \cdot \left(OD2\_035^2 - ID2\_035^2\right)}{4} \qquad twt2\_35 := 4 \cdot 2 \cdot wt2\_35$$

$$S\_065 := \frac{L}{2 \cdot A\_065} \qquad SF\_065 := \frac{S\_6061}{S\_065}$$

$$S\_049 := \frac{L}{2 \cdot A\_049} \qquad SF\_049 := \frac{S\_6061}{S\_049}$$

$$S\_035 := \frac{L}{2 \cdot A\_035} \qquad SF\_035 := \frac{S\_6061}{S\_035}$$

$$S2\_035 := \frac{L}{2 \cdot A2\_035} \qquad SF2\_035 := \frac{S\_6061}{S2\_035}$$

Results:
A_065 = 0.497	A_049 = 0.377
S_065 = 603.336	S_049 = 795.119
SF_065 = 58.011	SF_049 = 44.019
twt_65 = 4.678	twt_49 = 3.552
A2_035 = 0.189	A_035 = 0.271
$S2\_035 = 1.591 \times 10^3$	$S\_035 = 1.107 \times 10^3$
SF2_035 = 22	SF_035 = 31.621
twt2_35 = 1.811	twt_35 = 2.603

In the above results the S_065 (aluminum 2.5 od x 0.065 wall) has the least stress handling the subjective 600# weight imposed on it, and the greatest safety factor. On the negative side it weighs the most.

B. Leg Tubing Stress Analysis:
   Given: This analysis is to determine the strongest helicopter support structure able to withstand a ground impact of 4 g's (4 times the weight of the helicopter). Design lift off weight of the Airsport 254 is 525#. Listed below are the results of the various concepts to be evaluated. Note the

calculated stress values for each tube. Those calculations follow.

A. Tube: 1.75 od x 0.035 wall aluminum + 1.25 od x 0.035 wall aluminum + composite center, wt = 2.97# each, Stress on this tube 50,310 psi.

B. Tube: 2.50 od x 0.065 wall aluminum, wt 2.27, stress 71,180 psi.

C. Tube: 1.75 od x 0.035 wall aluminum + 1.00 od x 0.035 wall aluminum + composite, wt 3.5#, stress 43,370 psi.

D. Tube: 2.50 od, x 0.035 wall aluminum + composite, wt = 2.24#, stress 39,810 psi.

E. Tube: 1.75 od x 1.0 id composite, wt = 2.40#, stress 44,680 psi.

F. Tube: 4130 Steel 1.75 od x 0.049 wall, wt = 3.45#, stress = 193,900 psi.

G. Tube: 2 od x 1.25 id Kevlar tube, wt = 2.04#, stress = 39,110 psi.

Conclusion: All the calculations follow these remarks. As you can readily see all but one tube exceed the materials maximum yield strength (Kevlar yield at 43,000 psi) and that was a composites, so it was excluded. To solve the problem and keep the weight down I decided to use the same material as the mast and cross brace the legs as shown in the diagram. The point on the leg where the braces were located was such that the induced landing stress was less than the 35,000 psi yield. In the future this design area must be revisited so something better may be designed.

CAD Dwg 14.7-1

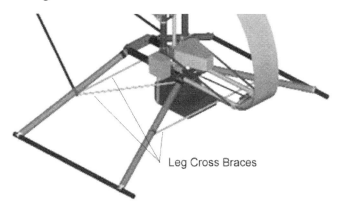

Leg Cross Braces

Definition of terms:

F_leg = Force on leg; 1/4 (4G's x 500# load); 525 gwt rounded off to 500#
A_leg = leg length = 42 inches
M_leg = Bending moment at joint = F_leg x A_leg
S_leg = Stress on leg joint = (M_leg x c) / I
S_mat = Aluminum 6061T6, 35,000psi yield, Kevlar49 yield 43000 psi
wt_b250 = 6061-T6, 2.5 od x 0.035 wall, wt/ft = 0.3254
wt_b175 = 6061-T6, 1.75 od x 0.035 wall, wt/ft = 0.2264
wt_a087 = 4130,   0.875 od x 0.065 wall, wt/ft = 0.5623

A tube terms:

I_175 = Moment of inertia- 1.75 od alum tube, 0.035 wall,
I_125 = Moment of inertia- 1.25 od alum tube, 0.035 wall

I_c175 = Moment of inertia composite- Kevlar /resin filler
wta1   = weight per lineal inch of 1.75 od tube plus 1.25 od tube
wta2   = weight per lineal inch of Kevlar filler material
wt   = total weight of 46.5 inch leg
I_175   = Moment of Inertia of 1.75 od tube
I_125   = Moment of Inertia of 1.25 od tube
I_c175   = Moment of Inertia of Kevlar center
I_leg   = Moment of Inertia of leg
c_leg   = leg centroid distance (1.75 od leg)
M_leg   = Bending moment at joint = F_leg x A_leg
S_leg   = calculated stress on 1.75 od leg
   B tube terms:
wt_a250 = 6061-T6, 2.5 od x 0.065 wall, wt/ft = 0.5847
wt2   = weight of leg = A_leg x wt_a250
I_250   = Moment of Inertia – 2.5 od x 0.065 wall tube
c_leg2   = leg centroid distance (2.5 od leg)
S_leg2   = calculated stress on 2.5 leg
   C tube terms:
I_175   = Moment of inertia- 1.75 od alum tube, 0.035 wall,
I_100   = Moment of inertia- 1.00 od alum tube, 0.035 wall
I_ca175 = Moment of inertia composite- Kevlar /resin filler
wt3a1   = weight per lineal inch of 1.75 od tube plus 1.25 od tube
wt3a2   = weight per lineal inch of Kevlar filler material
wt3   = total weight of 46.5 inch leg
I_175   = Moment of Inertia of 1.75 od tube
I_100   = Moment of Inertia of 1.00 od tube
I_ca175 = Moment of Inertia of Kevlar center
I_leg3   = Moment of Inertia of leg
c_leg   = leg centroid distance (1.75 od leg)
M_leg   = Bending moment at joint = F_leg x A_leg
S_leg3   = calculated stress on 1.75 od leg
   D tube terms:
wt4   = total weight of aluminum 2.5 od x 0.035 wall tube + composite
c_leg2   = leg centroid distance (2.5 od leg)
I_250b   = Moment of inertia- 2.50 od alum tube, 0.035 wall
Ic_250   = Moment of Inertia- solid Kevlar filler
I_leg4   = Moment of Inertia of tube plus filler
S_leg4   = calculated stress on 2.5 od +composite leg
   E tube terms:
wt5   = total weight of 1.75 od x 1.00 id composite
c_leg   = leg centroid distance (1.75 od tube)
I_cc175 = Moment of Inertia of Kevlar
S_leg5   = calculated stress on composite 1.75 od leg
   F tube terms:
wt6   = total weight of 0.875 od x 0.065 wall steel
c_leg6   = leg centroid distance (0.875 od tube)
I_leg6 = Moment of Inertia of 0.875 od steel tube

S_leg6 = calculated stress on steel 0.875 od leg
  G tube terms:
Wt7   = total weight of 2.0 od x 1.5 id Kevlar tube
c_leg7   = leg centroid distance (2.0 od tube)
I_leg7   = Moment of Inertia of 2.0 od x 1.5 id Kevlar tube
S_leg7   = calculated stress on composite 2.0 od x 1.5 id leg
Calculation 14.7-2

LEG STRESS ANALYSIS

Variable Assignment:

$F\_leg := 0.25 \cdot (4 \cdot 500)$ $A\_leg := 42$ $S\_mat := 35,000$

B.TUBE: 2.50od x 0.065wall

Formula

$$I\_250 := \frac{\pi}{4} \cdot \left[\left(\frac{2.50}{2}\right)^4 - \left(\frac{2.37}{2}\right)^4 \right]$$ $$c\_leg2 := \frac{2.5}{2}$$

$$S\_leg2 := M\_leg \cdot \frac{c\_leg2}{I\_250}$$ $$wt2 := \frac{46.5}{12} \cdot 0.585$$

$I\_250 = 0.369$ $wt2 = 2.267$

Results

$S\_leg2 = 7.118 \times 10^4$ WILL FAIL value greater than 35,000

A TUBE: 1.75od x 0.035wall +1.25od x 0.035wall + Composite Center
    $S\_leg = 5.031 \times 10^4$ $wt = 2.969$ NG > 35,000 psi

C TUBE: 1.75od x 0.035wall +1.00odx 0.035wall + Composite
    $S\_leg3 = 4.337 \times 10^4$ $wt3 = 3.498$ NG > 35,000 psi

D Tube 2.50od x 0.035wall + composite
    $S\_leg4 = 3.981 \times 10^4$ $wt4 = 2.243$ NG>35,000 psi

E. Tube 1.75od x 1.0id composite
    $S\_leg5 = 4.468 \times 10^4$ $wt5 = 2.409$ NG > 35,000 psi

F.Tube: 1.75od x 0.049 Wall, 4130 Steel (75,000yield)
    $S\_leg6 = 1.939 \times 10^5$ $wt6 = 3.449$ NG >75,000psi

G.Tube: 2.0od x 1.25id Composite
    $S\_leg7 = 3.911 \times 10^4$ $wt7 = 2.044$ possible < 43,000 psi

C. Tail Tubing Stress Analysis:

Given: This exercise is to determine how to support the tail rotor assembly and its counter-rotational force. Most helicopters utilize a single tail boom either of composite material or a lattice truss like assembly. I decided to use a tandem tube design mainly to support the tail belt drive pulley.

Definition of Terms:

E_6061t6	= Modulus of elasticity for aluminum = 10E6
c_leg2	= distance from tube outside diameter to neutral axis
A_tail	= tail moment arm (from rear engine support to tail rotor) = 113.6
A2_tail	= distance from rear tail support to tail rotor = 51.58
F_t	= resultant force on tail without tail support (use 50#)
F_v	= weight of tail rotor & pulley + horizontal tail + tail tube
M_tail	= tail moment without tail support
M2_tail	= tail moment with tail support
I_250	= Moment of Inertia 2.5od x 0.065 wall alum tube
I_250b	= Moment of Inertia 2.5od x 0.035 wall alum tube
S_tail	= stress on unsupported 2.5od x 0.065wall tube
S2_tail	= stress on unsupported 2.5od x 0.035wall tube
Sb_tail	= stress on unsupported 2.5od x 0.065wall tube
d_tail	= deflection at tail rotor assembly using tandem 2.5od x 0.065wall tubes
S2b_tail	= stress on unsupported 2.5od x 0.035wall tube
d_tail2	= deflection at tail rotor assembly using tandem 2.5od x 0.035wall tubes

Calculation 14.7-3

Tail Structure Design Calculations

Variables Assigned:

$$F\_v := 13.2 \qquad F\_t := 50 \qquad A\_tail := 113.6 \qquad A2\_tail := 51.6$$

$$E\_6061t6 := 10 \cdot 10^6 \qquad\qquad c\_leg2 := \frac{2.5}{2}$$

Formulas:

$$F\_tail := \sqrt{F\_v^2 + F\_t^2}$$

$$M\_tail := \frac{F\_tail}{2} \cdot A\_tail \qquad\qquad M2\_tail := \frac{F\_tail}{2} \cdot A2\_tail$$

a. Stress on 2.5od x 0.065wall tube with no tail support

$$S\_tail := M\_tail \cdot \frac{c\_leg2}{I\_250}$$

$$S\_tail = 9.956 \times 10^3 \quad \text{psi}$$

b. Stress on 2.5od x 0.035wall tube with no tail support

$$S2\_tail := M\_tail \cdot \frac{c\_leg2}{I\_250b}$$

$$S2\_tail = 1.783 \times 10^4 \quad psi$$

c. Stress on 2.5od x 0.065wall tube with tail support

$$Sb\_tail := M2\_tail \cdot \frac{c\_leg2}{I\_250}$$

$$d\_tail := \frac{\left[\dfrac{F\_v}{2}(A2\_tail)^3\right]}{3 \cdot E\_6061t6 \cdot I\_250}$$

$$Sb\_tail = 4.522 \times 10^3 \quad psi$$

$$d\_tail = 0.082 \quad inches$$

d. Stress on 2.5od x 0.035wall tube with tail support

$$S2b\_tail := M2\_tail \cdot \frac{c\_leg2}{I\_250b}$$

$$d\_tail2 := \frac{\left[\dfrac{F\_v}{2}(A2\_tail)^3\right]}{3 \cdot E\_6061t6 \cdot I\_250b}$$

$$S2b\_tail = 8.1 \times 10^3 \quad psi$$

$$d\_tail2 = 0.147 \quad inches$$

Conclusion: Use tandem 2.5 od x 0.065 wall aluminum tubes for tail structure with mid support. I decided to stick with the same material as used for the mast and legs as stated before to keep inventory simple. As it turns out during ground testing of the helicopter even with these supported 2.5 od x 0.065 tubes, the tail rotor really makes some dynamic gyrations. Whether or not this motion would occur in the air I'm not sure but I would never consider any structure less stiff in this area for any future design.

2. This section of the Main Frame discussion deals with packaging everything to make the helicopter fly. I found it easier to breakdown the total frame into three subassemblies in order to make the BOM easier to follow. Those subassemblies are called the Cockpit, Tail Section, and the Substructure. In the Cockpit subassembly all the ergonomic issues are addressed along with developing the vertical structure supporting the main rotor shaft and cooling system. The Tail Section includes the structure and all the supports required to package the Tail Rotor and its drive train. The substructure anchors the other two subassemblies, supports the engine, the main rotor drive, the landing skids and the fuel tank. Shown below in prints 14.7-1 thru 14.7-4 is my approach to helicopter frame design.

Not shown on the prints is my addition of a 3/16 steel cable between the rear two legs and between the front two legs located just above the skid tubes. This was done as an added safety measure to minimize leg stress on a landing exceeding 4 g's.

Print 14.7-1

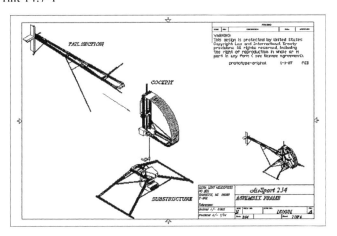

Print 14.7-2

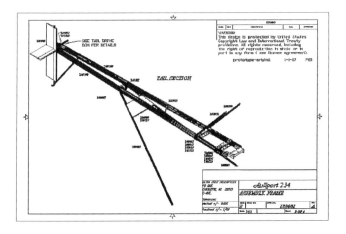

Print 14.7-3

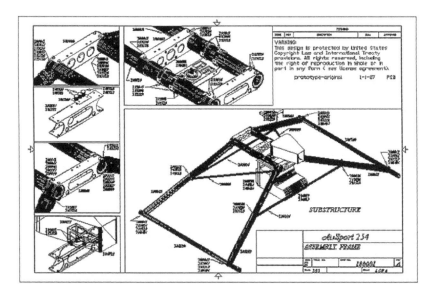

Print 14.7-4

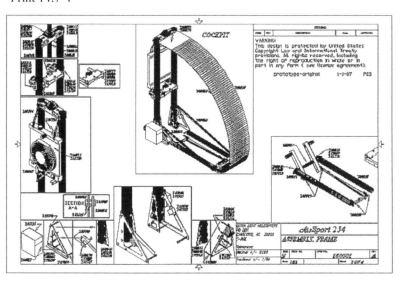

14.8 Drive train

Basic Function Description: Transfer engine power to drive the Main and Tail Rotors and the water pump.

Design Issues: 1. Select for each referenced subsystem the best drive to transfer the power required at the speed needed with the lowest weight.
Calculations:
1. Gearbox strength calculations:
2. Drive shaft stress calculations:
3. Belt and Chain selection as per drive:

1. Gearbox strength calculations:

Procedure: Based upon the relationship, horsepower = torque times rpm divided by a constant, one can readily see that for a given horsepower if rpm decreases torque must increase. Excessive torque will destroy a gearbox even though it could last for years if the transmitted horsepower stayed the same but the operational rpm was higher. That's the whole point of these calculations, to determine what that ratio of transmitted Torque versus Rpm should be for the gear set within the gearbox to have a useful life. If our calculations are correct we should be able to avoid the most common problems associated with a gearbox's load carrying capabilities such as excessive heat of operation, breaking of gear teeth, excessive wear of gear-tooth surfaces along with excessive noise in operation. We start this work by disassembling the gearbox to measure the pinion and gear surfaces. Solving the equations comes next.

In the first set of calculations (A) we will simplify the spiral bevel gear setup by modeling the gearbox as if spur gears were used. In this case the Buckingham equation is used to run some quick numbers to get close to the Torque versus rpm operational relationship for the horsepower we want to transmit. The second set of calculations (B) fine-tunes the load carrying capability of the gear set using the torque versus speed ratio determined above in the actual Spiral Bevel gear set equations. Note that hardened gears max pv is around 3,000,000 and hardened steel does not appear to have any definite endurance limit. Gears that are heavily loaded at high speed show burnt discoloration. When using hardened steel pinion & gear sets as in our gearbox you must use an extreme-pressure lubricant.

Gearbox Tear Down Data:

Definition of Variables:

N_p = pinion teeth = 11
N_{p1} = pinion inside diameter = 1.348" calculated
N_{p2} = pinion outside diameter = 1.884" calculated
N_{pr} = nominal pinion pitch radius = 1.616/ 2 =0.808", 20.5mm calculated
N_g = ring teeth = 32
N_{g1} = ring inside diameter = 3.92" (circumference=12.3", 313 mm)
N_{g2} = ring outside diameter = 5.48" (circumference =17.2", 437 mm)
N_{gr} = nominal ring pitch radius = ((5.48-3.92)/2+ 3.92)= 2.35", 60 mm

w   = tooth face = 1.10", 28 mm
p1   = pitch id = 0.385, 9.77 mm
p2   = pitch od = 0.538, 13.66 mm
rpmr11 = rpm of 11t pinion = 1680 rpm x 32/ 11 = 4887 rpm
rpmr32 = rpm of 32t ring gear = 455 rpm x 48/13 = 1680 rpm
cfmp1 = contact fpm = 2 x r32r x rpmr32/ 12
cfmp2 = contact fpm = 2 x p11r x rpmr11/ 12

A. Calculations Using the Buckingham Equation:
Fdyn = Ft + [(0.05 vr (w x C + Ft)) / (0.05vr + (w x C + Ft)^0.5)
F   = Hp x 33000 / vr
Fdyn   = dynamic force on the root of the gear tooth taking into effect the
            kinetic loading effects determined by the Barth speed factor
vr   = tangential velocity at the pitch circle = pi x rpm x d11(P) / 12
rpm   = pinion rpm = 4538
d11(P)   = pitch circle diameter as a function of pitch P
d32(P)   = pitch circle diameter as a function of pitch P
GR   = gear ratio
P   = possible pitch values ranging from 2 thru 6 to determine which
            one matches the measured gearbox data turns out P = 6
Fallow   = Load on tooth if it is assumed to carry the entire tangential load
            as a cantilevered beam. The maximum load using the Lewis beam
            strength equation is based on the maximum allowable bending
            stress = S (w x Y14) / P
C   = maximum error in action between gears, for a class 2 gear set
            with a 14 ½ tooth form and 6 pitch the error is 0.001 and C = 1600
Y14   = is the form factor for use in the Lewis equation, for a 11 tooth
            spur gear with a 14 ½ full depth gear tooth the constant = 0.21
Sst   = allowable gear stress for a 360 BHN hardened gear = 90,000 psi
S   = stress on gear (Fdyn x P) / (kd x w x Y14)
kd   = Barth speed factor = a (a + vr)
a   = constant used in the Barth equation = to 1200 for accurately cut
            gears with a tangential velocity less than 300 fpm
Hp   = transmitted horsepower = 50
Hp2   = allowable calculated hp

B. Calculations Using the Spiral Bevel Gear Equations:
W = limiting load for wear lbs. = (0.75 Dvp x w x K x Q) / (cos (As) x cos (As)
Definition of Variables:
K   = load stress factor for 0.040 case hardened steel worm (s =150000 psi)
        = 446
P   = diametrical pitch (must be the same in all meshing gears) = 6
w   = face width = 1.1
Np   = number of teeth on pinion = 11
Ng   = number of teeth on gear = 32

Apcp = pitch-cone angle of spiral bevel pinion = arc tan (Dp / Dg) in
       radians. Note: spiral angles generally employed are between 30 &
       35 degrees
Apcg = pitch-cone angle of spiral bevel gear = arc tan (Dg / Dp) in radians
As   = normal pressure angle at middle gear face = 14 degrees, = 14 / 57.3
        radians. NOTE: this pressure angle is used for 11 tooth
        pinions when the gear has 25 teeth or more.
Dp   = pitch diameter of spiral bevel pinion at large end, in. = Np / P
Dg   = pitch diameter of spiral bevel gear at large end, in. = Ng / P
Dvp   = virtual pitch diameter of spiral bevel pinion at middle of gear face,
        in = (Dp – w sin (Apcp)) / cos (Apcp)
Nvp   = virtual number of teeth in spiral bevel pinion = Np / cos (Apcp)
Nvg   = virtual number of teeth in spiral bevel gear = Ng / cos (Apcg)
Q   = ratio factor = 2 Nvg / (Nvp + Nvg)
Hp3   = Horsepower capable with this spiral bevel gearbox = torque x rpm /
        5250 where torque =W x diametrical pitch radius (ft)

Calculation 14.8-1

1. GEARBOX DESIGN CALCULATION

A. Using the Buckingham Equation:

Variables Assigned:

$P := 2,3..6$ $rpm := 4538$ $Y14 := 0.210$ $w := 1.1$

$C := 1600$ $Hp := 50$ $Sst := 90000$

$$d11(P) := \frac{11}{P} \qquad d32(P) := \frac{32}{P} \qquad GR := \frac{11}{32}$$

Formulas:

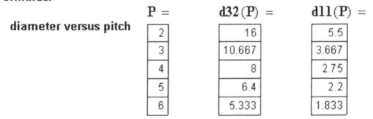

diameter versus pitch	$P =$	$d32(P) =$	$d11(P) =$
	2	16	5.5
	3	10.667	3.667
	4	8	2.75
	5	6.4	2.2
	6	5.333	1.833

Pitch value established from above tabulation: $P := 6$

$$vr := \frac{(rpm \cdot \pi \cdot d11(P))}{12} \qquad Ft := \frac{Hp \cdot 33000}{vr}$$

$$Fallow := \frac{(Sst \cdot w \cdot Y14)}{P} \qquad \text{Lewis eq.}$$

$$Fdyn := Ft + \left[\frac{0.05 \cdot vr \cdot (w \cdot C + Ft)}{0.05 vr \cdot \sqrt{(w \cdot C + Ft)}} \right] \qquad \begin{array}{l} \text{Buckingham} \\ \text{eq.} \end{array}$$

$$Hp2 := \left(\frac{rpm}{5250} \right) \cdot \left(Fdyn \cdot \frac{d11(6)}{2 \cdot 12} \right)$$

Results:

$$Fallow = 3.465 \times 10^3 \quad \text{lbs}$$

$$Fdyn = 807.722 \quad \text{lbs}$$

$$Hp2 = 53.333$$

B. Calculations using the Spiral Gear equations:

$$Apcg := atan\left(\frac{Dg}{Dp} \right) \qquad Nvp := \frac{Np}{\cos(Apcp)}$$

$$Q := \frac{2 \cdot Nvg}{(Nvp + Nvg)} \qquad Nvg := \frac{Ng}{\cos(Apcg)}$$

$$W := \frac{0.75 \cdot Dvp \cdot Fc \cdot K \cdot Q}{\cos(As)^2} \qquad Hp3 := \left(\frac{rpm}{5250} \right) \cdot \left(W \cdot \frac{d11(6)}{2 \cdot 12} \right)$$

Results:

$$W = 1.195 \times 10^3 \quad \text{Allowable working Load on teeth}$$

$$Hp3 = 78.892$$

The gearbox can handle 78.9 hp at our design rpm therefore it is capable of handling the drive requirements.

2. Drive Shaft Calculations:

If you consult Mark's Handbook on shaft design four possible scenarios are given. They are as follows: Case 1. The shaft is subject to pure torque (no bending involved) in which a steady torque has an alternating torque applied during its rotation. Case 2. The shaft is subject to a steady bending moment to which an alternating bending moment is applied (no torsion involved). Case 3. The rotating shaft is subject to not only a steady torque but also a completely reversed bending moment, generating both a steady shear stress and an alternating bending stress. Case 4. Everything is included such as steady torsion stress with added alternating torsion and steady bending stress with imposed alternating bending stress.

a. Based on the above possible types of shaft loading I consider case 3 to be the most appropriate assumption to determine the maximum and allowable torsional shear stress for this helicopter's main rotor shaft. The two most often used equations for this case are:

Maximum torsional shear stress (TSSm) = (16 x ((Km x M)^2 + (Kt x T)^2)^0.5) / (pi x d^3); where Km is a moment loading correction factor, M the moment, Kt the torsion loading correction factor, T the applied torque, pi a constant, and d the shaft diameter or shaft "equivalent diameter" if a tubular shaft is used.

b. Allowable torsional shear stress (TSSa) = (0.577 x yield stress) / SF; where the number "0.577" comes from Distortion Energy Theory. This theory agrees closely with historical part failure analysis in that you can use up to 57.7 % of steel's yield stress in a design and be relatively safe that the part will not fail. SF is the safety factor.

Before we go on let me say these two formulas I will use are not necessarily universal in their use. Some companies have developed their own codes, government agencies its' own procedures and adding to that there are other solutions available by different theoretical methods.

A. Main Rotor Shaft Calculations:

After numerous iterations and calculations I arrived at a weight saving 1.5-inch diameter tubular shaft with a 0.12-inch wall to satisfy the power transfer conditions. Shown below are the specific stress calculations for this shaft.

Procedure: Use the formula stated under Ssmax to calculate the stress on the shaft then determine the Safety Factor.

Definition of Terms:

Syeild   = 4130 steel yield = 90000 psi
Km = derating factor (rotating shaft gradually applied or steady load) = 1.5
Kt   = derating factor (rotating shaft gradually applied or steady load) = 1.0
SF   = safety factor
Ssmax   = maximum shear stress, =
          psi = [16 x {((Km x M)^2) + (Kt x T)^2)^0.5}] / (pi x d^3)
Ss_allow = allowable stress based upon the distortion energy theory =
          (0.577) Syeild / SF

Ten   = tensioning force on Goodyear tail drive belt = 50 lbs
P   = design horsepower = 50
RPM   = rpm of rotor shaft = 455 (note: final design rpm = 438)
M   = moment, in-lbs
T   = torque, in-lbs = P x 63000/RPM
d   = equivalent diameter of 1.5" diameter tube with 0.12 wall
d1   = shaft outside diameter, in = 1.5"
d2   = shaft inside diameter, in = 1.25"
L1   = distance to Goodyear pulley = 27"
L2   = distance between shaft bearing supports = 54"

Calculation 14.8-2

1. Main Rotor Shaft Calculations

Variables Assigned:

$$Km := 1.5 \quad Kt := 1.0 \quad P := 50 \quad L1 := 27$$

$$d1 := 1.5 \quad d2 := 1.25$$

$$Syield := 90000 \quad Ten := 50 \quad RPM := 455$$

Formulas:

$$M := Ten \cdot L1 \qquad T := P \cdot \frac{63000}{RPM} \qquad d := \sqrt[3]{d1^3 - d2^3}$$

$$Ss\_allow := 0.577 \cdot Syield$$

$$Ssmax := \frac{16 \cdot \sqrt{(Km \cdot M)^2 + (Kt \cdot T)^2}}{\pi \cdot d^3}$$

$$SF := \frac{Ss\_allow}{Ssmax}$$

Results:

$$M = 1.35 \times 10^3 \ in \cdot lb \quad T = 6.923 \times 10^3 \ in \cdot lb \quad d = 1.124 \ in$$

$$Ss\_allow = 5.193 \times 10^4 \ \frac{lbs}{in^2} \qquad Ssmax = 2.584 \times 10^4 \ \frac{lbs}{in^2}$$

$$SF = 2.01 \quad Safety \ Factor$$

B. Gearbox Output Shaft Calculations:

These calculations were easier to accomplish since the maximum shaft size that would fit thru the gearbox housing was 0.75-inch. Using this size shaft and the equations defined above for a case 3 loading along with an additional shaft derate factor for having a keyway cut on the outside diameter the specific stress calculations are shown below.

Procedure: Using the same steel shaft material, Km, Kt and engine horsepower numbers as in the previous calculations determine the Safety Factor of this 3 / 4 inch shaft with keyway.

Definition of Terms:

Smaxgb = maximum gearbox output shaft shear stress, psi
Ssk_allow = allowable stress on shaft after keyway derate
SFgb = safety factor for the gearbox output shaft
RPMgb = rpm of gearbox output shaft (455x48/13) = 1680
Dgb = 3/4" diameter gearbox shaft
L1gb = distance between shaft bearing supports = 5"
L2gb = cantilever distance to gearbox 13t sprocket = 0.859"
Tengb = total chain tension on 13t sprocket = 25 lbs
Mgb = gearbox output shaft moment, in-lbs
Tgb = gearbox output shaft torque, in-lbs = HP x 63000/RPMgb
K = derating factor for keyway in shaft = 0.75

Calculation 14.8-3

Gearbox Output Shaft Calculations

Variables Assigned:

$HP := 50$ $RPMgb := 1680$ $Syield := 90000$ $Tengb := 25$
$Km := 1.5$ $Kt := 1.0$ $K := 0.75$ $L1gb := 5$ $L2gb := 0.859$
$Dgb := 0.75$

Formulas:

$$Tgb := HP \cdot \frac{63000}{RPMgb} \qquad Mgb := Tengb \cdot L2gb$$

$$Ssk\_allow := 0.577 \cdot Syield \cdot K$$

$$Smaxgb := \frac{16 \cdot \sqrt{(Km \cdot Mgb)^2 + (Kt \cdot Tgb)^2}}{\pi \cdot Dgb^3}$$

$$SFgb := \frac{Ssk\_allow}{Smaxgb}$$

Results:

$$Tgb = 1.875 \times 10^3 \quad \text{in-lb} \qquad Mgb = 21.475 \quad \text{in-lbs}$$

$$Ssk\_allow = 3.895 \times 10^4 \quad \frac{lb}{in^2}$$

$$Smaxgb = 2.264 \times 10^4 \quad \frac{lb}{in^2}$$

$$SFgb = 1.72$$

B2. Gearbox Output Shaft Key Stress Calculation (cam and gearbox key are the same size)

Procedure: The above calculation gives us the torque on the shaft dividing that number by the shaft radius gives us the force acting on the shear area of the key resulting in the shear stress on the key. Comparing this number with the allowable shear stress of the material gives us the Safety Factor.

Definition of Terms:

Fkey   = force on key
Akey   = key shear area (0.188 x 1.625) = 0.305
Skey   = shear stress on key
Dgb   = gearbox shaft diameter
Ss_allow = allowable stress on key = 90,000 psi x 0.577

Calculation 14.8-4

2a Gearbox Output Shaft Key Stress Calculation

Variables Assigned:

$$Tgb = 1.875 \times 10^3 \quad Dgb := 0.75 \qquad Akey := 0.305$$

Formulas:

$$Fkey := \frac{Tgb}{\frac{Dgb}{2}} \qquad Skey := \frac{Fkey}{Akey} \qquad SFkey := \frac{Ss\_allow}{Skey}$$

Results:

$$Skey = 1.639 \times 10^4 \quad \frac{lb}{in^2} \qquad Ss\_allow = 5.193 \times 10^4 \quad \frac{lb}{in^2}$$

$$SFkey = 3.168$$

B3. Gearbox Output Shaft Critical Frequency Calculation:
Procedure: The purpose of this calculation is to determine possible vibration issues with this cantilevered output shaft. If the critical frequency is below our design shaft rpm we will have to add a physical bearing support to the free end. Adding the support will reduce shaft deflection and subsequently increase its critical frequency. As it turns out the shaft support is required but I have included both the cantilevered and fix support calculations for additional clarity.

Tengb = chain tension on shaft = 25 lbs
L1gb   = distance between shaft bearing supports = 5"
L2gb   = cantilever distance to gearbox 13t sprocket = 0.859"
I   = Moment of inertia = $\pi(Dgb)^4 / 4$
E   = Modulus of elasticity (steel) = 30 x 10^6 psi
dfmax = maximum shaft deflection = from beam deflection formulas the static deflection of the cantilever portion of the shaft will be = {wt x (L1gb^2) (L1gb+L2gb)}/ (3 x E x I)
fcrit = cantilevered shaft critical frequency = [(g/dfmax)^0.5] / (2 x pi)
a   = distance to applied chain tension = 0.859 inches
b   = applied chain tension to outboard shaft support = 4.141 inches
g   = (32 ft / sec^2) = 386 in / sec^2
L   = shaft length beyond gearbox = 5 inches
Df   = maximum shaft deflection = from beam deflection formulas the static deflection for a supported shaft will be = {wt x (a^3 + b^3)}/ (3 x E x I x L^3)
fcrit2 = critical frequency for gearbox output shaft with end support = [(g/dfmax2)^0.5] / 2 x pi
g   = gravitational constant = 386 inches / sec^2
  Calculation 14.8-5

Gearbox Output Shaft Critical Frequency Calculation

Variables Assigned

$$g := 386 \quad E := 30 \cdot 10^6 \quad Dgb = 0.75 \quad RPMgb = 1.68 \times 10^3$$

Formulas:

$$I := \pi \cdot \frac{Dgb^4}{4} \qquad dfmax := (Tengb) \cdot (L2gb)^2 \cdot \frac{(L1gb + L2gb)}{3E \cdot I}$$

$$fcrit := \frac{1}{2\pi} \sqrt{\frac{g}{dfmax}}$$

Results:

$$I = 0.249$$

$$fcrit = 1.422 \times 10^3 \quad \text{Add support since design rpm is 1680}$$

Recalculation Using Support on Shaft End

Formula for deflection for a shaft with fixed supports:

$$a := 0.859 \quad b := 4.141 \quad L := 5$$

$$df := (Tengb) \cdot \frac{\left[(a)^3 + b^3\right]}{3 \cdot E \cdot I \cdot L^3} \qquad fcrit2 := \frac{1}{2\pi} \cdot \sqrt{\frac{g}{df}}$$

$$fcrit2 = 3.907 \times 10^3 \qquad \text{This value is above operating rpm therefore acceptable}$$

Summary: The critical frequency (fcrit) of the cantilevered output shaft is 1422 rpm. The actual design speed of this shaft is 1680 rpm so we need to add a shaft support to the free end. Recalculating the critical frequency (fcrit2) of that shaft arrangement gives a value of 3900 rpm, which is acceptable.

C. Tail Rotor Forward Drive Shaft Calculation:

Procedure: Using the same steel shaft material, Km, Kt we will utilize the same shaft shear stress calculations as before but only 5 horsepower will transfer to the tail rotor and a hollow steel shaft will be utilized to save weight.

SFtr   = safety factor
HP_tr   = design horsepower tail rotor = 5
RPM_tr   = rpm of tail rotor shaft = 2700
d_tr   = equivalent diameter of 0.625 diameter tube with 0.095 wall
d1_tr   = shaft outside diameter, in = 0.625
d2_tr   = shaft inside diameter, in = 0.435
L_tr   = distance between shaft bearing supports = 20-3/8"
L_trs   = tail rotor drive shaft length = 23-1/4"
wt_63   = shaft wt/ft 5/8od x 0.095 wall = .54# per ft
T_tr   = torque, in-lbs = HP x 63000/RPM
M_tr   = bending moment = 0 inch pounds
Tenfs   = chain tension = 13#
b2 = distance from applied chain tension to outboard shaft support = 1"
Ssmax_ts = maximum shaft shear stress
Ss_allow = allowable stress based upon the distortion energy theory = (0.577) Syeild/ SF

Calculation 14.8-6

Tail Rotor Forward Driveshaft Calculation:

Variables Assigned:

$Km = 1.5$ $Kt = 1.0$ $HP\_tr = 5$ $L\_trs = 23.75$

$d1\_tr = 0.625$ $d2\_tr = 0.435$ $L\_tr = 20.375$ $wt\_63 = 0.54$

$M\_tr = 0$ $Ss\_allow = 5.193 \times 10^4$ $RPM\_tr = 2700$

Formulas:

Torque developed on tail rotor forward drive shaft

$$T\_tr = HP\_tr \cdot \frac{5250 \cdot 12}{RPM\_tr}$$

$$T\_tr = 116.667$$

Calculating equivalent shaft diameter of shaft selected:

$$d\_tr = \sqrt[3]{d1\_tr^3 - d2\_tr^3}$$

$$d\_tr = 0.545$$

a Calculation of maximum shaft shear stress:

$$Ssmax\_tr = \frac{16 \cdot \sqrt{(Km \cdot M\_tr)^2 + (Kt \cdot T\_tr)^2}}{\pi \cdot d\_tr^3}$$

$$Ssmax\_tr = 3.672 \times 10^3$$

$$SF\_tr = \frac{Ss\_allow}{Ssmax\_tr}$$

$$SF\_tr = 14.143 \qquad \text{Safety Factor}$$

Based upon this safety factor this shaft is more that adequate indicating a smaller size could be used, however the stock bore for the drive components used on this shaft lend itself to this diameter.

b. Determine weight of this shaft:

$$Wt\_63 := wt\_63 \cdot \frac{L\_trs}{12}$$

$$Wt\_63 = 1.069$$

c Tail Rotor Forward Shaft Critical Frequency Calculation

$$I2 := \pi \cdot \frac{d\_tr^4}{4} \qquad E = 3 \times 10^7$$

$$Tenfs := 13 \qquad b2 := 1 \qquad d\_tr = 0.545$$

$$df2 := (Tenfs) \cdot (b2)^2 \cdot \frac{(L\_trs + b2)}{3E \cdot I2} \qquad df2 = 5.162 \times 10^{-5}$$

$$fcrit2 := \frac{1}{2\pi} \cdot \sqrt{\frac{g}{df2}}$$

$$fcrit2 = 435.234$$

Summary: This forward shaft was added after the original helicopter was field tested. The modification was made for two reasons. One to eliminate the alignment problems associated with the tail rotor belt drive and secondly to address the Yaw control problems. This change accomplished both goals. The problem with the arrangement is the low critical frequency number calculated above. What this meant is the shaft would operate in the second or third harmonic. So far this has not been an issue. This will be improved on after field testing is complete.

D. Tail Rotor Shaft Calculation:
Procedure: First we determine the shaft stress as we have done previously. If this Safety Factor is acceptable we then calculate the critical frequency of the shaft using the weight of the tail rotor assembly as the overhung load.

d1ts   = outside diameter of steel shaft tubing = 0.75
d2ts   = inside diameter of steel shaft tubing = 0.51
dts   = equivalent diameter of steel tube
Smaxts = maximum shear stress on tail rotor shaft
Wtr   = overhung weight of tail rotor on shaft
shaft L = shaft length
a4   = overhung distance of tail rotor
b4   = shaft distance between support bearings
Mts   = moment generated by tail rotor weight and overhung distance
T_tr   = torque on shaft, in-lbs
SFtr   = shaft shear stress Safety Factor
I4   = shaft Moment of Inertia
df4   = shaft deflection
fcrit4 = critical frequency

Calculation 14.8-7

Tail Rotor Shaft Calculation:

Variables Assigned:

$$d1ts := 0.75 \quad d2ts := 0.51 \quad Wtr := 2.8 \quad shaftL := 10.875$$

$$T\_tr = 116.667 \quad a4 := 5.396 \quad b4 := 5.479$$

$$Ss\_allow = 5.193 \times 10^4$$

Formulas:

$$Mts := a4 \cdot Wtr \qquad dts := \sqrt[3]{d1ts^3 - d2ts^3}$$

$$Smaxts := \frac{16 \cdot \sqrt{(Km \cdot Mts)^2 + (Kt \cdot T\_tr)^2}}{\pi \cdot dts^3}$$

$$SFtr := \frac{Ss\_allow}{Smaxts}$$

a Calculation of maximum Shear Stress

Results:

$$Smaxts = 2.093 \times 10^3$$

$$SFtr = 24.814 \qquad \text{Safety Factor}$$

b Tail Rotor Shaft Critical Frequency Calculation:

$$I4 := \pi \cdot \frac{dts^4}{4} \qquad df4 := (Wtr) \cdot (a4)^2 \cdot \frac{(b4 + a4)}{3E \cdot I4}$$

$$df4 = 6.558 \times 10^{-5}$$

$$fcrit4 := \frac{1}{2\pi} \cdot \sqrt{\frac{g}{df4}}$$

$$fcrit4 = 386.133 \quad \text{rpm}$$

Summary: Due to this low critical frequency number yet high inherent strength of the shaft. This shaft will stay as is for the present time and operate in the second or third harmonic.

3. Belt and Chain selection as per drive:
Sample center distance calculation shown below.
Calculation 14.8-8

TAIL SPROCKET CENTER DISTANCE CALCULATION:

L = NUMBER OF CHAIN PITCHES
n = TEETH ON SMALL SPROCKET
N = TEETH ON LARGE SPROCKET
P = CHAIN PITCH
C = SPROCKET CENTER DISTANCE (INCHES)

Variables Assigned:

$$L := 126 \quad n := 16 \quad N := 60 \quad P := 0.375$$

Formula

$$C := \frac{P}{8}\left[2L - N - n + \sqrt{(2L - N - n)^2 - 0.81 \cdot (N - n)^2} \right]$$

Results

$$C = 16.288$$

Print 14.8-1

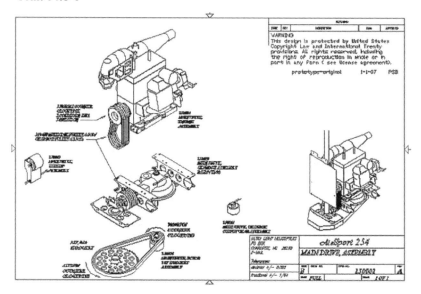

Print 14.8-2

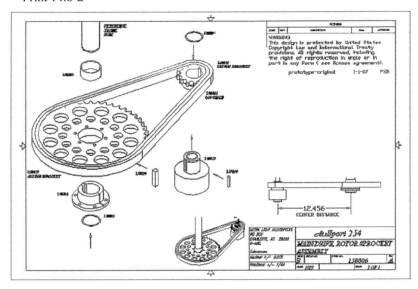

Print 14.8-3

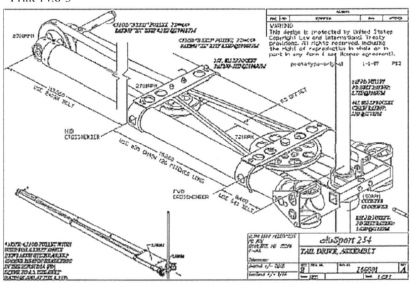

14.9 Electrical Systems:

Basic Function Description: On the Airsport 254 the electrical systems are separated into three types: the first to start and run the engine, the second to maintain proper cooling water temperature, and the third the instrumentation.

Shown below is the completed helicopter wiring diagram.

Detail 14.9-1

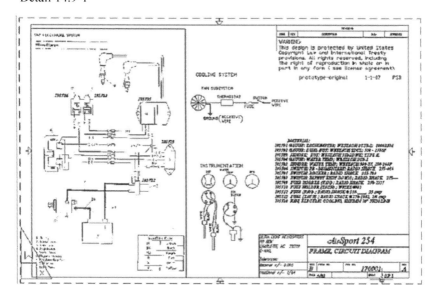

The design issues resolved in the above wiring diagram were:

1. Determine how to refigure the Jet Ski engine wiring to package intact on the helicopter frame.

2. Integrate the cooling fan and thermostatically controlled sensor into the main wiring circuit.

3. Following the instrument manufacturers' recommendations include all switches and fuses as they pertain to the device on the completed electrical system-wiring diagram including assigned wire numbers.

Note: For wiring signal current to the instruments I purchased a pre-packaged electrical wire stand from Harbor Freight that included ten different wire spools covering 24 gage on up to 10 gage. It was a big mistake! The Chinese wire supplied had the insulation thickness three times the size of an equivalent UL listed wire. Their wire worked but is very heavy, hard to route neatly and ugly to look at.

Calculations: none

14.9.1 Engine Electrical

The engine manufacturer's electrical wiring circuitry follows a certain coding and since I basically lifted the complete engine intact with all support hardware, their electrical wiring format may interest you. It is described as follows: "All electrical wires are either single-color or two-color and, with only a few exceptions, must be connected to wires of the same color. On any of the two-color wires there is a greater amount of one color and a lesser amount of a second color, so a two-color wire is identified by first the primary color then the secondary color. For example, a yellow wire with a thin red stripe is referred to as a "yellow/red" wire. It would be a "red/

yellow" wire if the colors were reversed to make red the main color.

As for the electrical components used on this engine they are listed below:

• Starter: Anybody who has ever started a two-stroke 30cc chain saw engine by way of a rope pull handle knows there is some repeated pulling involved. Now scale that engine up to 650cc and that pull rope task becomes a nightmare (by the way the 305cc Rotax engine which was considered for this project uses the hand start method). Fortunately, this engine has an electric starter that is more than capable to start this engine even with all the downstream drive train pulleys and sprockets attached. Originally, I designed a manually operated clutch to disengage the engine belt drive so the engine could spin faster but because the starter worked so well I removed it.

• Battery: 12 volt 19 amp hour. This size battery has never failed me in open water on my Kawasaki 650 Jet Ski the same engine as in the helicopter. In helicopter operation this 13-pound battery looses charge because the generator coil output is less than the heavy amperage demands of the electric cooling fan. In one hour a fully charged battery 12.7 volts will drop to 11.7 volts. Even at this charge level I have not experienced any engine performance issues. This is a good thing since one hour is the maximum flight time possible with the allowable 5 gallons of fuel on board. This battery voltage drop issue would be eliminated if the water-cooled test bed engine is replaced with an air-cooled engine.

• CDI Igniter: This device increases the 12 volt line voltage to 20,000 plus volts to enable a electronic spark to jump the gap between the spark plug center electrode and its grounded tip. This timed spark delivery, when the air/ fuel mix is at the ideal piston compression point, initiates the gasoline explosion that violently causes the downward piston motion creating the torque to turn the crankshaft.

• Magneto Flywheel: You probably know that your car uses a belt driven alternator to produce the electricity to run your car. In the old days (1960's and earlier) a generator / regulator set was used. The main difference between the two is the alternator produces ac (alternating current) then converts it back to 12 volt dc (direct current) while the generator / regulator setup produces straight 12 volt dc. The reason I was told that the auto industry changed over to the alternator was that it could produce the required current at a lower rpm making it more efficient.

As for the magneto / flywheel setup, its history goes back to the Ford Model "T" days (early 1900's). Electricity is produced when the permanent magnets attached to the rotating flywheel speeds by two coils fix mounted to the engine block. The magnetic induced fields generated by this arrangement force electrons to flow thru the coils to produce a current whose voltage is a function of the flywheel speed. This is basically the same method used by the generator. The subtle difference between the two is, in the generator its permanent magnets are fixed (they are attached to the inside of the cylindrical generator housing) and the coils rotate (they are integrated into the rotating center mass called the stator) just the reverse of the magneto.

The advantage of the magneto is it is compact, cost efficient, sets up engine timing, (without the need of a spark plug firing distribution device such as a distributor (older cars) or an engine management computer coupled to a flywheel crank angle sensor (newer cars)) and is direct driven. Unfortunately, where an alternator or generator are easily replaced the magneto is internal to the engine and when a coil fails it's a major job to fix. In fact on the Jet Ski I mentioned earlier the entire engine must be removed to even access any of the parts. The output of this 650 Kawasaki magneto / flywheel is 4.0 amps / 13 volts @ 6000 rpm.

• Regulator / Rectifier: this device limits the output voltage of the magneto to 13 volts in order to protect the battery from exploding and downstream electrical components from melting, frying, or generating excess heat.

• Spark Plug: NGK BR7ES, gap 0.7 – 0.8 mm. how could anyone come up with anything to say about something as simple looking as a spark plug? In fact today I doubt any kid owning a car has ever even seen a spark plug.

So much for me even writing all this, but as a teenager when Holly carburetors ruled engine fuel delivery and not fuel injectors you religiously changed your points and spark plugs every 6000 miles or your engine wouldn't run. When you went to the auto parts store and asked for the cars recommended plugs if they didn't have them you took what they had and installed them. Today's cars on the other hand do not require you to replace the spark plugs with car mileage under 50,000 miles.

Prior to getting involved with this project I never gave spark plugs differences much thought. Sure I could understand how Bosch would try to out do Autolite by adding platinum tips or adding three firing points versus one. As to whether or not one plug was better than another in increasing performance I figured it to be only sales hype. My attitude changed radically as to the significance of spark plugs when I was given a different plug than recommended for my 50cc Yamaha 2-stroke motorcycle. I was told at the parts counter that they didn't have the plug I wanted but the plug they gave me would work fine because the operating heat range was only slightly different. Well, it didn't work fine and the old plug worked better. The point is having the correct plug for your engine is important and there is science and engineering involved in their design. In addition the main reason the FAA has never certified a 2-stroke engine for aircraft use is due to the tendency of the spark plugs to foul at low rpm (become so soaked with oil they short out the intended spark) so make sure you always use the right plug.

• Timing: 15 degrees BTDC @ 6000 rpm (meaning that the magneto plate must be rotated 15 degrees on center from the crankshaft bottom position of the number one piston). Correct timing of a spark plug spark required to ignite the fuel in the cylinder is rather obvious especially when operating at 6000 rpm. The piston is only in that correct position to maximize horsepower every 1/100th of a second.

14.9.2 Engine Cooling System

As discussed previously if you use the air-cooled Hirth helicopter engine the cooling system is integral so you really have no additional electrical wiring to do.

For this engine there is a wiring circuit to design. Reiterating, the cooling system I added utilized an engine driven water pump to circulate water thru the engine and the radiator. To optimize engine efficiency I decided the engine water temperature should be around 190 degrees F like my Jeep thermostat setting (water cooled engines are held to tighter piston / cylinder tolerances so higher compression ratios can be obtained generating more horsepower than an equivalent sized air cooled engine). To accomplish that goal I purchased a Hayden 14 inch diameter fan with the optional adjustable electrostatic thermostat. This thermostat is operated by attaching a capillary type sensor to the radiator fins to remotely turn on and off the cooling fan to maintain water temperature set point.

14.9.3 Gages:

Shown below is the helicopter instrument panel used on the first test flight. Photo 14.9.3-1

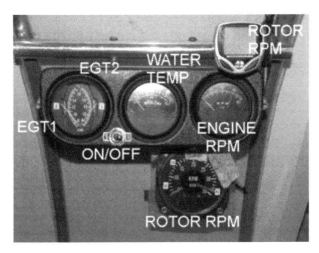

I purchased all Westach aircraft grade instruments. Let me just say they have impressed me by just surviving this long.

Initially the panel vibration was so severe that my eyes couldn't focus on what was displayed on the instrument cluster. To reduce instrument shake, I changed the way the panel was mounted to the cross tube. Instead of hard mounting the two together I added longer bolts that enabled me to add a light duty compression spring to the bolt shank between the cross tube and the instrument cluster. Then on the lower surface of the cluster I added a brace that tied back to the windshield. The brace was a formed piece of Lexan made wide enough to house the 3-inch main rotor tachometer. In one last effort towards vibration reduction I didn't hard mount the windshield end of the Lexan brace to the windshield but glued in a foamed strip between the two.

This modification was a vast improvement in reading all the 2-inch gages but just not good enough. I purchased a 3-inch tachometer for the main rotor rpm. If I were to do this console over again I wouldn't order any gage less than 3 inches. There's just too much vibration to read a small gage.

One more point, on the upper right hand corner of the console you will notice a small rectangular flat screen display approximately 2" by 3". This is a battery operated Schwinn bicycle speedometer digital display costing around $10 and comes with a remote wheel/frame mounted magnetic sending unit. You can calibrate this device to read in either km/hr or mph and adjust for your particular bicycle wheel diameter. I bought this thing on a whim never expecting to actually use it on the helicopter. Much to my surprise when I started to work with the calibration numbers I realized I could obtain a 44.0 readout to match the main rotor design rpm. This fact peaked my curiosity so I mounted the display as shown and the sending unit adjacent to the main rotor shaft with magnet attached. I never expected the device to keep up with a 440 rpm rotor shaft but it did and when cross checked with my optical tachometer it's readings were quite accurate so I kept it as a backup unit even though the display is quite small (note a 44.0 reading equals 440 rotor rpm).

The instruments used are as follows:

• EGT: This displays the exhaust gas temperature of both the #1 & #2 cylinder. I had to modify the exhaust manifold to use it, and when I ordered the probe to use with the gage I got the temperature range right but should have ordered the one with the "threaded mount" instead of "probe only". At the time I didn't think I could drill that much material out of the manifold to install the "threaded mount" type but after all the trouble I've had trying to keep the existing probe in position I think I should have tried to make the other type work.

The main reason I wanted this EGT information was to verify my calculations on heat transfer used in my engine water-cooling sizing.

• Water Temperature: This gage was used to give the water temperature "exiting the engine" or "exiting the radiator" via a two-position selector switch mounted on the joystick. The selector switch sent a signal from either the engine thermocouple or the radiator thermocouple to the gage. The single gage was to save instrument space. The purpose of the two readings was to determine how effective the engine cooling system was under various ambient air conditions while hovering.

• Engine RPM: This instrument in conjunction with the main rotor rpm gage indicates if the drive train is functioning properly. At ~ 5500 rpm the main rotor should be rotating at ~ 440 rpm. If it is not then the engine belt drive may be slipping or something more serious happening.

• Rotor RPM: This is the most critical gage to watch since you must be at

439 rpm before attempting to raise the collective to hover. The entire rotor hub assembly was designed to handle the stresses developed at this rotational speed. To make sure you're at the right spot on the gage it has been red lined (you are at the right rpm if the gage needle aligns with the red mark on the instrument face).

14.10 Hydraulic Systems:
None on this helicopter.

14.11 Stability Augmentations System:
Currently mechanical and electronic control is under investigation.

14.12 Autopilot:
Maybe some day I'll take that on but for the time being it is not high on the list.

PART 5- THE HELICOPTER TAKES SHAPE
Chapter 15- Fabrication

15.1 General Information
In Chapter 11 all the fabrication machinery and how each was used was discussed so this chapter focuses on the integration of all the equipment required when making a particular part.

The first step in this process is getting the material you specified. In Chapter 9, under the aluminum materials section, we noted that a 6061-type alloy could vary in tensile strength from a 12,000 to 35,000 pounds per square inch depending on its heat treatment. You need to make sure you're starting part fabrication with the type stated on the print and what you based your calculations on. Fortunately, ordering aircraft grade aluminum from a reputable supplier is the only way to go since the materials they send you are dye marked as to what type of material it is. These vendors, for a price, will also send you certification sheets if you ask.

The next step is setting aside a space to be used solely for your aircraft grade aluminum storage. As I noted above the vendor supplied material comes in dye marked, but the spacing of the dye nomenclature is such that any residual material left over after the band saw operation will not be marked. You should use an indelible marker to note the type before saving it. NEVER, NEVER throw any scrap away since you might be able to use it in the future and it is expensive.

Personally, I didn't have the time to look thru all the prints to see where this cutoff piece of scrap could be used but simply looked at my stock before ordering any new raw material. I do remember working on a 50 million dollar NAVY ELCAS (3000ft elevated causeway) project where the sole function of one of the purchasing agents was to maximize material usage of every steel plate ordered by going print by print to see how they could squeeze out the most parts from every sheet.

Now that you have a print and raw material in front of you, look at the lower left hand corner of the print to the material listing. There you will find the description of the material including its cutoff length, which under most circumstances is 1/8 of an inch longer than the final part length. This extra length is to allow for machining off any band saw cutoff variations.

After the band saw cuts the piece you have to decide what machining operations you want to do next. Looking at the print below titled "Frame Cable Gusset" you quickly realize three operations have to be done to produce this part, those being bending, milling and drilling. In this case I would bend the flange first then mill off the flange material not required. If you did any more work to this part first and the flange snaps off during the bending operation you just wasted a lot of effort. To save time and money it is a wise idea to mentally go over how to best produce a part before turning any machine on.

Print 15.1-1

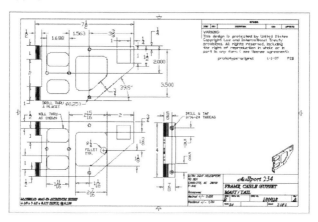

Print 15.1-2 Shown below is a part requiring milling, drilling and lathe work in order to create.

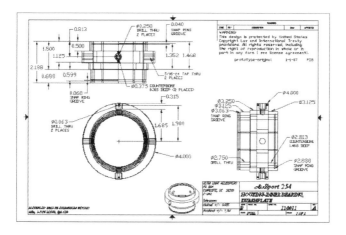

15.2 Process Sheets

There comes a time when you just can't figure out how to make the part shown on the print (see the print below titled "Bearing, Modified Spherical, Swashplate").

Print 15.2-1

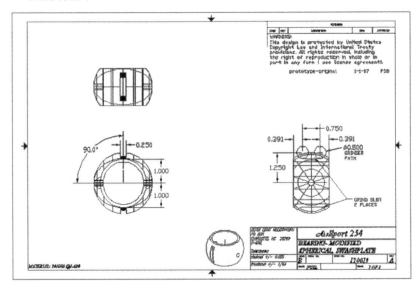

Sure in my case I "now" know how to produce it because of all the trial and error that came before it but you don't. To alleviate possible problems making parts someone came up with the concept of "process sheets" to accompany the design print. These process sheets did two things. It broke down times for every individual machining operation and associated a cost to it. It also gave the process the machinist needed to follow to make the part.

Some companies make process sheets for all the parts for costing purposes. This is helpful if you were generating a new model year sales price for a standard machine where for example, costs to mill parts increased 20%. In this case only the milling operations would have the multiplier added and the new cost for the machine quickly totaled. This pricing strategy is beneficial when your competition is very aggressive and a less analytical pricing approach could lose you market share.

Back to the part at hand, which is hardened alloy steel spherical bearing to which I had to add a slot to the surface. After print review I knew machining the part having a Rockwell hardness of 62 or better was out because none of my tool bits had a higher number. Then I decided to purchase diamond grit tool bits. The problem with that solution was they only worked at 20,000 rpm at which none of my equipment operated. Along with these bits I had to buy a handheld Dremel rotary tool. The process required a water-cooling system to keep the diamond-cutting surface cool. Needless to say "water flow" and holding a "120-volt" electrical device was not a sound option so I tried cutting without water and fried the diamond bit in less than 10 seconds! At this point I had only two options left send the spherical bearing to a heat treatment shop to reduce the heat treatment, then machine the groove then send it back for heat treatment to harden it, or attempt to grind the groove with my cutoff grinder.

Years ago when I had close contact with heat-treating shops I would have gone that route. Unfortunately most of those shops have gone out of business so the only real option left that I could afford was to try the grinder approach. After modifying the hand held grinder to mount to the milling machine spindle assembly I was able to perform the slot cutting operation within the specified print tolerance. Needless to say the write up for this particular process sheet would be extensive.

15.3 Part Finish

Although mentioned before most parts require more than one operation to complete but what hasn't been discussed is the way you want your part to look aesthetically. The 6061 and the 2024 aluminum you will be ordering both have a dull finish. They polish up to a mirror finish if you want to take the time. I polished every part because it just made me feel good looking at them.

I used two methods for polishing, the 3/8-hand drill and the bench grinder depending on the part size. The polishing compound itself comes in stick form in either red rouge or white. To me the red polished quicker. One thing you must know is that on large parts it helps to work only one area completely before inching forward since the polish works best warm. Keep in mind polishing is an abrasive act that generates a substantial amount of heat. In fact within a short time you'll be putting gloves on to hold the part. Just for your greater knowledge on the subject of "finish" or correctly stated as "surface-texture", there are written specifications (MIL-STD-10A and SAE AS 291C) as to what to include on your print. This surface-texture number will get the look you want if perhaps you sent the print to some machine shop to have the part made. At one point in my career a vendor actually gave me a key ring attached to a small piece of metal in which all the typical surface-textures were physically etched onto it along with the corresponding number. Here are some of the typical surface-texture numbers: 250 (rough machining), 125 (threads, surfaces for soft gaskets), 63 (gear shafts and bores), 32 (press fit parts, gear teeth, precision parts), 16 (spline shafts, camshaft lobes), 13 (engine cylinder bores, crankshaft bearings), 8 (lapped antifriction bearings), 4 (ball bearing races), 2 (shop-gage faces), 1 (gages and mirrors, micrometer anvils). The specifications stated above also give values for surface waviness limits for the textured surface.

At the company I worked for the 125 surface-texture number was the number we used most commonly for the machine parts we fabricated. When parts were to be chrome plated that number dropped into the two-digit range.

Chapter 16- Assembly

16.1 Assembly

My father was a very colorful character. When he was young, he and his rough neck friends played motorcycle polo. A game more daring than when played on horseback but also cheaper. Needless to say in order to play the season, good motorcycle repair skills were a given. Surprisingly enough with all his mechanical knowledge, he only gave me two rules to follow when working with tools. Never walk around with a screwdriver in your pocket with the tip pointed up because if you trip and fall the screwdriver may puncture your kidney. The second was never push forward on a wrench because if it slips off your knuckles will inadvertently smash into some metal part and be cut. If pulling toward you if the wrench slips off your hand will simply bounce off your chest.

In addition to my dad's advice I suggest on any assembly requiring multiple fasteners put them all in first before tightening them. Since this helicopter has two sides that have parts which are mirror images of one another it is recommended to place all bolts in both mating halves (for example the right and left tail sections) before tightening. Before moving on in this discussion lets look at some pictures.

Photo 16.1-1

What better way to find out if your helicopter CG is located correctly than to just lift it from the main rotor shaft. Shown in the picture is a standard size 2-ton engine hoist lifting the helicopter about 3 inches off the ground. In this pilot's position you can really get a feel for how sensitive this helicopter's balance is. Just a simple joystick movement will instantly have a skid touch the pavement; note the forward left skid contact.

I must say I never expected this engine hoist lift experiment to accomplish more than to verify that the helicopter CG was correct. As it turned out I gained a lot of seat of the pants (literally seat of the pants) experience on how to sit in it without having the skids hit the ground. Here's a thing you can try to get a feel for what I am talking about. Get a piece of wood about the same size as a skateboard deck and place it on top of a tennis ball then try to balance on it. Now figure out the type of control you need without being able to use your feet.

Photo 16.1-2

Photo 16.1-3

Photo 16.1-4

The previous two photos are at the grass strip airport. Here I am mounting the rotor blades to the rotor hub. The two critical pieces in this operation are the 8-foot stepladder and the 3-foot ladder. Let me just say getting the blades aligned can take hours.

Some of the things you can see from the photos are the aluminum ramp extensions inserted into the lowered trailer. What isn't too visible are the rubber wheels installed into the front and rear of the helicopter skids. Obviously they were added to facilitate placement on or off the trailer.

The other transportation aid was the eight-foot long pipe with pneumatic tires on each end. This wheeled axle assembly would slip under skids by the front legs and be secured. Then with very little effort the entire helicopter could be moved across the field via lifting the tail and pushing to the test area.

Chapter 17- Programming

17. 1 Automated Control

Currently there are no automated controls on this helicopter. I do plan on adding engine speed control before the next testing cycle because it is very difficult to do manually with all the other coordinated control efforts required just to hover. The engine speed control would require a programmable controller, a digital tachometer direct connected to the main rotor shaft and a servo-drive to operate the carburetor linkage.

PART 6- THE FINAL PHASE

Chapter 18- Testing

You've come a long way to get to this point and now its time to field test your helicopter. Just for the record, it's a good idea to get some formal pilot training. In fact some kit manufacturers will not sell you their product unless you are a qualified pilot with helicopter certification. Needless to say I didn't go that route of getting my helicopter pilot license before going to the airfield and just decided to see what would happen. Just to be on the safe side I decided to limit any flight motion to only six inches of helicopter travel. This was accomplished by leaving only that much nylon strap slack between the extreme skid corners and the anchoring 100-pound weights. Here's what happened.

18.1 Flight line Evaluation

During the first several trips to the airfield engine cooling was a major issue. Basically, in helicopter operation maximum power is required at zero speed therefore helicopter engines must be force air-cooled for them to perform at peak performance. In my case engine cooling finally came under control when I completely shrouded the radiator and centered an 1100 cfm fan on it.

Once cooling was under control additional airfield testing revealed the engine just wasn't delivering the power I needed. What happened next was bizarre. One afternoon while running flat out on my Kawasaki 650 standup jet ski (actually it's mate is the source of my helicopter engine) the retaining coupling holding the two halves of the exhaust manifold separated. Instantaneously the Jet Ski stalled out. It was then I realized the exhaust system I modified to use on my helicopter was flow limited due to the development of sonic shock waves at the exit point. Adding backpressure to the choke point eliminated the sonic shock condition. That of course sounds technical but my first pass at solving this problem was to add a long section of pipe to the existing exhaust pipe. It worked and I was able to get the helicopter off the ground. The next step was to design an exhaust pipe extension that would maximize engine power at the operational rpm.

Now that lift off was possible the next step was to find out how controllable the helicopter was. Well, that didn't take long. As soon as the helicopter was lifting off it pivoted clockwise in spite of my effort of Foot Pedal control. My first thought was the Foot Pedal adjustment was off but it was also possible that the Tail Rotor was not developing the force required. Ultimately the tail rotor drive had to be redesigned to increase its relative speed to the main rotor rpm and that corrected the problem.

Solving the yaw controllability problem set the stage for the next issue to surface, that being Joy Stick control sensitivity. Using the word "sensitive" really doesn't even come close to describing how difficult it was to control helicopter motion with the joystick. Apparently this is a typical flaw associated with lightweight helicopters with inherent low inertia rotor blades. The following pictures were taken from that first minute and a half hovering flight.

Photo 18.1-1

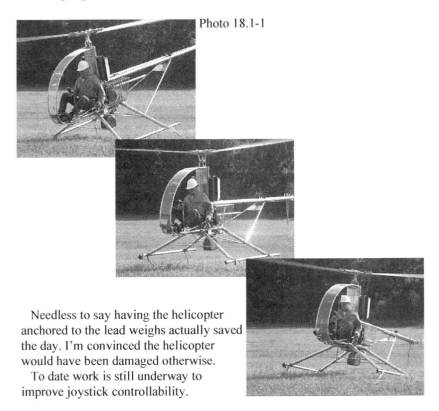

Needless to say having the helicopter anchored to the lead weighs actually saved the day. I'm convinced the helicopter would have been damaged otherwise.

To date work is still underway to improve joystick controllability.

18.2 Data Collection

Flight data collection is probably not as critical when building a prefab kit or when buying a new or used helicopter but when you design your own, be prepared to record a lot of data. In fact when you are finished with your project you will have notebooks full of the stuff. The main point here is to record all your real time numbers because as soon as the helicopter shuts down you will inadvertently suffer memory loss about what you just experienced. Furthermore with time, all the things you wanted to remember will be merged into one big glob of worthless information. Documentation is especially important if you want to compare specific hardware changes with recorded performance data.

The first thing you should do is purchase a clipboard to write on because all your data collection will either be in your shop or in the field and it will be illegible at times. Just look at me with clipboard in hand sitting in the running helicopter trying to record the numbers. Then look at the clipboard paper I was writing on at the time.

Photo 18.2-1

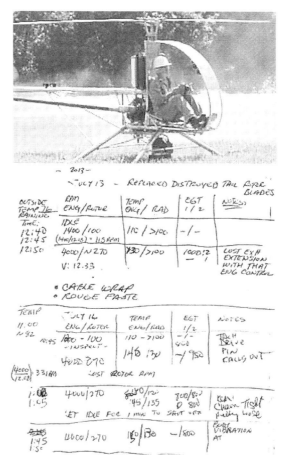

As you can see these clipboard records need to be rewritten, preferably in a bound notebook.

In addition to my recording flight line data some of the other records I generated dealt with engine water pump flow rates versus engine speed, tail rotor thrust versus foot pedal travel, main and tail rotor drive chain stretch, indicated instrument rpm versus hand held optical tachometer readings, etc. Then adding to that type of information I recorded how to best do something. This came about when I realized that the way I thought was the easiest to put something together or take apart was not necessarily the most efficient. Here are some of the areas I developed procedures for: engine drive belt replacement, main rotor blade alignment, main rotor drive chain slack adjustment, tail rotor drive belt adjustment, etc.

Documentation, as you can tell by now covers a lot of territory, and for me I found setting up a binder with individual section titles worked the best to organize everything. When I had to change the tail drive ratios I first set up a binder section title for it. Then I included in it all my previous calculations for the drive I was going to replace, the new calculations for the revised drive, the bill-of-materials for the parts I needed, included a sketch of where the parts went, and a copy of the parts ordered from each vendor involved. After the parts arrived and were installed I added a summary of the improvements if any, over the drive I replaced based upon previously recorded data. Eventually, I wound up with a binder with a very comprehensive developmental history of my work.

One other point I may add concerns "mind lapse"! After a couple of years have gone by, when someone asks why you did something a certain way and you can't quite remember all the nuance for the response, opening the binder to help refresh your memory is extremely beneficial. Or, if you think you can improve a particular area even further, say on the tail drive, just go to that section in your binder and instantly you will see the exact history of your prior work in that area.

Keep Records!!!!

18.3 Failure Analysis

This is an interesting chapter because after all the work you have completed to date by either just reading this book, building your purchased kit or getting ready to fly your own design I bring you the bad news. I'll start by quoting something I remember hearing a long time ago which went something like this "The best laid plans of mice and men often go asunder"!

This chapter starts out with parts that failed on my helicopter and the reasons behind the failures. The chapter ends with a cover copy of the National Transportation Safety Board's special investigation report into the Robinson Helicopter Company R22 loss of main rotor control accidents. You can review the complete report via the internet, which is an interesting read since this rather successful small helicopter had some major flaws as shown in the included Accident Summary page (4 deaths per 100,000 flight hours).

18.3.1 Airsport 254 Part Failures

1)   Engine Pulley Failure: I needed a five-groove pulley for the amount of power being transferred. Unfortunately this was not a stock size so I fabricated one on my lathe from 2024-T3 aluminum. Then to save machining time I utilized the existing engine coupling to mount my pulley to it via a calculated light press interference fit. I'm not exactly sure which contributed more to the failure of the press fit, engine vibration or elevated crankshaft temperature but nevertheless the two pieces fell apart. The fix was simple I bolted the two parts together and the fix has worked fine ever since. The point here is press-fitting parts together is a rather routine process and I have specified this method many times over the years and never had problems, so this failure definitely blind-sided me.

2)   Goodyear Eagle Synchronous Belt Wear: This drive transferred power from the main rotor shaft to the first of two tail rotor power transfer drives. The cause of the problem was pulley dynamic alignment aberrations.

Typically in the machine fabrication process the side plates that tie the machine together are drilled in pairs. This insures that cross-members, tie-bars, bearing mounted idler rolls and shafts are parallel since theoretically all this stuff would mount to identically drilled side-plates. In my case the shaft on the tail cross-member and the main rotor shaft were not necessarily parallel and getting them parallel in three-dimensional space without the aid of side-plates was difficult but possible with guide fixtures. What was missed was the slight deflection of the main rotor shaft while at operational rpm and rotor load. This resulted in the main rotor shaft pulley being skewed with respect to the other pulley causing the belt to try to walk up the face of the other pulley fraying the belt on one side. The fix was to add a bearing support to the main rotor shaft adjacent to the pulley to eliminate any shaft deflection at that point.

I am still not satisfied with the degree of difficulty it takes to align this drive so this area will be revised at a later date.

3)   Engine Bolt Failure:
Photo 18.3.1-1

First lets fill in the background of the bolt's location. Originally this helicopter frame layout was designed to support a three-cylinder "Two Stroke International" engine. That manufacturer went out of business when the helicopter frame was already completed. At this point I decided to use the Kawasaki engine to prove out the helicopters capabilities using the as built helicopter frame. The modifications seemed simple enough. Just add a pair of aluminum support pieces between the existing frame cross members to allow any engine under consideration to mount to them. These support members were then mounted on 50 durameter rubber isolators and secured by the bolt shown above. The head of the bolt contacted the top of the aluminum support piece while the nut secured the above to the frame cross member. Repetitive bolt head contact with the steel spacer between it and the aluminum frame while engine was at maximum power caused the fracture. Additional rubber isolation was added under the head of the bolt to eliminate the problem.

4)   The Main Cross Member Failure: This cross member housed the lower main rotor shaft radial bearing. The bearing was centered between two

1 x 2" rectangular channels and these fastened together with four aircraft grade bolts. The failure occurred when I didn't completely bend the cotter pin ends over the castle nuts. During testing the cotter pins worked themselves out, which in turn allowed the castle nuts to loosen. The result of these loose bolts was a buckling of the cross member closest to the engine. Looking at the subject part the extent of the structural deformation is evident as is the crack running from the bearing bore to the lightening hole (lightening hole was made to reduce overall piece part weight). When the cross member shifted it contacted the 5500 rpm engine pulley which virtually wore down both contacting surfaces. The fix: remade this cross-member without lightening holes and made sure all cotter pins were correctly installed. See the following pictures:

Photo 18.3.1-2

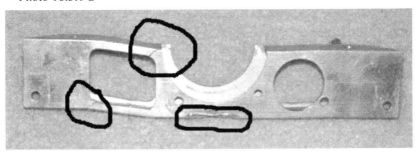

Photo 18.3.1-3

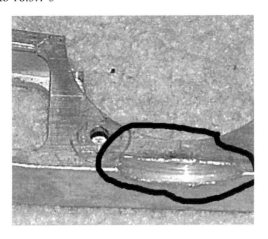

5)   Tail Rotor Failure: This failure was initiated when one of the two stiffeners supporting the "Tail Rotor Ground Strike" structural support failed. When this occurred one support tube moved into the path of the rotating tail rotor, damaging both. This failure occurred because two of the rivets retaining one of the structural tube inserts sheared. The fix: redesigned connection to use bolts instead of rivets.

Photo 18.3.1-4

Photo 18.3.1-5

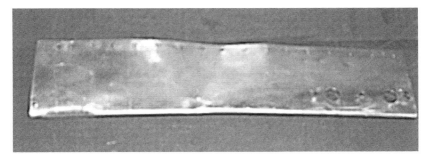

6)   Foot Pedal Linkage Rod Failure:
This failure was a direct result of pilot error, meaning specifically my
error. While attempting hover during an early test run I couldn't maintain
Yaw control (keep the helicopter straight) within the range of motion of the
foot pedals. During this rapid directional change I forgot to ease off the
opposite foot pedal. The amount of force I exerted buckled the left foot pedal
linkage rod. Needless to say I doubt this type of failure would happen again
but to be on the safe side both linkage rods will be replaced in the future
with stiffer ones.

Summary: As these failures occurred more attention was paid to adding
redundancy to the fastened parts. For example any pulley mounted on a
vertical shaft was not only retained by keyway and setscrews but also locked
into proper position with snap rings. Any belt after properly tensioned had
the moveable section dowel pinned. A pre-transportation checklist was
generated to insure all cotter pins were properly installed, gearbox oil level

checked, guards in proper position, etc. Then at the airfield a pre-flight checklist was prepared so I could double check the craft prior to startup. I also generated a startup procedure so I could review everything I needed to do for the flight-testing. This checklist came about after forgetting a couple of times to switch on the cooling fan circuit.

18.3.2 National Transportation Safety Board report PB96-917003

Table 1-U.S. Loss of control[1] (LOC), non-loss of control (non-LOC), and all fatal helicopter accidents, flight hours, and corresponding accident rates for the years 1981-1994, by helicopter model.

Helicopter model[a]	Fatal Accidents			Flight hours[b]	Fatal Accidents per 100,000 flight hours		
	LOC	Non-LOC	All		LOC	Non-LOC	All
Bell 206	2	119	121	13,369,702	0.015	0.890	0.905
Hughes 369	2	38	40	3,002,236	0.067	1.267	1.333
Hiller UH12	1	13	14	987,796	0.101	1.316	1.417
Enstrom F28	1	16	17	845,032	0.118	1.893	2.012
MBB BO 105	1	12	13	806,750	0.124	1.487	1.611
Bell 212	1	3	4	497,129	0.201	0.603	0.805
Hughes 269	5	28	33	1,992,301	0.251	1.405	1.656
Bell 47	6	44	50	2,343,215	0.256	1.878	2.134
Bell 204	1	2	3	227.683	0.439	0.878	1.318
Robinson R22	23	39	62	1,524,483	1.509	2.558	4.067
Totals	43	314	357				

18.4 Vibration Analysis

Vibration management is a major design consideration when developing a helicopter, especially a lightweight one. On this helicopter the engine represents close to 30% of the base weight so when running at 5500 rpm the operational vibration created is strongly transmitted throughout the airframe even when mounted on rubber isolators. Add to that all the vibrations induced by slightly eccentric drive shafts (machining tolerance 0.005 TIR), shaft movements due to allowable clearances within the bearings themselves, chain and belt pulsations, main and tail rotor load irregularities, etc. Analysis of all these varying amplitudes of energy inputs is complex, so much so that I decided to address problem areas after getting the helicopter up and running. Upon startup two major areas of concern readily became apparent: 1) anything fastened together loosened up 2) cockpit mounted instruments could not be read. Addressing the first issue, all rivets were upgraded to ones with steel pin centers. All set screws, bolts, and nuts were coated with Locktite blue before being torqued. Dowel pins, where possible, were positively retained with set screws. The only thing I didn't try was adding an epoxy bond prior to mechanically fastening parts together. This is still under consideration.

The instrument shake issue was a much harder problem to resolve .The best results occurred by spring mounting the instrument cluster to a small diameter aluminum tube whose ends were glued to the inside of a foam one-inch wide donut. The circumference of this donut was in turn glued to the inside of an aluminum sleeve bolted to the cockpit framework. Although this helped significantly to reduce small amplitude vibrations it did not solve the problem completely. I decided to replace the two inch diameter Main Rotor RPM gage with a three inch diameter one, as for the remaining gages I used an idea I learned in the Air Force which was to "Red Line" them. Basically, you put a physical "Red Line" on the surface of the gage face as to where the gage should read under normal operating conditions. With just a simple glance you could easily notice whether or not the gage needle was in the proper position. This was the instrument panel mounting setup in place during the first tethered flight.

18.5 Trouble Shooting

Fortunately this helicopter is rather straightforward in that there are no hydraulic or electrically assisted subsystems. Our helicopter control is thru direct mechanical linkage and as a result troubleshooting is confined to only electrical and mechanical systems.

8.5.1 Mechanical

Troubleshooting a mechanical system is simple because mechanical problems are rarely intermittent and will progressively degrade until failure. The key here is to develop an intimate harmonic relationship between you and your helicopter straight up.

Now for another personal experience to drive home the point. When I was seventeen I had a 1949 Dodge tow truck. I used that truck to deliver old junk cars given to me to take to the Newark scrap yards. The truck was an antique, even then, and the cars of course ready for the bone yard so on any particular trip something mechanical would go wrong. Some of these mechanical problems could quickly endanger other drivers if not addressed.

For example one time I missed a loose hood on an Oldsmobile I was towing from the rear bumper (front steering wheel tied to the vent window structure so the front wheels would remain straight) and I lost the hood in heavy traffic. Another time the rusty lug nuts on the car I was towing snapped, sending the tire into the fast lane where a speeding car hit it and launched it 20 feet into the air. Another time a car I was towing backwards had the tied steering wheel come off because someone had removed the retaining nut, which I didn't notice. The result was the car abruptly swerved to the curb where it crashed into a high-tension telephone pole.

These were the mechanical things I missed and fortunately no one was killed or hurt. That launched tire barely missed landing in a convertible car traveling in the opposite lane with six teenagers in it, and the car that hit the telephone pole missed by mere yards kids playing near it. As for those potentially dangerous situations I did avoid, I heard them. To this day I find myself analyzing an unfamiliar sound to determine exactly what caused it. The point here is to develop your hearing sensitivity and acumen so any unfamiliar mechanical sound can be acted upon in order to prevent it turning into a costly repair or a dangerous condition. It's similar to the advice given when exercising. If you feel a muscle hurt in the middle of a set you should stop, in order to prevent doing extensive damage.

So far in this helicopter development process mechanical problems or potential problems have been isolated and corrected by concentrating on one area at a time during a run-up ground-testing schedule. This procedure I highly recommend following regardless of whether or not you build your helicopter or a kit one. Also having an observer at the airfield to signal you if something begins to go wrong can be a big plus! You have many things to keep track of and another set of eyes and ears can only help.

18.5.2 Electrical

Trouble shooting electrical problems is just tedious, mainly because of the erratic nature of the cause. Just recently I had a rear signal light on my Jeep Wrangler go awry. At first I thought it was a bad ground because every time I tapped on the light housing the light would work properly. I didn't get a chance to add the extra ground until several weeks later and to my surprise the next time I checked that light it was once again out, until I shook the housing. At that point I removed the bulb and replaced it. That corrected the problem. Upon close inspection of one of the two internal bulb filaments I noticed one end of the top filament was cracked thru. I surmise that when the vehicle was moving that filament would be jarred enough to no longer make contact with its mating piece. When the vehicle was stopped and I tapped the housing that filament would snap back into operational position. As I said before these type of erratic and intermittent electrical problems are indeed tedious to resolve.

Our helicopter has only two main "external" electrical systems (radiator fan and the starter), one "internal" engine charging circuit, and the "engine instruments" to worry about. To date only the starter circuit has not been a problem, as for the others here's the breakdown.

On one of the engine modifications I decided to replace the rear crankshaft driven oil injection pump with a water pump. Unfortunately the mounting screw was a tad bit longer than necessary and it eventually severed one of the charging circuit magneto wires. This in turn caused a battery voltage drop that affected cooling fan performance.

The instruments have been dependable in spite of the excessive cockpit vibration but the EGT senders have a tendency not to display correctly until the engine is running above 5000 rpm. I believe this is due to the fact that the inherent nature of two-stoke engine operation is to coat everything in the flow path with oil. Upon startup this oil buildup on the EGT thermocouple probes must burn off before the sensors can respond and this I believe happens at around 5000 rpm. Although I have not experienced instrument function problems, shown below is a typical vendor-troubleshooting guide if one were to occur.

TROUBLE SHOOTING

CHT or EGT Senders:
- Make sure all connections are clean and secure.
- Reverse connections.
- Disconnect sender from gauge, check the resistance of the sender. A cold probe on the sender should be approx. 1.5 ohms for CHT and 2.5 ohms for EGT.

CHT or EGT Gauges: Use OHM meter ONLY for this test!
- Disconnect the gauge from the sender; check the resistance of the gauge. Depending on the scale the resistance will be between 8-15 ohms. You will also notice a needle deflection on the unit (only w/ analog ohm meters).

Temperature, Pressure & Fuel level Senders (Water, Oil, Air, and Carb. Temp):
- Make sure all connections are clean and secure.
- For best results all grounds should be on one common point.
- Temperature senders: Disconnect the sender leads from the gauge, cool or heat the probe to a known temperature then check the resistance. See chart below:
    - 32 deg. F — 9800 ohms
    - 70 deg. F — 3570 ohms
    - 100 deg. F — 17400 ohms
    - 212 deg. F — 212 ohms
- Pressure and Fuel level senders: Disconnect sender leads from the gauge, use the resistance values below for 240 ohm pressure and fuel level senders:
    - E or 0 scale — 230-240 ohms
    - Mid scale — 90-100 ohms
    - Full scale — 27-37 ohms
- For 3 wire capacitance type fuel senders it is recommended to return the unit to the factory for recheck, or check with a known good gauge.
- Do not use water as a medium to calibrate the capacitance type fuel level probes as this will make the gauge read full scale and will need to be flushed out.
- Do not use a battery charger to bench test, as this will cause an erratic reading at best on the gauge.

Temperature Gauges (Water, Oil, Air, and Carb. Temp):
- To test the operation of the unit, disconnect the sender leads from the gauge and substitute the sender with a resistor of the appropriate value that you wish to check from the list above. Such as 615 ohms is 150.
- If the sender lead is shorted to ground, the meter should peg full scale.
- If the sender lead is open (broken) or sender is open internally, meter should peg hard to the left. Slight needle movement to the left is normal when the temperature is below the starting point of the gauge.
- If Pin # 5 is ungrounded with sender attached correctly, meter will read full scale.

KV & 4 wire capacitance type fuel senders:
- OP produces .5-4.5 across the psi range with 5vdc input.
- Capacitance senders produce 0-5vdc with 12/24 vdc input.

Chapter 19 The FAA

19.1 Federal Aviation Regulations (FAR103)
FAR part 103 states the legal rules relating to Ultralights. They are as follows:

Adopted July 30, 1982, effective on October 4 that same year, Federal Aviation Regulation Part 103 formally established what truly is recreational flight.
Part 103 established limits on size, performance, and configuration and also established that people flying them needed no certificate or medical qualification.

FEDERAL AVIATION REGULATION PART 103—ULTRALIGHT VEHICLES
Authority: 49 U.S.C. 106(g), 40103–40104, 40113, 44701.
Source: Docket No. 21631, 47 FR 38776, Sept. 2, 1982, unless otherwise noted.
Subpart A—General
§ 103.1   Applicability.
This part prescribes rules governing the operation of ultralight vehicles in the United States. For the purposes of this part, an ultralight vehicle is a vehicle that:
(a) Is used or intended to be used for manned operation in the air by a single occupant;
(b) Is used or intended to be used for recreation or sport purposes only;
(c) Does not have any U.S. or foreign airworthiness certificate; and
(d) If unpowered, weighs less than 155 pounds; or
(e) If powered:
(1) Weighs less than 254 pounds empty weight, excluding floats and safety devices which are intended for deployment in a potentially catastrophic situation;
(2) Has a fuel capacity not exceeding 5 U.S. gallons;
(3) Is not capable of more than 55 knots calibrated airspeed at full power in level flight; and
(4) Has a power-off stall speed which does not exceed 24 knots calibrated airspeed.
§ 103.3   Inspection requirements.
(a) Any person operating an ultralight vehicle under this part shall, upon request, allow the Administrator, or his designee, to inspect the vehicle to determine the applicability of this part.
(b) The pilot or operator of an ultralight vehicle must, upon request of the Administrator, furnish satisfactory evidence that the vehicle is subject only to the provisions of this part.
§ 103.5   Waivers.
No person may conduct operations that require a deviation from this part

except under a written waiver issued by the Administrator.

§ 103.7   Certification and registration.

(a) Notwithstanding any other section pertaining to certification of aircraft or their parts or equipment, ultralight vehicles and their component parts and equipment are not required to meet the airworthiness certification standards specified for aircraft or to have certificates of airworthiness.

(b) Notwithstanding any other section pertaining to airman certification, operators of ultralight vehicles are not required to meet any aeronautical knowledge, age, or experience requirements to operate those vehicles or to have airman or medical certificates.

(c) Notwithstanding any other section pertaining to registration and marking of aircraft, ultralight vehicles are not required to be registered or to bear markings of any type.

Subpart B—Operating Rules

§ 103.9   Hazardous operations.

(a) No person may operate any ultralight vehicle in a manner that creates a hazard to other persons or property.

(b) No person may allow an object to be dropped from an ultralight vehicle if such action creates a hazard to other persons or property.

§ 103.11   Daylight operations.

(a) No person may operate an ultralight vehicle except between the hours of sunrise and sunset. (b) Notwithstanding paragraph (a) of this section, ultralight vehicles may be operated during the twilight periods 30 minutes before official sunrise and 30 minutes after official sunset or, in Alaska, during the period of civil twilight as defined in the Air Almanac, if:

(1) The vehicle is equipped with an operating anticollision light visible for at least 3 statute miles; and

(2) All operations are conducted in uncontrolled airspace.

§ 103.13   Operation near aircraft; right-of-way rules.

(a) Each person operating an ultralight vehicle shall maintain vigilance so as to see and avoid aircraft and shall yield the right-of-way to all aircraft.

(b) No person may operate an ultralight vehicle in a manner that creates a collision hazard with respect to any aircraft.

(c) Powered ultralights shall yield the right-of-way to unpowered ultralights.

§ 103.15   Operations over congested areas.

No person may operate an ultralight vehicle over any congested area of a city, town, or settlement, or over any open air assembly of persons.

§ 103.17   Operations in certain airspace.

No person may operate an ultralight vehicle within Class A, Class B, Class C, or Class D airspace or within the lateral boundaries of the surface area of Class E airspace designated for an airport unless that person has prior authorization from the ATC facility having jurisdiction over that airspace. [Amdt. 103–17, 56 FR 65662, Dec. 17, 1991]

§ 103.19   Operations in prohibited or restricted areas.

No person may operate an ultralight vehicle in prohibited or restricted areas unless that person has permission from the using or controlling agency, as appropriate.

§ 103.20   Flight restrictions in the proximity of certain areas designated by notice to airmen.

No person may operate an ultralight vehicle in areas designated in a Notice to Airmen under §91.137, §91.138, §91.141, §91.143 or §91.145 of this chapter, unless authorized by:

(a) Air Traffic Control (ATC); or

(b) A Flight Standards Certificate of Waiver or Authorization issued for the demonstration or event.

[Doc. No. FAA–2000–8274, 66 FR 47378, Sept. 11, 2001]

§ 103.21   Visual reference with the surface.

No person may operate an ultralight vehicle except by visual reference with the surface.

§ 103.23   Flight visibility and cloud clearance requirements.

No person may operate an ultralight vehicle when the flight visibility or distance from clouds is less than that in the table found below. All operations in Class A, Class B, Class C, and Class D airspace or Class E airspace designated for an airport must receive prior ATC authorization as required in §103.17 of this part.

Airspace	Flight visibility	Distance from clouds
Class A	Not applicable	Not applicable
Class B	3 statute miles	Clear of Clouds
Class C	3 statute miles	500 ft below, 1000ft above, 2,000 ft horizontal
Class D	3 statute miles	500 ft below, 1000ft above, 2,000 ft horizontal
Class E; Less than 10,000ft MSL	3 statute miles	500 ft below, 1000ft above, 2,000 ft horizontal
Class E; At or above 10,000ft MSL	5 statute miles	1000ft below or above, 1 statute mile horizontal
Class G; 1,200ft or less above the surface (regardless of MSL altitude	1 statute mile	Clear of clouds
Class G; More than 1,200ft above the surface and at or above 10,000ft MSL	1 statute mile	500 ft below, 1000ft above, 2,000 ft horizontal
Class G; More than 1,200ft above the surface but less than 10,000ft MSL	5 statute mile	1000ft below or above, 1 statute mile horizontal

The FAA does grant training exemptions to a limited few organizations that have two place training aircraft that are similar to and can provide instruction for some ultralight types. This FAA exemption limits the two-place powered ultralight to have a maximum empty weight of 496 pounds, a maximum fuel capacity of not more than 10 US gallons, not be capable of more than 75 knot calibrated airspeed at full power in level flight, and have a power-off stall speed which does not exceed 35 knots calibrated airspeed. Consult the FAA document to review the complete details of the limitations and conditions concerning this aircraft trainer.

19.2 The Rotorcraft Flight Manual

This chapter addresses the compilation of the "Rotorcraft Flight Manual" in accordance with the format standardized by the document "Specifications for Pilot's Operating Handbook" written by the General Aviation Manufacturers Association (GAMA). The "Rotorcraft Flight Manual" includes operating limitations, which a pilot must comply with in accordance with Title 14 of the Code of Federal Regulations (14 CFR) part 91. The following Rotorcraft Flight Manual is for the Airsport 254 (serial number 2011001). It should be noted that although this particular manual is not required for an ultralight helicopter all the standardized title headings are in accordance with the standard. I set mine up in accordance with these rules. One additional point is that the FAA will allow the main title of the document to be called the "Pilot's Operating Handbook" provided it states on the title page that this is the FAA approved "Rotorcraft Flight Manual".

19.3.1 Preliminary Pages:

Each and every helicopter has its own specific "Rotorcraft Flight Manual" and as such on the title page the serial number and registration for which the manual applies must be listed. If it is not specifically stated then that Rotorcraft Flight Manual is relegated as generic material for general use.

It is in this section a table of contents is listed as well as a statement of how the manual is paginated, for example section number followed by the page number 1-1, 1-2, 1-3 etc. If loose-leaf bound then tabs indicating the title of that section along with the number may also separate sections. Adding to the ease of use the tab for the Emergency Procedures may be red.

19.3.2 General Information:

This section gives the general description of the helicopter and the powerplant. Included here would be a front, side and top view of the aircraft with overall dimensions along with the rotor diameter. This is the first place to start to learn about the basics of a specific helicopter.

Complimenting the above terminology, symbology, abbreviations and any other unique features germane to this manual are defined. At the manufacturer discretion metric and conversion tables could be located here.

19.3.3 Operating Limitations:

This section states the regulatory limitations as well as those stated by the manufacturer for safe operation of the helicopter and its subsystems. In addition to Airspeed, Altitude, Rotor, Powerplant and Flight limitations topics such as Weight and Loading Distribution, Instrument markings/color coding, Placards, as well as Fuel and Oil requirements are also detailed.

19.3.3.1 Airspeed Limitations: The airspeed limitations of a helicopter are indicated by graphs and placards and in the cockpit along with visuals on the airspeed gage itself. The airspeed gage visuals include a redline superimposed on the airspeed which is the aircraft's extreme limit and called the "never to exceed speed" or Vne. If the Vne is exceeded structural damage may occur. A blue line is superimposed on the gage indicating maximum safe autorotation speed. A green arc placed adjacent to the gage speed range indicates normal operation

19.3.3.1.1 Airsport 254 Airspeed Limitations: The Airsport 254 although structurally capable of safe operation at airspeeds greater than 50 knots it is power output limited to that maximum speed in accordance with FAR 305 governing Ultralight operation.

19.3.3.2 Altitude Limitations: The maximum altitude attainable by a helicopter is based upon load and available power. Charts included in this section may indicate the maximum operating density altitude as a function of rotorcraft takeoff weight.

19.3.3.2.1 Airsport 254 Altitude Limitations as per FAR103

19.3.3.3 Rotor Limitations: A rotor hub and blade system is engineered to withstand the dynamic loads of the rotating blades in achieving the required Lift. This design lift occurs at an optimal rotational speed that enhances the rotor blade efficiency and minimizes power requirements. Below this minimum rotor RPM limit sufficient lift is not produced. Rotor RPM can only go so high before high stresses can cause catastrophic damage to hub bearings and rotor blade integrity as well as rotor blade instability as a result rotor RPM has operational limits.

Reciprocating engines develop peak torque over a rather narrow RPM range. It is around this engine torque band that the optimized rotor RPM is geared. On a typical dual-needle tachometer a superimposed red line at the critical RPM highlights the maximum and minimum operating points of the engine with a green arc connecting those two points depicting the normal operating range. The rotor portion of the tachometer is setup in similar fashion with red line indicating maximum and minimum RPM limits connected by a green band. The rotor limits change during Power-on and power-off. This being whether the engine is powering the rotor or the engine is off and the rotor is in autorotation mode.

19.3.3.3.1 Airsport 254 Rotor Limitations: The Power-on operating range of the rotor is 439- 450 RPM power-off range 420-455 RPM.

19.3.3.4 Powerplant Limitations; This section defines the operating parameters for the powerplant such as the RPM range, maximum power rating, operating temperatures and pressures, type of oil used and fuel type requirements. The typical monitoring instrument being the tachometer. Reciprocating engines include gages for EGT (exhaust gas temperature), manifold pressure, and water temperature all of which are redlined at their respective limits. Turbine engines gages in addition to indicating a maximum continuous power rating (the green arc band on the gage) will state a time limited maximum peak power rating (the yellow arc band with a red line positioned at the power band limit). The turbine outlet temperature gage is set up similarly with the green band zone for normal operation followed by a yellow band the upper envelope of the temperature range terminating with the redline not to exceed temperature markings. A placard in close proximity to the gage states maximum values as a function of specific situation.

19.3 .3.4.1 Airsport 254 Powerplant Limitations: The engine operational range is 5800-6000 RPM. Engine redline (or not to exceed RPM) is 6500. Water temperature-200 F, EGT-1450 F, Manifold Pressure-27 "hg, 2-cycle oil-Kawasaki marine, Fuel- 87 octane gasoline.

19.3.3.5 Weight and Loading Distribution: The Weight and Balance Section of the FAA- Approved Rotorcraft Manual gives the methodology in performing weight and balance calculations critical to accurate placement of load and personnel within the allowable rotorcraft "Center of Gravity" operating range. This section states the maximum certified weights, the actual allowable CG (center of gravity) range and the x-y-z Datum points from which to calculate the revised CG after all takeoff loads have been added.

19.3.3.5.1 Airsport 254 Weight and Loading Distribution: This rotorcraft has a base certified weight of 254 pounds. 25 pounds of instrumentation and safety equipment, a 180 pound pilot in the cockpit and 35 pounds of fuel with a CG at 0,0, 20. Where 0, 0 in the x-y plane is the position of the main rotor shaft.

19.3.3.6 Flight Limitations: This section states maneuvers and flight conditions which are prohibited, for example certain acrobatics or those boundary conditions causing system component icing. If the rotorcraft is limited to VFR flight it is stated here as well as minimum crew, and specific pilot seating position during solo flight.

19.3.3.6.1 Airsport 254 is limited to 50 knots, one hour flight time, daylight only operation, compliance with all in flight operating rules stated in FAA Far 103 subpart b and all aerial acrobatics are prohibited.

19.3.3.7 Placards: This section addresses the critical visual aids strategically located in the cockpit that enable safe operation of the rotorcraft. They are listed and explained here. Although placards vary in number and type per aircraft, all helicopters have the chart placard "Vne as a function of altitude".

19.3.3.7.1 Airsport 254 has the following placards, "Vne as a function of altitude", Startup checklist, cockpit gage allowable limits and collective settings.

19.3.4 Emergency Procedures: The exact actions the pilot must take to survive emergencies such as fires, system failures, engine failure, tail rotor failure etc. are presented in two checklists here. The first checklist highlights the procedural sequence in abbreviated detail. This should be memorized. If an emergency occurs and time permits this checklist would be used to verify all actions were addressed. The second checklist elaborates in detail the procedures and enables clarification of the action required. This checklist should be reviewed in order to understand the logic supporting the action.

Typically, manufacturers of larger helicopters will include a section titled "Abnormal Procedures" for those malfunctions not deemed of a critical nature.

19.3.4.1 Airsport 254 Emergency Procedures are as follows:
In the normal course of flight whether in hover or a translating type flight if any unrecognizable vibration occurs land immediately! Should the engine rpm inexplicably drop instantly lower the Collective lever and setup for an autorotation landing.

19.3.5 Normal Procedures: This section is the most referenced since it concentrates on normal flight operating conditions. Typically the first part is dedicated to listing airspeeds for basic maneuvers, which enhance safe flight. Following this part are the checklists such as, preflight, pre-startup, startup, how to engage the rotor, and those directed at the flight itself spanning takeoff, approach, landing and possibly autorotation. Other areas covered would be ground checks and shutdown. Never perform any of the above without consulting the checklist if one is available. It is simple to do and minimizes chance for error.

   19.3.5.1 Airsport 254 Normal procedures are as follows: If the airfield you are operating from has a tower notify them as to your intention to land. Then at the designated landing area reduce flight speed and begin lowering the Collective while maintaining proper engine rpm. Continue this procedure until safely contacting the ground. If taxiing is required then continue your ground flight path until your designated spot is reached. At that point record any last minute data then turn off the engine and wait until the main rotor stops.

   19.3.6 Performance: This section deals with the regulations and manufacturer performance data that will aid in operating the helicopter. Charts included here are as follows: "airspeed conversion of indicated versus calibrated", "hovering ceiling versus gross aircraft weight", and a "velocity versus height diagram".

   19.3.6.1 Airsport 254 Performance data: see Chapter 8.6 for details.

   19.3.7 Weight and Balance: This section contains everything dictated by FAA regulation to calculate Weight and Balance to establish a takeoff CG and most often will include a sample problem.

   19.3.7.1 Airsport 254 Weight and Balance is very simple in that only pilot weight must be factored into the takeoff CG calculation; any added fuel is automatically positioned at the ideal takeoff CG since the fuel tank is positioned at the craft's CG. This helicopter has its takeoff X-Y coordinates located exactly at the main rotor shaft centerline (the ideal takeoff CG) if the pilot weighs 180 pounds and sits in the seat positioned close to its rearmost setting. Any pilot weight variation from 180 pounds the calculation is as follows:

Definition of terms:
  NSPD = normal seat position distance = 12"
  StdP = standard pilot weight = 180 pounds
  Pwt = weight of actual pilot (in pounds)
  NewD = new seat position distance (in inches) move seat forward if positive, rearward if negative.

Formula:
        (NSPD) (StdP) / (Pwt)-NSPD = NewD

Example:
Pwt = the pilot of the craft weighs 170 pound
NewD = for this pilot the seat should be moved the following distance

(12" x 180# / 170#) –12" = 0.7 ". In this example since the 0.7" is a positive distance we would move the seat forward that amount.

19.3.8 Aircraft and Systems Description:
This section was established to provide a pilot an opportunity to review and learn about this helicopter. In clear concise language all the systems and subsystems are explained at the technical level pilots can grasp. Larger more complex helicopters of course will add to the level of difficultly in the detailing of the descriptions but not beyond the capability of the pilots who are authorized to fly them.

19.3.8.1 Airsport 254 Aircraft and Systems Description: these descriptions have been covered in great detail in section 3.2 so I'll just give you the reference numbers in an effort to spare you the redundancy.

a) Main Frame: aircraft grade aluminum construction, see section 3.2.1.

b) Main Rotor: see section 3.2.2.

c) Flight Controls: all mechanical linkage setup in accordance with military specification functionality and operation, see section 3.2.3.

d) Swashplate: see section 3.2.4.

e) Tail Rotor: see section 3.2.5.

f) Engine: 650 cc Kawasaki, see section 3.2.6.

g) Transmission: hypoid gear set, see section 3.2.9.

19.3.9 Handling, Service, and Maintenance:
An aircraft in general is a very critical piece of machinery due to its requirement to transport humans in flight. The FAA regulates its airworthiness at its initial registration continuing with follow-up service, maintenance and any mandatory upgrades as presented in their Airworthiness Directives. All the above is in addition to those maintenance and inspection procedures recommended by the manufacturer. The manufacturer will also include direction for the pilot/operator as to how to insure any work performed conforms to accepted procedure and verification.

Certified pilots are also informed in this section what preventative maintenance procedures they can perform. Additionally stated here are instructions by the manufacturer as to proper tie down procedures, hangering precautions and storage guidelines.

19.3.9.1 Airsport 254 Handling, Service, and Maintenance:
Handling: Needless to say the most important advice in this regard is "Never Let Anyone Touch Your Helicopter, PERIOD"! I can remember being at an air show and I started to get too close to this fellow's helicopter tail rotor. I was sternly told to move back. Now that I have my helicopter I fully understand the logic. A helicopter is a critical piece of machinery whose perfect balance of drive train, shafts, rotor blades, etc. must be maintained for optimal performance. A misplaced force to any part of a helicopter could easily cause concern.

In addition this helicopter is a showpiece of highly polished aluminum. Fingerprints are very time consuming to buff out. To keep spectators from causing you a lot of extra work it is wise to keep them a safe distance away via roped off barriers. That way no one leaves marks on the structure.

Whenever I move the helicopter or install the rotor blades I work with soft white cotton gloves. To date I haven't relied on help in this work but if I ever do the first thing I would insist on is the person's use of the white gloves. Typically, any motive force to maneuver the helicopter to the trailer loading ramps is done at the tail support tubes. The spot where the tail tubes are braced underneath back to the engine base is very rigid and can easily support the manhandling. Once the helicopter is aligned with the trailer loading ramps a manual nylon strap ratchet winch can be attached to the center of the engine support and the helicopter winched on board. Tie down straps at the four skid corners is all that is required to secure the helicopter to the trailer.

Service: I haven't operated the helicopter long enough to determine an adequate service schedule but after every five hours of flight testing I tear down the main rotor and check for bearing wear. The other inspections are visual and cover the entire belt and chain drives. As for the engine itself so far its power appears to be constant so all I verify is compression after each flight.

Maintenance: I guess service and maintenance pretty much go hand in hand. Whenever I teardown the main rotor I automatically repack all the bearings with marine high-pressure grease. I also top off the transmission with 90-weight high-pressure gear oil after every flight. Included in the post flight maintenance is checking belt tension and viewing the frame structure to see if the belts are contacting the aluminum and leaving any trace amounts of black rubber marks at the contact points. Initially, excessive belt play was marking the tail tubes but that condition was corrected by the addition of several belt guides. Next is to check chain drive wear by measuring chain stretch. If any stretch is detected the take-up idler is adjusted to bring the chain tension back to specification. Last but not least is a comprehensive inspection of all bolted connections, cable termination points, and control system linkage continuity.

19.3.10 Supplements:

Any optional equipment installed on the aircraft but not covered in the "Aircraft and Systems Description" section would be included here. To date anything I add to make this helicopter more functional is automatically added to the basic system and its description included there. It should be noted that instrument weight by FAA ruling must be included in an ultralights total empty weight so it behooves you to not get too carried away with adding too many instruments.

19.3.11 Safety and Operational Tips:

Main safety tip is to shut down immediately if any unrecognizable vibration is felt. So far I have had forced shutdowns due to a failed set-screw on a tail drive sprocket, a fractured bolt on the tail rotor ground support tube and a failed engine mounting bolt, all of which could be felt seat-of-pants. The most damage occurring to date was from the bolt fracture on the tail rotor ground support tube that caused both tail rotor blades to be damaged. This could have been prevented if the shut down occurred quicker. In fact I am seriously considering adding a rear view mirror so I can see more of what is going on aft of the cockpit.

19.3.12 Index:

None written to date.

Chapter 20- Legal Paperwork

20.1 *Registration*

EAA Ultralight Information:

The EAA maintains an ultralight vehicle and pilot registration program for your benefit. Although the federal regulations that pertain to ultralights do not require you to register yourself or your vehicle, insurance companies often require it. If you are a member of EAA you can register yourself and your ultralight vehicle gratis. If not a member the cost is $15 for each registration card or $30 for you and your ultralight vehicle.

It should be noted that the FAA has allowed ultralight rules to be simpler and less restrictive based upon the assumption the industry would self regulate, which is the function EAA does. If you want to know what is required from the FAA if you exceed the 254 pound limit you can check out the details of registering an "Experimental Aircraft" on the FAA website. You will find that the paperwork is extensive.

Chapter 21- Costs

21.1 *Drafting software*

Unquestionably AutoCAD and SolidWorks 3D drafting packages are the ones to use as a professional engineer. In addition to being able to model in 3 dimensional space they provide engineering data such as "moments of inertia", part volumes and in the case of SolidWorks even a part stress analysis. Plan on spending in excess of $6000 for a single use license. Fortunately you can obtain basic 3-D CAD functionality at a fraction of the cost using either TurboCAD Deluxe ($130) or DesignCAD 3D Max ($100). I currently have TurboCAD Deluxe, which has some exceptional presentation graphics. DesignCAD 3D Max gives you the capability to do animations.

21.2 *Manufacturing Equipment*

Band Saw: WT 4-1/2 motorized horizontal / vertical..........$319
Lathe: WT 7" x 10" precision mini table lathe.................$569
Milling Machine: WT tabletop vertical mill $1375
Drill Press: WT floor model drill press..........................$399
Bench Grinder with stand...$98
Hydraulic Press: WT 12-ton.....................................$189
Welder: HF 180 amp Mig /Flux wire feed......................$350
Rotary Table: WT...$129
Mill Setup Package WT package #2 (Bridgeport clamps)......$168
Arbor Press: WT 1-ton $65
Hand Tools:
   1/2-inch Electric Drill: HF variable speed heavy duty........$40
   3/8-inch Electric Drill: HF variable speed reversible drill...$25
   Crescent 170 pc. Mechanic's tool set (Metric & SAE) $119

Metric & SAE Tap & Die Set: $129
Die Grinder: HF...$35
Hand Rivet Tool: WT...$6
Brake Rivet Tool:...$30
Hand Files: Nicholson round, half round and flat.............$50
Snap Ring Tool: $24
Milling Vise WT precision machining........................$119
Vise: HF 6-inch Swivel vise with anvil........................$80
C-Clamps: 4, 3-inch; 4, 4-inch; 4, 6-inch;....................$60
Air Compressor: WT 6-gallon....................................$119
Electronic Scale: WT $169
Drill Bit Sharpener: WT drill doctor DD750X................$149
Calipers: WT 0-12-inch...$49
Shop vacuum..$150
Consumables:
End mill set: HF..$80
Drill set: HF 115 pc cobalt drill set........................$130
Drill set: WT 1/2 shank size to 1-1/2..........................$107
Grinding discs: HF 10 pc. 4-inch cut off wheels for metal...$10
Welding wire: HF 0.030-inch E71T-GS, 2#....................$25
Bandsaw blades: HF 64-1/2 inch bimetal supercut............$28
Safety Items:
Welding helmet: HF auto-darkening............................$70
Gloves: HF split leather 5 pr.....................................$10
Glasses: HF impact resistant......................................$4
Respirator: HF dual cartridge – medium..........................$25
Welding apron: HF split leather.................................$15

21.3 The Helicopter

Photo 21.3-1

Estimated material cost to build this helicopter using an air-cooled engine versus the water-cooled Kawasaki jet-ski engine I used.

- Swashplate_____$100
- Main Rotor Assembly____$1500
- Controls_____$50
- Tail Rotor Assembly_____$250
- Engine (w/3:1gearbox)__$4800
- Main Frame_____$500
- Drive Train_____$500
- Electrical (w/instruments)_$600
- Miscellaneous_____$800

Estimated Material Grand Total (2013 dollars) = $9000
Labor hours - over 1000 with majority being machining time.

21.4 Insurance

AVEMCO Insurance, one of the worlds leading aviation insurers, has been insuring ultralights for over a decade. As EAA's approved insurer, they will insure most ultralights and ultralight pilots that participate in EAA's registration program.

21.5 Airport

Photo 21.5-1

Finding an airport to fly your ultralight from is not as easy as it sounds. The local airports in the Charlotte area do not allow non-FAA registered aircraft to use them, period! When I reached a dead end at the last small airport I visited I decided to have a coke at a picnic table that was adjacent to the tower. There I managed to join in a conversation with some old timers and they told me of a private grass strip still in use that might allow me to do some flight-testing. I got lucky. The manager there was an FAA certified mechanic who was willing to accept the fact I had an unregistered plane and an ultralight at that. I was able to rent some airstrip space a safe distance from the landing strip itself and there I started debugging the helicopter.

21.6 Flight testing

I mentioned previously that the person overseeing the grass strip was an FAA certified mechanic. He was an aviation purest and had the attitude if it flew he was supportive. Soon after I rented the spot I transported the helicopter to the field to show him. Much to my surprise he took about an hour out of his schedule to fully inspect my craft. Out of the whole inspection he had only two criticisms, both dealing with the way I safety wired the bolts retaining the main and tail rotor blades. His help gave me some reassurance I had done a good job in the details.

Another source of help would have been thru the EAA organization. Later in this project I went to a local EAA chapter meeting and there I met several highly experienced members who volunteered help in either the construction or the testing. One fellow even offered me hanger space to work in. The members that can help you the most are referred to as EAA Technical Advisors and EAA Flight Advisors and more about their programs can be found online.

Obviously, at this flight testing stage of the project I knew everything there was to know about the details of my helicopter but I was only vaguely familiar with the tower / pilot interface. This is where the outside help becomes important. At one particular EAA meeting the topic was communication radios and GPS devices. Apparently the state-of-the-art equipment available now can give you in-flight locations of every plane flying in your immediate vicinity. This type of information provided by end users is great to know since the purchase of this communications gear is very expensive.

As for flight-testing itself, in my case it was very unnerving. Here I had close to ten years invested in this helicopter, between the engineering, equipment, part fabrication and assembly and it all could be destroyed in a heartbeat with bad pilot input. I know I mentioned some of this before but I'd just like to re-iterate some of the highlights. To minimize risk I replaced the 5 gallon fuel tank with a 1.5 gallon one and tethered the craft at all four corners with 1 foot long straps tied to 100 pound weights. The first round of testing involved me getting the engine cooling under control. Adding a larger radiator, 1100 cfm fan and a complete radiator shroud eventually solved this.

Next was yaw control. I didn't have it. This was corrected by changing the tail drive train speed ratio to increase tail rotor rpm. This fix worked perfectly.

Currently I am experiencing control stick sensitivity. This means the main rotor response to a small joystick input is too drastic, causing the entire fuselage to either pitch uncontrollably forward or rearward depending on joystick direction. This I believe is an inherent problem with small helicopters based upon the information obtained on the "Robinson Helicopter Company R22 loss of main rotor control accidents". I am engineering a fix for this at the present time.

21.7 Training:

I have yet to do any of the following but probably will once the helicopter's hovering capabilities are completely easy to perform and totally predictable. First up would be to take one of those helicopter training courses that prepare you for getting your helicopter rating. Next schedule a helicopter flight with an instructor preferably in a small helicopter like a Robinson R22.

Here comes the long-range possibility: The EAA maintains an FAA exemption for flight instruction using two-place trainers. Currently, as of this writing, the EAA administers its training programs in cooperation with the Aero Sports Connection (ASC). The three types of EAA instructors are Basic Flight Instructors (BFI), Advanced Flight Instructors (AFI) and Certified Flight Instructors (CFI). All three can provide you with instruction to be an ultralight pilot. Problem is there are no two place helicopters like mine!

In view of EAA's FAA exemption I could redesign the Airsport 254 to be a two-place ultralight helicopter trainer. The restrictions for this trainer would be it could not have a maximum empty weight of more than 496 pounds, have a fuel capacity of more than 10 U.S. gallons, would not be capable of more than 75 knots calibrated airspeed at full power in level flight, and have a power-off stall speed which does not exceed 35 knots calibrated airspeed. There are other restrictions but they are similar to those of ultralights in general. Once this trainer aircraft was built you could turn it over to Aero Sports Connection (ASC) to enable a formal training program to be developed for the craft. When that was established you could probably start marketing your ultralight helicopter. How's that for a long-range plan!

One last point about training deals with the insurance implications here's the question and answers to that one.

Q: Is there any type of additional training that would lower my premiums?
A: Avemco's Safety Rewards is a two-part program, knowledge and flight training, which reward pilots for participating in specialized training to improve their skills as pilots. Participation can save up to 10% on your annual insurance premium. To be eligible for the ground credit, complete any FAA Safety Team (FAASTeam) WINGS Knowledge course and receive a 5% annual premium credit. Take an Avemco-recognized flight-training course, including WINGS-approved training and become eligible for another 5% annual premium credit. Or complete both to obtain an upgraded certificate or Instrument Rating and you become eligible for the full 10% savings. Completion of any of these approved programs qualifies you to receive the savings from Avemco immediately upon notification.

Should you decide to skip the ultralight path to helicopter flying, part of what will be expected of you during your formal flight training program is covered under Far part 61 section c, "For a helicopter rating: Except as provided in paragraph i (section i gives credit for flight simulator training), a person who applies for a private pilot certificate with rotorcraft category and helicopter class rating must log at least 40 hours of flight time that includes at least 20 hours of flight training from an authorized instructor and 10 hours of solo flight training in the areas of operation listed in FAR 61.107 (b)(3) (this section is included below), and the training must include at least-3 hours of cross-country flight training in a helicopter

Except as provided in FAR 61.110 (this section addresses the State of Alaska's specific night flying requirements) 3 hours of night flight training in a helicopter that includes: One cross-country flight over 50 nautical mile total distance; and 10 takeoffs and 10 landings to a full stop (with each landing involving a flight in the traffic pattern) at an airport.

3 hours of flight training in preparation for the practical test in a helicopter, which must have been performed within 60 days preceding the date of the test; and

10 hours of solo flight time in a helicopter, consisting of at least- 3 hours cross-country time, - one solo cross-country flight of at least 75 nautical miles total distance, with landings at a minimum of three points, and one segment of the flight being a straight-line distance of at least 25 nautical mines between the takeoff and landing location; and -Three takeoffs and three landings to a full stop (with each landing involving a flight in the traffic pattern) at an airport with an operating control tower.

Far 61.107 a) General. A person who applies for a private pilot certificate must receive and log ground and flight training from an authorized instructor on the areas of operation of this section that apply to the aircraft category and class rating sought.

b) Areas of operation:

For a rotorcraft category rating with a helicopter class rating:
Preflight preparation;
Preflight procedures;
Airport and heliport operations;
Hovering maneuvers;
Takeoffs, landings, and go-arounds;
Performance maneuvers:
Navigation;
Emergency operations;
Night operations, except as provided in 61.110 and
Post flight procedures

21.8 Operational

First up, the helicopter has to be registered and a tail number assigned. Next you need a two-way radio in order to communicate with the tower. This being an ultralight aircraft with only 5 gallons of fuel on board an a maximum speed of 50 knots you probably aren't flying too far from your takeoff point so knowing what's going on at that airport is a good idea or else you might wind up landing on someone's lawn.

21.9 Storage

I designed the aircraft so the rotor blades are removable and can be stored on holding brackets located on the tail. In this configuration the helicopter will easily store in a standard bay of a two-car garage.

If you intend to fly often then rent some covered space. Fortunately, the grass strip where I flight-tested had a row of covered bays and I was able to rent one. At first I thought it was a good idea just to protect the craft from any rain. As it turned out the protection it provided from the July and August sun while adjusting blades or doing maintenance was more than worth the extra cost of rent. The greatest benefit of renting the bay was not having to go through the elaborate procedures of installing and aligning the main rotor blades every time you wanted to test fly.

I also had the opportunity to rent enclosed hanger space. This is the way to go if you can afford it.

Chapter 22- Transportation

22.1 Trucking versus Trailering:

The least expensive way to go is to trailer it. I bought a used trailer, added some structure to protect the tail then mounted a huge "always on" tail light. Lowering the nose of the trailer with a floor jack and then connecting two 5-foot long loading ramps accomplished loading and unloading the helicopter. The helicopter is just pushed either on or off. Wheels mounted on the helicopter skids make moving this 254-pound craft manageable without requiring the need of a winch.

Now for the down side. This helicopter represents a huge investment in time and energy and as of this writing is not insured. In this day of cell phones and texting, automobile drivers are definitely scary. You are absolutely taking a big risk traveling anywhere. Point in fact I was driving home from a morning of jet skiing with this very same trailer at 2 in the afternoon on a deserted 4 lane highway (interstate 485) when I saw in my rearview mirror a GMC Yukon quickly approaching me in my right-hand lane. I was traveling at 55 mph and he obviously at 70. There was absolutely nothing I could do but lean on the horn up until he crashed into the back of the trailer which sent the Jeep careening barely under control. The guy in his early forties admitted to falling asleep and was truly contrite. Since that incident whenever I transport the helicopter to the field, which is about 15 miles away, my wife follows closely behind me.

The safest way to transport the helicopter to the field is by car retriever but unfortunately this is not cheap!

Photo 22.1-1

Photo 22.1-2

Chapter 23 The End

I'm really not sure why I stuck with this project so long especially since its still not finished to my satisfaction. Having reached this point in the project my wife insisted I should pass all this acquired information to others having the desire to build a helicopter but who often wind up squandering their money on useless Internet helicopter plans. Hopefully I have provided some insight into the real complexity of such a project and more detailed information for those desiring to know about ultralight helicopters beyond just the available FAA manuals. It isn't a project for everyone but it has definitely kept me occupied and intrigued for years!

APPENDIX

C1 Drill Chart

Drill #	Dia. in.	Drill #	Dia. in.	Drill #	Dia. in.	Drill #	Dia. in.	Drill #	Dia. in
1	0.2280	17	0.1730	33	0.1130	49	0.0730	65	0.0350
2	0.2210	18	0.1695	34	0.1110	50	0.0700	66	0.0330
3	0.2130	19	0.1660	35	0.1100	51	0.0670	67	0.0320
4	0.2090	20	0.1610	36	0.1065	52	0.0635	68	0.0310
5	0.2055	21	0.1590	37	0.1040	53	0.0595	69	0.02925
6	0.2040	22	0.1570	38	0.1015	54	0.0550	70	0.0280
7	0.2010	23	0.1540	39	0.0995	55	0.0520	71	0.0260
8	0.1990	24	0.1520	40	0.0980	56	0.0465	72	0.0250
9	0.1960	25	0.1495	41	0.0960	57	0.0430	73	0.0240
10	0.1935	26	0.1470	42	0.0935	58	0.0420	74	0.0225
11	0.1910	27	0.1440	43	0.0890	59	0.0410	75	0.0210
12	0.1890	28	0.1405	44	0.0860	60	0.0400	76	0.0200
13	0.1850	29	0.1360	45	0.0820	61	0.0390	77	0.0180
14	0.1820	30	0.1285	46	0.0810	62	0.0380	78	0.0160
15	0.1800	31	0.1200	47	0.0785	63	0.0370	79	0.0145
16	0.1770	32	0.1160	48	0.0760	64	0.0360	80	0.0135

C2 Material Gage Chart

C2 Material Gage Chart

Gage No.	Steel Sheets		Aluminum Sheets		Gage No.	Steel Sheets		Aluminum Sheets	
	Wt. #/sq.ft	Tk. (in)	Wt. #/sq.ft	Tk. (in)		Wt. #/sq.ft	Tk. (in)	Wt. #/sq.ft	Tk. (in)
3	10.000	0.2391	3.235	0.2294	21	1.375	0.0329	0.402	0.0285
4	9.375	0.2242	2.881	0.2043	22	1.250	0.0299	0.357	0.0253
5	8.750	0.2092	2.565	0.1819	23	1.125	0.0269	0.319	0.0226
6	8.125	0.1943	2.284	0.1620	24	1.000	0.0239	0.283	0.0201
7	7.500	0.1793	2.035	0.1443	25	0.875	0.0209	0.252	0.0179
8	6.875	0.1644	1.812	0.1285	26	0.750	0.0179	0.224	0.0159
9	6.250	0.1495	1.613	0.1144	27	0.688	0.0164	0.200	0.0142
10	5.625	0.1345	1.437	0.1019	28	0.625	0.0149	0.178	0.0126
11	5.000	0.1196	1.279	0.0907	29	0.563	0.0135	0.159	0.0113
12	4.375	0.1046	1.139	0.0808	30	0.500	0.0120	0.141	0.0100
13	3.750	0.0897	1.015	0.0720	31	0.438	0.0105	0.126	0.0089
14	3.125	0.0747	0.904	0.0641	32	0.406	0.0097	0.112	0.0080
15	2.813	0.0673	0.805	0.0571	33	0.395	0.0090	0.100	0.0071
16	2.500	0.0598	0.716	0.0508	34	0.344	0.0082	0.089	0.0063
17	2.250	0.0538	0.639	0.0453	35	0.313	0.0075	0.079	0.0056
18	2.000	0.0478	0.568	0.0403	36	0.281	0.0067	0.071	0.0050
19	1.750	0.0418	0.506	0.0359	37	0.266	0.0064	0.063	0.0045
20	1.500	0.0359	0.451	0.0320	38	0.250	0.0060	0.056	0.0040

C3 Aluminum Sheet Standard Types

ALUMINUM SHEET STANDARD TYPES							
Alloy & Temper	Surface Finish	Thickness (in.)	Wt #/sq.ft.	Alloy & Temper	Surface Finish	Thickness (in.)	Wt #/sq.ft.
2024T3	Alclad	0.016	0.230	6061T6	Bare	0.016	0.228
		0.020	0.288			0.020	0.285
		0.025	0.360			0.025	0.353
		0.032	0.461			0.032	0.452
		0.040	0.576			0.040	0.570
		0.050	0.720			0.050	0.713
		0.063	0.907			0.063	0.889
		0.071	1.02			0.090	1.27
		0.080	1.15			0.125	1.76
		0.090	1.30			0.190	2.68
		0.125	1.80	6061-0	Bare	0.020	0.282
		0.190	2.74			0.040	0.564
		0.250	3.60			0.063	0.889
2024-0	Alclad	0.032	0.461			0.080	1.13
		0.040	0.576	7075T6	Alclad	0.025	0.364
6061T4	Bare	0.025	0.353			0.032	0.466
		0.032	0.452			0.040	0.582
		0.040	0.564			0.050	0.727
		0.050	0.706			0.063	0.916
		0.063	0.889			0.125	1.82
		0.080	1.13	7075-0	Alclad	0.025	0.364
		0.125	1.76			0.032	0.466
						0.040	0.582
						0.063	0.916

C4 Steel Sheet Thickness Chart

4130 Steel Sheet @ 0.28395 #/ cu. in.							
Tk. (in.)	Weight (#/sq. ft)	Tk. (in.)	Weight (#/sq. ft)	Tk. (in.)	Weight (#/sq. ft)	Tk. (in.)	Weight (#/sq. ft)
0.025	1.022	0.050	2.044	0.080	3.271	0.125	5.111
0.032	1.308	0.063	2.576	0.090	3.680	0.190	7.769
0.040	1.636	0.071	2.903	0.100	4.089	0.250	10.222

C5 4130 Tubing Sizes

4130 ALLOY STEEL ROUND SEAMLESS TUBING								
OD	Wall Tk.	Wt Per Ft.	OD	Wall Tk.	Wt Per Ft.	OD	Wall Tk.	Wt Per Ft.
3/16	0.028	0.0478	3/4	0.028	0.2159	1-3/4	0.049	0.8902
				0.035	0.2673		0.058	1.048
				0.049	0.3668		0.065	1.170
				0.058	0.4287		0.095	1.679
				0.065	0.4755		0.120	2.089
				0.095	0.6646		0.188	3.136
				0.120	0.8074			
				0.156	0.9897			
				0.188	1.1280			
1/4	0.028	0.0672	7/8	0.028	0.2533	1-7/8	0.250	2.249
	0.035	0.0804		0.035	0.3140			
	0.049	0.1052		0.049	0.4323			
	0.058	0.1189		0.058	0.5061			
	0.065	0.1284		0.065	0.5623			
				0.083	0.7021			
				0.095	0.7914			
				0.120	0.9676			
				0.188	1.3790			
5/16	0.028	0.0852	1	0.028	0.2907	2	0.049	1.021
	0.035	0.1039		0.035	0.3607		0.065	1.343
	0.049	0.1382		0.049	0.4977		0.095	1.933
	0.058	0.1580		0.058	0.5835		0.120	2.409
	0.065	0.1722		0.065	0.6491			
	0.095	0.2212		0.083	0.9182			
				0.095	1.1280			
				0.120	1.4060			
				0.188	1.6300			
3/8	0.028	0.1038	1-1/8	0.035	0.4074	2-1/8	0.065	1.430
	0.035	0.1271		0.049	0.5631			
	0.049	0.1706		0.058	0.6609			
	0.058	0.1964		0.065	0.7359			
	0.065	0.2152		0.095	1.0450			
	0.083	0.2588		0.120	1.2880			
	0.095	0.2841						

4130 ALLOY STEEL ROUND SEAMLESS TUBING

OD	Wall Tk.	Wt Per Ft.	OD	Wall Tk.	Wt Per Ft.	OD	Wall Tk.	Wt Per Ft.
7/16	0.035	0.1506	1-1/4	0.035	0.4542	2-1/4	0.120	2.730
	0.049	0.2036		0.049	0.6285			
	0.065	0.2589		0.058	0.7384			
	0.095	0.3480		0.065	0.8226			
	0.120	0.4075		0.083	1.1320			
				0.095	1.1720			
				0.120	1.4480			
				0.156	1.8230			
1/2	0.028	0.1411	1-3/8	0.035	0.5009	2-1/2	0.095	2.440
	0.035	0.1738		0.049	0.6939		0.120	3.050
	0.049	0.2360		0.058	0.8158		0.250	6.008
	0.058	0.2738		0.065	0.9094			
	0.065	0.3020		0.095	1.2990			
	0.083	0.3696		0.120	1.6090			
	0.095	0.4109		0.188	2.3830			
	0.120	0.4870						
9/16	0.035	0.1974	1-1/2	0.035	0.5476	2-3/4	0.120	3.370
	0.049	0.2690		0.049	0.7593			
	0.060	0.3457		0.058	0.8932			
	0.120	0.5677		0.065	0.9962			
				0.083	1.2560			
				0.095	1.4260			
				0.120	1.7690			
				0.188	2.6340			
5/8	0.028	0.1785	1-5/8	0.058	0.9707			
	0.035	0.2205		0.065	1.0830			
	0.049	0.3014		0.083	1.100			
	0.058	0.3512		0.095	1.552			
	0.065	0.3888		0.120	1.929			
	0.095	0.5377		0.156	2.447			
	0.120	0.6472		0.188	2.885			
	0.156	0.7814						

Psychrometric Chart:

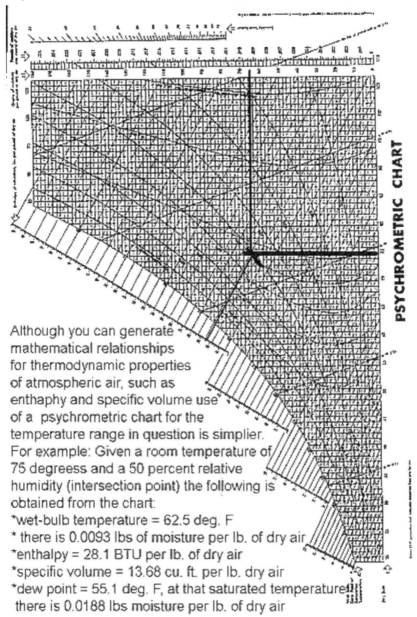

Although you can generate mathematical relationships for thermodynamic properties of atmospheric air, such as enthaphy and specific volume use of a psychrometric chart for the temperature range in question is simplier. For example: Given a room temperature of 75 degreess and a 50 percent relative humidity (intersection point) the following is obtained from the chart:
*wet-bulb temperature = 62.5 deg. F
* there is 0.0093 lbs of moisture per lb. of dry air
*enthalpy = 28.1 BTU per lb. of dry air
*specific volume = 13.68 cu. ft. per lb. dry air
*dew point = 55.1 deg. F, at that saturated temperature there is 0.0188 lbs moisture per lb. of dry air

Typically I am comfortable sitting in a room with a temperature set at 75 °F having a 50% relative humidity, looking at the chart at that intersection point of those two curves (75 °F, 50% rh) there is approximately 0.009 pounds moisture per pound dry air. It would appear if I am comfortable in an environment where the moisture per pound of dry air is 0.009 then looking at the chart again this moisture to dry air value is also satisfied when the temperature is at 90 °F with 30% relative humidity or at 103 °F with a 20% relative humidity. Yes this is an over simplification of the definition of "comfort" and if you want to get into the "Cylindrical Model of Thermal Interaction of the Human Body and the Environment" which delves into the formulas for "steady-state energy balance and human thermal comfort", check out the ASHRAE Fundamentals Handbook where you can obtain data as significant as garment insulating values of clothing down to such items as a bra (0.01 ft ^{2} x h x °F/ Btu) or panties (0.03 ft^2 x h x °F/ Btu). The point is when people from Arizona living in 100F weather say the temperature is fine because it's a "dry heat" at first you may think they are crazy but based upon the chart 103 °F and 20% relative humidity seems like it might work.

Federal Aviation Advisory Circular

U.S. Department of Transportation; Federal Aviation Administration
Date: 1/30/84; Initiated by: AFD-820; AC No: AC 103-7
1. PURPOSE. This advisory circular provides guidance to the operators of ultralights in the United States. It discusses the elements that make up the definition of ultralight vehicles for the purposes of operating under Federal Aviation Regulation (FAR) Part 103. It also discusses when an ultralight must be operated as an aircraft under the regulations applicable to certificated aircraft.
2. BACKGROUND.
    a. The number of ultralight vehicles and participants in the various aspects of this sport has increased dramatically in recent years. All indications are that this growth will continue. The presence of these vehicles in the national airspace has become a factor to be considered in assuring the safety of all users of the airspace.
    b. On October 4, 1982, a new regulation (Part 103) applicable to the operation of ultralight vehicles became effective. 1b is a regulation defining those vehicles which may be operated as "ultralight vehicles" and provides operating rules which parallel those applicable to certificated aircraft. The Federal Aviation Regulations regarding aircraft certification, pilot certification, and aircraft registration are not applicable to ultralight vehicles or their operators.
    c. Ultralight vehicle operations may only be conducted as sport or recreational activity. The operators of these vehicles are responsible for assessing the risks involved and assuring their own personal safety. The rules in Part 103 are intended to assure the safety of those not involved in the sport, including persons and property on the surface and other users of the airspace. The ultralight community is encouraged to adopt good operating practices and programs in order to avoid more extensive regulation by the Federal Aviation Administration (FAA).
3. DEFINITIONS. For the purpose of this advisory circular, the following definitions apply:
    a. Ultralight Vehicle. This term refers to ultralights meeting the

applicability for operations under Part 103.

b. Recognized Technical Standards Committee. This term refers to a group of at least three persons technically qualified to determine whether a given ultralight meets the requirements for operations under Part 103, as follows:

(1) It is recognized by a national pilot representative organization,

(2) It is comprised of persons not directly associated with the manufacture and/or sale of the make of ultralight being inspected, and

(3) It conducts its review and documents the findings in accordance with the guidance provided in this circular.

4. WHAT DOES THIS MEAN FOR THE PERSON WHO WANTS TO FLY UNDER PART 103?

a. You are Responsible for Your Personal Safety. Certificated aircraft are designed, flight tested, manufactured, maintained, and operated under Federal regulations intended to provide an aircraft of consistent performance, controllability, structural integrity, and maintenance. An ultralight vehicle is not subject to Federal aircraft certification and maintenance standards. This means that the costs of purchasing and maintaining an ultralight vehicle may be considerably less than the purchase of a certificated aircraft. There is no assurance that a particular ultralight vehicle will have consistent performance, controllability, structural integrity, or maintenance. Your safety, and potentially that of others, depends on your adherence to good operation and maintenance practices. This includes proper preflight techniques, operation of the vehicle within the manufacturer's recommended flight envelope, operation only in safe weather conditions, and providing safety devices in anticipation of emergencies. Part 103 is based on the assumption that any individual who elects to fly an ultralight vehicle has "assessed the dangers involved and assumes personal responsibility for his/her safety.

b. You are Limited to Single-Occupant Operations. Part 103 is based on the single-occupant concept meaning operation by an individual who has assumed all responsibility for his/her personal safety. Pilots of ultralight vehicles subject to Part 103 are not required to have training or previous experience prior to the operation of these vehicles. You should consider receiving adequate training prior to participation.

c. You are Limited to Recreation and Sport Purposes. Operations for any other purpose are not authorized under the applicability of Part 103.

d. You are Limited as Necessary for the Safety of Other Persons and Property. Part 103 consists of operating rules that were determined necessary for the safety of other users of the airspace and persons on the surface. These rules were developed in consideration of the capabilities of the vehicles and their pilots. It is your responsibility to know, understand and comply with these rules. Ignorance of the regulations pertaining to the activities you pursue is not an acceptable excuse for violating those regulations.

e. You are Responsible for the Future Direction the Federal Government Takes With Respect to Ultralight Vehicles. The actions of the ultralight community will affect the direction Government takes in future regulations. The safety record of ultralight vehicles will be the foremost factor in determining the need for further regulations.

5. FAA CONTACT POINTS. The FAA will provide clarification of particular subject areas, information, and assistance pertaining to the operations of ultralight vehicles through the following contacts:

a. Flight Standards Field Offices. Flight Standards District Offices (FSDOs), General Aviation District Offices (GADOs), and Manufacturing and Inspection District Offices (MUDs) are the FAA field offices where information and assistance are available regarding the operation of ultralight vehicles, acceptable methods of complying with Part 103 requirements, and compliance with other regulations should it become necessary to operate an ultralight as a certificated aircraft.

b. Air Traffic Control Facilities. FAA Air Traffic Control facilities are located throughout the United States and maintain jurisdiction over the use of the controlled airspace in their particular area. To obtain authorization to operate from or into the airspace designated in § 103.17, contact must be made with the controlling facilities.

c. Flight Service Stations. These facilities provide operational information to pilots, such as weather briefings, advisory information regarding the status of facilities, etc., and are the most accessible of the FAA points of contact. They can provide additional information regarding how to reach the other points of contacts mentioned here.

d. Airports District Offices. These offices inspect airports certificated under Part 139 of the FARs to determine whether an airport is safe for public use. Persons wanting to establish new airports or flight parks, or operate ultralight vehicles from federally funded airports, may contact these offices for assistance.

6. –9. Reserved

SECTION 1. WHAT IS AN ULTRALIGHT VEHICLE?

10. SCOPE AND CONTENTS. This section discusses the elements contained in § 103.1 which make up the definition of an "ultralight vehicle" and the proper way to assure that Part 103 applies.

11. APPLICABILITY OF PART 103.

a. Probably the single most critical determination which must be made is whether or not your vehicle and the operations you have planned are permitted under Part 103. The fact that you are operating a vehicle which is called or advertised as a "powered ultralight, hang glider, or hang balloon", is not an assurance that it can be operated as an ultralight vehicle under Part 103. There are a number of elements contained in 1 3.1 which make up the definition of the "ultralight vehicle. Ii If you fail to meet any one of the elements, you may not operate "under Part 103. Any operations conducted without meeting all of the elements are subject to all aircraft certification, pilot certification, equipment requirements, and aircraft operating rules applicable to the particular operation.   .

b. The FAA realizes that it is possible to design an ultralight which, on paper meets the requirements of § 103.1, but in reality does not. However, the designers, manufacturers of the kits, and builders are not responsible to the FAA for meeting those requirements. Operators of ultralights should bear in mind that they are responsible for meeting § 103.1 during each flight. The FAA will hold the operator of a given flight responsible if it is later determined that the ultralight did not meet the applicability for operations under Part 103. Be wary of any designs that are advertised as meeting the requirements for use as an ultralight vehicle, yet provide for performance or other design innovations which are not in concert with any element of § 103.1. The FAA may inspect any ultralight which appears, by design or performance, to not comply with § 103.1.

c. If the FAA Determines Your Ultralight Was Not Eligible for Operation as an Ultralight Vehicle. If your ultralight does not meet 103.1, it must be operated in accordance with applicable aircraft regulations. You will be subject to enforcement action ($1000 civil penalty for each violation) for each operation of that aircraft.

12. ELEMENTS MAKING UP THE DEFINITION OF AN ULTRALIGHT VEHICLE.

a. Single Occupancy. An ultralight cannot be operated under Part 103 if there is more than one occupant or if it has provisions for more than one occupant.

b. Sport or Recreational Purposes Only. An ultralight cannot be operated under Part 103 if it is operated for purposes other than sport or recreation or if it is equipped for other uses.

c. No Airworthiness Certificate. An ultralight cannot be operated under Part 103 if it has been issued a current U. S. or foreign airworthiness certificate.

d. Unpowered Vehicles. An unpowered ultralight cannot be operated under Part 103 if it weighs 155 pounds or more. Balloons and gliders are unpowered vehicles.

e. Powered Vehicles. A powered ultralight cannot be operated under Part 103 when it has an empty weight of 254 pounds or more; has a fuel capacity exceeding 5 U.S. gallons; is capable of more than 55 knots airspeed at full power in level flight; and has a power-off stall speed which exceeds 24 knots.

13 SINGLE OCCUPANT.

a. The Rationale for Allowing Single-Occupant Operations Only. One aspect of the rationale for allowing ultralight vehicles to operate under special rules that do not require pilot and aircraft certification is the single-occupant limitation. The assumption is made that a person who elects to operate an uncertificated vehicle alone is aware of the risks involved. This assumption does not necessarily told true for a passenger. Because the pilot qualifications for ultralight vehicle operations are not Federally controlled or monitored, the single-occupant requirement is a necessary component to the continuation of the policies and regulations which allow the operation of Ultralight vehicles free from many of the restrictions imposed on the operation of certificated aircraft.

b. Guidelines Regarding Seating Arrangements Which Should be Considered when Purchasing or Operating an Ultralight Vehicle.

(1) Any provisions for more than one occupant automatically disqualify an ultralight for operations under Part 103.

(2) Some powered ultralights were originally manufactured with bench or "love" seats with only one seatbelt, but have been advertised as two-place in the ultralight periodicals. They are not eligible for operations under Part 103. While no maximum width standards for the size of a "single" seat have been established at this time, most manufacturers are providing seats that have a width of 18 to 22 inches. Any seat notably wider than 22 inches raises a question as to whether the ultralight is intended for single occupancy.

(3) An ultralight with provisions for More than one occupant can only be operated as a certificated aircraft, even when occupied by only one person. In addition to the previously stated aircraft certification and registration requirements, the pilot must hold a medical certificate and at least a student pilot certificate with the proper endorsements for solo operations. At least one occupant during two-occupant operations must hold at least a private pilot certificate.

c. Two-place Ultralight Operations under Part 103. The AOPA Air Safety Foundation, Experimental Aircraft Association, and the United States Hang Gliding Association have been granted exemptions from the applicable aircraft regulations to authorize use of two-place Ultralights under Part 103 for limited training purposes and for certain hang glider operations. Except as authorized by exemption, no person may operate an ultralight under Part 103 with more than one occupant.

14. RECREATION AND SPORT PURPOSES ONLY (103.1b)

a. The Rationale for Only Allowing Recreation and Sport Operations Under Part 103. In combination with the single-occupant requirement, the limitation to recreation and sport operations only is the basis for allowing ultralight vehicle operations under minimum regulations. The reason for allowing the operation of these vehicles without requiring aircraft and pilot certification is that this activity is a "sport" generally conducted away from concentrations of population and aircraft operations.

b. Determining Whether a Particular Operation is for Recreation and Sport Purposes. There are several considerations that are necessary in determining whether a given operation is conducted for recreation or sport purposes:

(1) Is the flight undertaken to accomplish some task, such as patrolling a fence line or advertising a product? If so, Part 103 is not applicable.

(2) Is the Ultralight equipped with attachments or modifications for the accomplishment of some task, such as banner towing or agricultural spraying? If so, Part 103 does not apply.

(3) Is the pilot advertising his/her services to perform any task using an ultralight? If so, Part 103 does not apply.

(4) Is the pilot receiving any form of compensation for the performance of a task using an ultralight vehicle? If so, Part 103 does not apply.

c. Examples of Operations which are Clearly Not for Sport or Recreational Purposes.

(1) Aerial Advertising. Part 103 does not apply to operations that include the towing of banners and the use of loudspeakers, programmed light chains,

smoke writing, dropping leaflets, and advertising on wings; nor does it apply to the use of interchangeable parts with different business advertisements or flying specific patterns to achieve maximum public visibility.

(2) Aerial Application. Part 103 does not apply to operations that include using an ultralight to perform aerial application of any substance intended for plant nourishment, soil treatment, propagation of plant life or pest control. An ultralight with an experimental certificate as an amateur aircraft could be used to perform this function under specific, limited circumstances. Paragraph 35b provides more detail on this subject.

(3) Aerial Surveying and Patrolling. Patrolling powerlines, waterways, highways, suburbs, etc., does not come under Part 103. The conduct of these activities in an ultralight must be in compliance with applicable aircraft regulations as outlined in paragraph 34. Local, state, or Federal government entities may operate an Ultralight as a "public aircraft." This is discussed in greater detail in paragraph 35a

(4) Carrying parcels for hire.

d, Examples of Situations Involving Money or Some Other Form of Compensation Allowable Under the Recreation and Sport Limitation.

(1) Rental of Ultralight Vehicles. Renting an ultralight vehicle to another person is permissible.

(2) Receiving a Purse or Prize. Persons participating in sport or competitive events involving the use of ultralights are not prohibited from receiving money or some other form of compensation in recognition of their performance.

(3) Authoring Books About Ultralights. Persons are not prohibited from flying ultralights and then authoring books about their experiences, for which they ultimately receive compensation.

(4) Receiving Discount on Purchase of an Ultralight. There is no prohibition which would prevent you from taking advantage of any discount on the price of an ultralight a company might offer where its logo or name appears on a portion of the vehicle. You cannot, however, enter into any agreement which might specify the location; number, or pattern of flights contingent on the receipt of that discount. Any operation under such an agreement could not be conducted under Part 103.

(5) Participation in Air shows and Events. You may participate in air shows and other special events where persons are charged for viewing those events, so long as you receive no compensation for your participation. This does not hold true where you stand to benefit directly from the proceeds as the organizer or producer of the event.

15. AIRWORTHINESS CERTIFICATE (103.1c).

a. If your Ultralight has been issued an airworthiness certificate you cannot operate it as an ultralight vehicle under Part 103. An ultralight cannot be operated interchangeably as a certificated aircraft and an ultralight vehicle.

b. If you want to operate your ultralight under Part 103, you must turn in, to the issuing authority, any airworthiness certificates currently issued for the craft.

c. You may operate an ultralight as a certificated aircraft if you obtain the proper certification. If you do not already hold an airworthiness certificate, you should consult paragraph 31 for further guidance.

d. An Ultralight is eligible for operation under Part 103, even where the same make and model is also being issued airworthiness certificates, so long as a all elements of the definition of an ultralight vehicle contained in § 103.1 are satisfied. As an example, assume that there is a model which would meet the definition of an ultralight vehicle being manufactured in Canada and is issued a Canadian airworthiness certificate. If you purchased one, you would have to turn in the airworthiness certificate to the Canadian authorities before operating it in the United States under Part 103.

16. UNPOWERED ULTRALIGHT VEHICLES.

a. Unpowered Ultralight Vehicles Eligible for Operation Under Part 103. All forms of gliders and free balloons weighing less than 155 pounds and meeting ill other requirements of § 103.1 are eligible for operation under Part 103.

b. Unpowered Ultralights eligible for operations under Part 103 are not required to be operated under that Part. In some cases, you can obtain certification of your glider or free balloon as an experimental aircraft.

c. Computing the Empty Weight of an Unpowered Ultralight Vehicle.

(1) Gliders. The fuselage, wings, structure, control surfaces, harnesses, and landing gear, etc., are included in this determination. Parachutes and all personal operating equipment and harnesses associated with their use are not included.

(2) Free Balloons. The envelope, lines, harnesses, gondola, burner, and fuel tank are included in this determination. Parachutes and all personal operating equipment and harnesses associated with their use are not included. The weight of the fuel, in the case of a hot-air balloon, or any logical amount of removable ballast, when intended for control of the buoyancy of a gas balloon, is not included in the weight specified in § 103.1 (d) .

d. Free Balloons are Considered "Unpowered." A balloon, for Part 103 eligibility, is considered an unpowered ultralight, regardless of whether it drops ballast to ascend or uses heated air. The burner on a hot-air balloon is used to raise the temperature of the air in the envelope allowing the balloon to rise. This can be compared to the glider's use of lifting air as a means of ascending. In both cases, no method of horizontal propulsion is employed and a loss of the lifting force will cause the vehicle to descend to the surface.

17. POWERED ULTRALIGHT VEHICLES.

a. "Powered" Ultralights Eligible For Operation Under Part 103. All ultralights with a means of horizontal propulsion which also meet the provisions of § 103.1 are eligible; this includes ultralight airships, helicopters, gyrocopters, and airplanes.

b. A Powered ultralight eligible for operation under Part 103 is not required to be operated under that Part. You may elect to certificate and operate it as an experimental aircraft. The applicable procedures and regulations are explained in Advisory Circular 20-27C, Certification and Operation of Amateur-Built Aircraft.

18. POWERED VEHICLE WEIGHT.
    a. Items Excluded From the Computation of the Empty weight of a Powered Ultralight Vehicle.
    (1) Safety Devices Which are Intended for Deployment in a Potentially Catastrophic Situation. Parachutes and some associated additional equipment necessary for their operation meet this criteria. Other devices, such as seatbelts, roll cages, instruments, or wheel brakes, are considered part of the airframe and are included in the empty weight.
    (i) Up to 24 pounds of weight associated with the parachute system may be excluded by the FAA without requiring a separate weighing of the system components.
    (ii) No weight allowance will be given for any component of the parachute system if, when it was operated, the parachute was not carried and attached to the ultralight at the reinforced points/fittings provided.
    (2) Floats Used For Landings On Water. Only the weight of the floats and any integral, external attachment points are excluded. All other items associated with attachment of the floats to the airframe are included in the vehicle's empty weight. Up to 30 pounds per float may be excluded by the FAA without requiring substantiation of the float's actual weight. This exclusion was allowed under the rationale that float-equipped ultralights would not usually be operated in the vicinity of airports and large concentrations of people and, thus, would be even less of a safety hazard than those which had conventional landing gear. While amphibious capability would appear to negate somewhat that rationale, some allowance for the "float" capability is made.
    (i) Amphibious Floats. Up to 30 pounds per float may be excluded by the FAA. The weight of all attached items associated with the installation and operation of the landing gear is included in the calculation of the dry, empty weight specified in § 103.1 (e) (1). Satisfactory evidence of the weight of those components must be available.
    (ii) Amphibious Fuselage. Where the fuselage is intended to function as a float during water landings, up to 30 pounds (the average weight of a single float) is allowed by the FAA to be excluded from the empty weight where the ultralight is capable of repeated water takeoffs and landings. (Operators may be required to demonstrate the water operational capability of their vehicle in order to receive an allowance for the added weight.) Up to 10 pounds per outrigger float and pylon is also allowed by the FAA.
    (iii) "Float" provisions not discussed here should be reviewed with FAA personnel at a Flight Standards field office.
    b. Acceptable Methods for Determining the Weight of an Ultralight. The completely assembled ultralight should be taken to a draftless location and placed on:
    (1) A Single Scale. A determination may be made on a calibrated scale which has sufficient weighing surface to accommodate the ultralight resting fully on that surface without any stabilizing assistance, or
    (2) Two or More Scales. A determination may be made on two or more calibrated scales if they are located at all points where the ultralight contacts

the surface when parked and it is resting fully on those scales without any stabilizing assistance. In this case, the sum of the scales will be used.

19. MAXIMUM FUEL CAPACITY OF A POWERED ULTRALIGHT VEHICLE. The maximum fuel capacity for a powered ultralight vehicle is 5 U.S. gallons. Any powered ultralight with fuel tank(s) exceeding this capacity is ineligible for operation as an ultralight vehicle

a. Determination of Fuel Capacity. The total volume, including all available space for usable and unusable fuel in the fuel tank or tanks on the vehicle is the total fuel capacity. The fuel in the lines, pump, strainer, and carburetor is not considered in a calculation of total volume.

b. Use of an Artificial Means to Control Capacity.

(1) Tanks which have a permanent standpipe or venting arrangement to control capacity are permitted, but may be subject to demonstration of the capacity if there is any reason to doubt that the arrangement is effective.

(2) A temporary, detachable, or voluntarily observed method for restricting fuel capacity, such as a "fill-to" line, is not acceptable.

20. MAXIMUM LEVEL FLIGHT SPEED OF A POWERED ULTRALI GHT VEHICLE. The maximum speed-of an ultralight vehicle at full power in level flight cannot exceed 55 knots.

a. The 55 knots specified in § 103. 1 (e) (3) is a performance limitation, not a speed limit. It is not a speed limit that a pilot has to observe. The vehicle, as configured (exposed drag areas, engine power output, and propeller efficiency), cannot be capable of driving through the air in level flight at full power faster than 55 knots. It is also not a structural never-exceed speed (Vne). The vehicle may well be structurally capable of higher airspeeds.

b. The use of "voluntarily observed" or arbitrarily specified maximum airspeeds, such as a red line on the airspeed indicator, is not acceptable-where the ultralight is capable of more than 55 knots in level flight.

c. Acceptable Methods of Determining the Maximum Level Flight Airspeed of an Ultralight.

(1) A calculation, using the information in Appendix 1, is an acceptable method for making this determination.

Note: The engine manufacturer's maximum horsepower rating will be used for all computations associated with maximum level flight speeds (unless the operator can provide documentation from the engine manufacturer that a method of derating an engine will result in a predictable reduction in horsepower).

(2) A series of three or more full-power level runs in both directions along a 1,000-foot course under specified conditions could be used by a recognized technical standards committee to make this determination. The average speed derived should be adjusted for atmospheric conditions other than sea level on a standard day.

NOTE: "While these guidelines contain provisions allowing flight testing to establish eligibility for operations under Part 103, the FAA has provided charts in Appendixes 1 and 2 which encompass most normal aircraft design factors without requiring flight testing. Any flight testing to establish eligibility for operations under Part 103 is done at the risk of the participants. (3) A calibrated radar gun may also be used. Again, a series of full-power level runs as described in subparagraph c(2) could be used by a recognized technical standards committee to make this determination.

d. Use of an Artificial Means to Limit the Maximum Level Flight Airspeed.

(1) An artificial means of restricting the total power output of an engine in order to lower the maximum level flight speed at full power would be acceptable if the method used to restrict the power available is one that cannot be modified, bypassed, or overridden in flight and the pilot or operator can provide the FAA, on request, satisfactory evidence that the vehicle meets the requirement of § 103.l(e)(3).

NOTE: Vehicles which require artificial restrictions to power or propeller arrangements may incur a substantial penalty in terms of takeoff, climb, and absolute performance. This factor should be considered when assessing the safety of ultralight vehicle operations, especially at high altitude locations.

(2) As a general guideline, a method is unacceptable if it can be modified, bypassed, or overridden in any way while sitting in the pilot seat so as to further increase the power. There may be some ultralights which could be operated as ultralight vehicles if such restrictions are employed to meet the requirements of § 103.l (e) (3). If you change or modify the restricting elements, your vehicle may be ineligible or use under Part 103.

(3) The use of voluntarily-observed restrictions, such as a lower power setting, instead of using all available power, is unacceptable.

e. Use of a Less Efficient Propeller/Shaft Arrangement. The use of a less efficient propeller shaft arrangement to lower the maximum level flight speed at full power is acceptable, if the operator or pilot can provide the FAA, on request, satisfactory evidence that the vehicle meets the requirements of § 103.l (e)(3). If you change or modify that arrangement to increase the efficiency, your vehicle may be ineligible for use under Part 103.

f. Use of an Aerodynamic Restriction. The use of an aerodynamic restriction, such as a limiting device to pitch control travel on a canard arrangement, automatically deployed speed brakes, or a strut installed for drag purposes only, is acceptable, provided a recognized technical standards committee has evaluated the resulting maximum full-power level flight speeds at a pilot weight of 170 pounds and determined that the vehicle is not capable of maintaining level flight above 55 knots. (Again, modification of that arrangement may render the vehicle ineligible for use under Part 103.)

NOTE: Vehicles using aerodynamic restrictions to limit maximum speed may have undesirable flight characteristics when operated near the controllability limits.

21. MAXIMUM POWER-OFF STALL SPEED OF A POWERED ULTRALIGHT VEHICLE. The maximum power-off stall speed of an ultralight vehicle cannot exceed 24 knots (28 mph).

a. Acceptable Methods of Determining the Power-Off Stall Speed of an Ultralight Vehicle.

(1) A calculation, using the information provided in Appendix 2, is an acceptable method of providing satisfactory evidence that your vehicle meets this requirement.

NOTE: For the purpose of all stall speed calculations, the pilot's weight will be considered to be 170 pounds and the fuel tank(s) filled (6 lbs./gal.).

(2) This speed can also be determined by a recognized technical standards committee which can take the average speed from a series of power-off stalls using existing flight test procedures.

b. Use of High-Lift Devices to Lower Stall Speed to 24 Knots. Slots, slats, flaps, and any other devices which would lower the stall speed are acceptable. A determination of the resulting average stall speed by a technical standards committee is acceptable evidence of compliance. .

22. DOCUMENTATION OF A TECHNICAL STANDARDS COMMITTEE'S FINDINGS. If an ultralight is found by a recognized technical standards committee to meet the requirements of § 103.1 with respect to the items specified in paragraphs 18 through 21, the committee should issue a document confirming its findings.

23. CONTENTS OF THE DOCUMENT. To be acceptable, the document will contain, as a minimum, the:

a. Name and address of the person requesting the determination.

b. Type/model and general description of the ultralight, including any installed equipment.

c. Empty weight of the ultralight, showing the allowances given for parachutes, floats, and fuel, and how it was determined.

d. Fuel capacity and how it was determined.

e. Maximum speed at full power in level flight and how it was determined, including a description of any method incorporated to limit the power or thrust output or the ability of the vehicle to fly in level flight at more than 55 knots. (This description should allow an inspector reviewing the document to determine that the limiting devices are still operational.)

f. Maximum power-off stall speed and how it was determined, including a description of any lift devices used.

g. Typed or printed names of the committee members, their signatures, and the name of the organization which recognizes their committee.

24. CONTACTS WITH FAA INSPECTORS. Most ultralight operators will probably only encounter FAA field inspectors during accident, incident, or public complaint investigation. On initial contact, the inspector will usually ask for your pilot certificate and the aircraft airworthiness certificate. You should inform the inspector that you are operating your ultralight under Part 103 and provide evidence that it meets the applicability of § 103.1.

a. Failure to Provide Satisfactory Evidence. If you cannot provide this evidence, or if the evidence provided is not satisfactory, your ultralight will be considered an aircraft subject to all applicable aircraft regulations and you will be subject to all requirements applicable to the operator. It is your responsibility to prove that your ultralight and any operations you may have conducted meet the applicability for operation under Part 103. Until you do, the FAA will proceed with any enforcement investigation resulting from your inability to provide that proof.

b. "Satisfactory Evidence."

(1) The use of the graphs provided in Appendixes 1 and 2 will be acceptable for determination of the maximum level flight speed and power-off stall speed if your ultralight has no special limitations to maximum speed or power and no special high-lift devices.

(2) An FAA-certificated aircraft mechanic or repair station may also / weigh your ultralight and provide a weight document similar to that provided for aircraft, listing the components and attachments of the ultralight when weighed. An FAA-certificated mechanic may also make the determinations in paragraphs 18 through 21 and issue the determinations outlined in paragraph 23, provided that the maximum speeds were determined through the use of the graphs provided in Appendixes 1 and 2 (not included in this book).

(3) A recognized technical standards committee's findings documented as provided in paragraph 23 will usually be considered acceptable. A committee may issue their findings in relation to a given model of ultralight which are then included by the manufacturer in the sale of the ultralight. The subsequent operators of that model of ultralight may use those findings without having another inspection made, provided that there are no changes or modifications to the configuration, components, engine, or propeller arrangements of the basic model originally reviewed by the committee and any artificial means of restricting maximum airspeed is installed and operational.

c. FAA Ultralight Inspection Authority. The FAA has the legal authority to inspect any ultralight, whether it is operated as an aircraft under Part 91 or as an ultralight vehicle under Part 103. In the case of an ultralight operated under Part 103, this authority will usually be exercised only when an inspector has reason to doubt the validity of the evidence provided by the operator or that the Ultralight still conforms to the findings contained in that evidence.

(1) Refusal to Allow the Inspection. Refusal to allow the inspector to inspect the ultralight would be a violation of the Federal Aviation Act of 1958, as amended, and the applicable FAR, and would result in enforcement action.

(2) Usual Content of the Inspection. The inspector may ask you to show compliance with § 103.1 by measuring the capacity of the fuel tank, weighing the vehicle, measuring the wing, stabilizing and control surface areas, and showing that any artificial means required to restrict the maximum airspeed are installed, operational, and cannot be bypassed. Further checks may be made in situations where the inspector has reason to doubt the effectiveness of any restriction to maximum airspeed.

25, -29, RESERVED.

SECTION 2. HOW TO CERTIFICATE AND OPERATE AN ULTRALIGHT AS AN AIRCRAFT

30. SCOPE AND CONTENTS. This section outlines the regulations which

are applicable to the operation of ultralights as certificated aircraft and provides general information regarding how to comply with the regulations.

31. AIRCRAFT CERTIFICATION. A person who chooses to operate an ultralight as a certificated aircraft has two options for airworthiness certification of the vehicle, depending primarily on the configuration of the vehicle or kit when purchased, as follows;

a. Completely Assembled at the Factory, or Assembled by the Purchaser From a "Bolt-together" Kit With Little or No Fabrication Operations. An ultralight in this category would be eligible for airworthiness certification only for the purpose of exhibition in the experimental classification. Application for an experimental certificate for exhibition may be made to the nearest Flight Standards field office.

b. Major Portion (Over 50%) Fabricated by the Builder/Purchaser, Either from Raw Materials to the Builder's Own Design or From a Partially Prefabricated Kit. A vehicle shown to meet the provisions of this category would be eligible for airworthiness certification as an amateur-built aircraft, in addition to eligibility for experimental exhibition. Detailed information pertaining to amateur-built aircraft requirements are in FAA Advisory Circular 20-27C, Certification and Operation of Amateur-Built Aircraft. Applications for such certification may be made to the nearest Flight Standards field office.

32. REGISTRATION. An ultralight that is to be certificated and operated as an aircraft is subject to the registration and marking requirements applicable to aircraft. The applicant should contact the nearest Flight Standards field office to obtain the required forms and information concerning the procedures to be followed. Advisory Circular 20-27C also contains information concerning registration and marking requirements as they apply to amateur-built aircraft.

33. PART 61 (CERTIFICATION: PILOTS AND FLIGHT INSTRUCTORS). Part 61 of the Federal Aviation Regulations contains the regulations which define the certificates and ratings which pilots must hold to function as a pilot of a certificated aircraft in the United States. It also outlines the minimum experience levels and standards to qualify for those certificates and ratings. The minimum levels of pilot currency for certain operations are also contained in Part 61.

34. PART 91 (GENERAL OPERATING AND FLIGHT RULES). Part 91 contains the general operating rules (Subpart A), flight rules (Subpart B), and maintenance rules (Subpart C) which are applicable to all certificated aircraft operations. Pilots of certificated Ultralight aircraft must comply with Part 91. No certificated aircraft can the operated under Part 103. The flight rules of Subpart B are the minimum standards for flight operations except where the operating limitations of the particular aircraft establish more stringent standards. The majority of the rules contained in Subpart A and Subpart C will not apply to operations of certificated ultralight aircraft; however, a thorough review of these regulations should be conducted to determine those applicable to a particular type of ultralight aircraft.

35. SPECIAL FLIGHT OPERATIONS. There are some special operations

of ultralight aircraft that are allowed under present regulations.

a. "Public" Aircraft. An ultralight may be used exclusively in the service of a Federal, state, or local government without an airworthiness certificate. (The pilots do not have to hold pilot certificates.)

(1) The ultralight must be properly registered with the FAA and display appropriate registration markings, and

(2) All operations must be conducted in accordance with the applicable operating and flight rules of Part 91.

b. Aerial Agricultural Application. A farmer owning an amateur-built experimentally certificated aircraft may use that aircraft for aerial agricultural applications over his/her own property, provided that,

(1) The ultralight is certificated as an amateur-built aircraft and does not have any operating limitations prohibiting agricultural operations;

(2) The pilot holds at least a private pilot certificate and successfully completes a knowledge and skill test as specified in § 137.19 (e); and

(3) The farmer holds at least a Private Agricultural Operator Certificate under Part 137 and all operations are conducted in accordance with that regulation.

36. –50. RESERVED.

Kenneth S. Hunt Director of Flight Operations

The Military Specification on Helicopter Control

MIL-H-8501A, 7 SEPTEMBER 1961;

SUPERSEDING MIL H-8501, 5 NOVEMBER 1952

MILITARY SPECIFICATION

HELICOPTER FLYING AND GROUND HANDLING QUALITIES; GENERAL REQUIREMENTS FOR

This specification has been approved by the Department of Defense and is mandatory for use by the Departments of the Army, the Navy and the Air Force

1. SCOPE

1.1 This specification covers the design requirements for flying and ground handling qualities of U.S. military helicopters.

2. APPLICABLE DOCUMENTS There are no applicable documents.

3. REQUIREMENTS

3.1 General.

3.1.1 With the exception of 3.6, section 3 contains the requirements (or the flying qualities, und for certain relevant ground-handling characteristics, of all helicopters procured by the Department of the Army, the Department of the Navy, and the Department of the Air Force, that are required to operate under visual flight conditions. Paragraph 3.6 applies to helicopters required to operate under instrument flight conditions. The required characteristics are those which are considered, on the basis of present knowledge, as tending to insure satisfactory handling qualities and are subject to modification as indicated by new information. Every effort shall be made by designers to provide additional desirable characteristics which have been omitted as specific requirements.

3.1.2 Unless otherwise specified, the requirements of section 3 shall apply at all normal service loadings over the operating rotor speed range and all operational altitudes and temperatures. For the purposes of section 3, normal service loadings shall include all combinations of gross weight and center of gravity location that could ordinarily be encountered in normal service operations.

3.2 Longitudinal characteristics.

3.2.1 It shall be possible to obtain steady, smooth flight over a speed range from at least 30 knots rearward to maximum forward speed as limited either by power available or by roughness due to blade aerodynamic limitations, but not by control power. This speed range shall be construed to include hovering and any other steady state flight condition, including steady climbs and steady descents. Throughout the specified speed range a sufficient margin of control power, and at least adequate control to produce 10 percent of the maximum attainable pitching moment in hovering shall be available at each end to control the effects of longitudinal disturbances. This requirement shall apply not only to powered flight, but also to autorotative flight at forward speeds between zero and the maximum forward speed for autorotation. Within the limits of speed specified in 3.2.1 and during the transitions between hovering and the specified extremes, the controls and the helicopter itself shall be free from objectionable shake, vibration, or roughness, as specified in 3.7.1

3.2.2 The helicopter shall be reasonably steady while hovering in still air (winds up to 3 knots), requiring a minimum movement of the cyclic controls to keep the machine over a given spot on the ground, for all terrain clearances up to the disappearance of ground effect. In any case, it shall he possible to accomplish this with less +/- 1.0 inch movement of the cyclic controls.

3.2.3 For all conditions and speeds specified in 3.2.3, it shall be possible in steady-state flight to trim steady, longitudinal control forces to zero. At these trim conditions, the controls shall exhibit positive self-centering characteristics. Stick "jump" when trim is actuated is undesirable.

Table 1. Power and speed conditions

Initial trim and power conditions ~ speed range of interest
Hovering~0 to 30 knots ~ 15 to 60 knots
Level flight at 35 knots ~ 60% Vmax-Vmax
Level flight at Vmax ~ 80% Vmax-Vlimit
Climb at best rate of climb ~ Vmax R/C +/- 15 knots
Partial power decent at 300 to 500 fpm ~ 15 to 60 knots
Autorotation with trim as in "Level flight at 80% Vmax" above ~ 60% Vmax- Vmax for autorotation
Autorotation at speed for minimum rate of descent ~ 15 knots (trim speed +20 knots)

Table 2. Limit control force values (pounds) (when measured in flight with adjustable friction off)

Control	~ Limit control force	~ Limit control force for breakout, including friction force
Longitudinal cyclic	~ 8.0 # ~	minimum 0.5 #, maximum 1.5#
Lateral cyclic	~ 7.0 # ~	minimum 0.5 #, maximum 1.5#
Collective	~ 7.0 # ~	minimum* 1.0 #, maximum 3.0#
Directional	~ 15.0 # ~	minimum* 3.0 #, maximum 7.0#

* May be measured with adjustable friction set.

3.2.4 At all trim conditions and speeds specified in 3.2.1, the longitudinal force gradient for the first inch of travel from trim shall be no less than 0.5 pound per inch and no more than 2.0 pounds per inch. In addition, however, the force produced for a 1-inch travel from trim by the gradient chosen shall not be less than the breakout force (including friction) exhibited in flight. There shall be no undesirable discontinuities in the force gradient, and the slope of the curve of stick force versus displacement shall be positive at all times with the slope for the first inch of travel from trim greater than or equal to the slope for the remaining stick travel.

3.2.5 With the helicopter trimmed in steady, level, horizontal flight at maximum forward speed, it shall be possible readily and safely to bring the machine to a quick stop and hover. With the helicopter trimmed in hovering flight, it shall be possible to accelerate rapidly to maximum forward speed, maintaining approximately constant altitude.

3.2.6 Without retrimming, the longitudinal control forces required to change from any trim and power condition to any other trim and power condition as specified in table I, or for performance of the maneuvers discussed in 3.2.5 and 3.5.4 or any other normal helicopter maneuvers, shall not exceed the values given in table II.

3.2.7 With the control trimmed for zero force, the breakout forces, including friction in the longitudinal control system, shall conform with the values given in table II when measured in flight.

3.2.8 The controls shall be free from objectionable transient forces in any direction following rapid longitudinal stick deflections. During and following rapid longitudinal displacement of the control stick from trim, the force acting in a direction to resist the displacement shall not at any time fall to zero. Longitudinal control displacement shall not produce lateral control forces in excess of 20 percent or pedal forces in excess of 75 percent of the associated longitudinal force. For helicopters employing power-boosted or power-operated controls, there shall be no lateral or directional control force developed.

3.2.9 There shall be no objectionable or excessive delay in the development of angular velocity in response to control displacement. The angular acceleration shall be in the proper direction within 0.2 second after longitudinal control displacement. This requirement shall apply for the speed range specified in 3.2.1.

3.2.10 The helicopter shall, at all forward speeds and at all trim and power conditions specified in table I, except as noted below, possess positive, static longitudinal control force, and control position stability with respect to speed. This stability shall be apparent in that at constant throttle and collective pitch control settings a rearward displacement of and pull force on the longitudinal-control stick shall be required to hold a decreased value of steady, forward speed, and a forward displacement and push force be required to hold an increased value of speed. In the speed range between 15 and 50 knots forward, and 10 to 30 knots rearward, the same characteristics are desired, but a moderate degree of instability may be permitted. However, the magnitude of the change in the unstable direction shall not exceed 0.5 inch for stick position or 1.0 pound for stick force.

3.2.10.1 The stability requirements of 3.2.10 are intended to cover all steady flight conditions in which the helicopter might be operated for more than a short time interval. As a guide for the conditions to be investigated, the tabulation of pertinent conditions in table I may be utilized, all referred to the most critical center of gravity position.

3.2.10.2 The helicopter shall not exhibit excessive longitudinal trim changes with variations of rate of climb or descent at constant airspeed. Specifically, when starting from trim, at any combination of power and airspeed within the flight envelope, it shall be possible to maintain longitudinal trim with a longitudinal control displacement of no more than 3 inches from the initial trim position as the engine power or collective pitch, or both, are varied throughout the available range. Generally, the airspeeds needing the most specific investigation of the above characteristics include Vmax and the speeds between zero and one-half the speed for minimum power.

3.2.11 The helicopter shall exhibit satisfactory dynamic stability characteristics following longitudinal disturbances in forward Fight. Specifically, the stability characteristics shall be unacceptable if the following are not met for a single disturbance in smooth air:

(a) Any oscillation having a period of less than 5 seconds shall damp to one - half amplitude in not more than 2 cycles, and there shall be no tendency for undamped small amplitude oscillations to persist.

(b) Any oscillation having a period greater than 5 seconds but less than 10 seconds shall be at least lightly damped.

(c) Any oscillation having a period greater than 10 seconds but less then 20 seconds shall not achieve double amplitude in less than 10 seconds.

3.2.11.1 The following is intended to insure acceptable maneuver stability characteristics. The normal acceleration stipulations are intended to cover all speeds above that for minimum power required; the angular velocity stipulations shall apply at all forward speeds, including hovering.

(a) After the longitudinal control stick is suddenly displaced rearward from trim a sufficient distance to generate a 0.2 radian/sec. pitching rate within 2 seconds, or a sufficient distance to develop a normal acceleration of 1.5 g within 3 seconds, or 1 inch, whichever is less, and then held fixed, the time-history of normal acceleration shall become concave downward within 2 seconds following the start of the maneuver, and remain concave downward until the attainment of maximum acceleration. Preferably, the time-history of normal acceleration shall be concave downward throughout the period between the start of the maneuver and the attainment of maximum acceleration. Figure 1 (a) is illustrative of the normal acceleration response considered acceptable.

(b) During this maneuver, the time-history of angular velocity shall become concave downward within 2.0 seconds following the start of the maneuver, and remain concave downward until the attainment of maximum angular velocity; with the exception that for this purpose, a faired curve may be drawn through any oscillations in angular velocity not in themselves objectionable to the pilot. Preferably, the time-history of angular velocity should be distinctly concave downward throughout the period between 0.2 second after the start of the maneuver and the attainment of maximum angular velocity. Figure 1 (b) is illustrative of the velocity response considered acceptable.

Figure 1. Typical normal acceleration and pitch rate response. (In this case the control input was limited by normal acceleration)

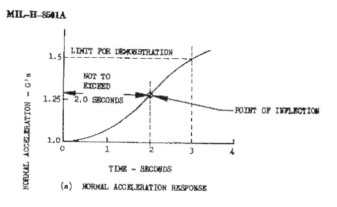

(a) NORMAL ACCELERATION RESPONSE

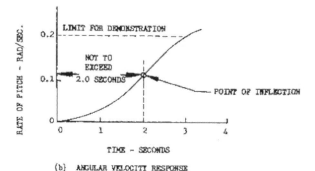

(b) ANGULAR VELOCITY RESPONSE

3.2.11.2 To insure that a. pilot has reasonable time for corrective action following moderate deviations from trim attitude (as for example, owing to a gust), the effect of an artificial disturbance shall be determined. When the longitudinal control stick is suddenly displaced rearward from the trim, the distance determined in 3.2.11.1 above, and held for at least 0.5 second, and then returned to and held at the initial trim position, the normal acceleration shall not increase by more than 0.25 g within 10 seconds from the start of the disturbance, except 0.25 g may be exceeded during the period of control application. Further, during the subsequent nose down motion (with the controls still fixed at trim) any acceleration drop below the trim value shall not exceed 0.25g within 10 seconds after passing through the initial trim value.

3.2.12 The response of the helicopter to motion or the longitudinal control shall be such that in the maneuver described in 3.2.11.1, the resulting normal acceleration always increases with time until the maximum acceleration is approached, except that a decrease not perceptible to the pilot may be permitted.

3.2.13 Longitudinal control power shall be such that when the helicopter is hovering in still air at the maximum overload gross weight or at the rated power, a rapid 1.0-inch step displacement from trim of the longitudinal control shall produce an angular displacement at the end of 1.0 second which is at least $45/(3 \times (W+1000)^{**}1/2$ degrees. When maximum available displacement from trim of the longitudinal control is rapidly applied, the angular displacement at the end of 1.0 second shall be at least $180/(3 \times (W+1000)^{**}1/2$ degrees. In both expressions W represents the maximum overload gross weight of the helicopter in pounds.

3.2.14 To insure satisfactory initial response characteristics following a longitudinal control input and to minimize the effects of external disturbances, the helicopter in hovering shall exhibit pitch angular velocity damping (that is, a moment tending to oppose the angular motion and proportional in magnitude to the angular velocity) of at least $8 \, (\mathrm{Iy})^{**}0.7$ ft-lb/rad/sec, where Iy is the moment of inertia about the pitch axis expressed in slug-ft$^{**}2$.

3.3 Directional and lateral characteristics.

3.3.1 Directional control shall be sufficiently powerful, in order that its use in conjunction with the other normal controls will permit easy execution of all normal taxiing maneuvers with wheel gear on land and float gear in water· using' normal rotor speeds. In particular, the following ground handling conditions shall be met:

(a) It shall be possible, without the use of brakes, to maintain a straight path in any direction in a wind of 35 knots.

(b) It shall be possible to make a complete turn in either direction by pivoting on one wheel in a wind of 35 knots.

3.3.2 From the hovering condition, it shall be possible to obtain steady, level, translational flight at a sideways velocity of 35 knots to both the right and the left. At the specified sidewise velocity and during the transition from hovering, the controls and the helicopter itself shall be free from objectionable shake, vibration, or roughness as specified in 3.7.1.

3.3.3 The requirements of 3.2.2 shall be applicable to lateral as well as to longitudinal control motions. It shall be possible to meet this requirement with less than+/- 1-inch movement of the directional control.

3.3.4 In all normal service loading conditions, including those resulting in asymmetrical lateral center of gravity locations and steady flight under the

conditions specified in 3.2.1 (including autorotation) and 3.3.2, a sufficient margin of control effectiveness, and at least adequate control to produce 10 percent of the attainable hovering rolling moment shall remain at each end.
3.3.5 Directional control power shall be such that when the helicopter is hovering in still air at the maximum overload gross weight or at rated takeoff power, a rapid 1.0-inch step displacement from trim of the directional control shall produce a yaw displacement at the end of 1.0 second which is at least $110/(3 \times (W+1000)^{**}1/2)$ degrees. When maximum available displacement from trim of the directional control is rapidly applied at the conditions specified above, the yaw angular displacement at the end of 1.0 second shall be at least $330/(3x (W+1000)^{**}0.5)$ degrees. In both equations W represents the maximum overload gross weight of the helicopter in pounds.
3.3.6 It shall be possible to execute a complete turn in each direction while hovering over a given spot at the maximum overload gross weight or at takeoff power (in and out of ground effect), ill a wind of at least 35 knots. To insure adequate margin of control during these maneuvers, sufficient control shall remain at the most critical azimuth angle relative to the wind, in order that, when starting at zero yawing velocity at this angle, the rapid application of full directional control in the critical direction results in a corresponding yaw displacement of at least $110/(3x(W+1000)^{**}0.5$ degrees in the first second, where W represents the maximum overload gross weight of the helicopter in pounds.
3.3.7 The response of the helicopter to directional-control deflection, as indicated by the maximum rate of yaw per inch of sudden pedal displacement from trim while hovering shall not be so high as to cause a tendency for the pilot to over control unintentionally. In any case, the sensitivity shall be considered excessive if the yaw displacement is greater than 50 degrees in the first second following a sudden pedal displacement of 1 inch from trim while hovering at the lightest normal service loading.
3.3.8 It shall be possible to make coordinated turns in each direction while in autorotation, at all autorotation speeds.
3.3.9 The helicopter shall possess positive, control fixed, directional stability, and effective dihedral in both powered and autorotative flight at all forward speeds above 50 knots, 0.5 Vmax or the speed for maximum rate of climb, whichever is the lowest. At these flight conditions with zero yawing and rolling velocity, the variations of pedal displacement and lateral control displacement with steady sideslip angle shall be stable (left pedal and right stick displacement for right sideslip) up to full pedal displacement in both directions, but not necessarily beyond a sideslip angle of 15 degrees at Vmax, 45 degrees at the low speed determined above, or beyond a sideslip angle determined by a linear variation with speed between these two angles. Between sideslip angles of +/- 15 degrees, the curve of pedal displacement and lateral control displacement plotted against sideslip angle shall be approximately linear. In all flight conditions specified above, a 10% margin of both lateral and longitudinal control effectiveness (as defined in 3.2.1 and 3.3.4) shall remain.

3.3.9.2 During pedal fixed rolling maneuver, there shall be no objectionable adverse yaw.

3.3.10 For all conditions and speed specified in 3.2.1 and 3.3.2, it shall be possible in steady flight to trim steady lateral and directional control forces to zero. At these trim conditions, the controls shall exhibit positive self-centering characteristics. Stick "jump" when trim control is actuated is undesirable

3.3.11 At all trim conditions and speeds specified in 3.3.10, the lateral force gradient for the first inch of travel from trim shall be no less than 0.5 pound per inch and no more than 2.0 pounds per inch. In addition, however, the force produced for a 1-inch travel from trim by gradient chosen shall not be less than the breakout force (including friction) exhibited in flight. The slope of the curve of stick force versus displacement shall be positive at all times and the slope for the first inch of travel from trim shall always be greater than or equal to the slope for the remaining stick travel. The directional control shall have a limit force of 15 pounds at maximum deflection with a linear force gradient from trim position. There shall be no undesirable discontinuities in either the lateral or directional force gradients.

3.3.12 From trimmed initial conditions, the lateral and directional control forces required for the performance of the maneuvers discussed in 3.2.6, 3.3.1, 3.3.2, 3.3.4, 3.3.5, 3.3.6, 3.3.8, and 3.3.9.1, shall conform with the values given in table II.

3.3.13 With the controls trimmed for zero force, the breakout forces including friction in the lateral and directional control systems shall conform with the values given in table II when measured in flight.

3.3.14 The controls shall be free from objectionable transient forces in any direction following rapid lateral stick or pedal deflections. During and following a rapid lateral displacement of the control stick from trim or a rapid pedal displacement from trim, the force acting in a direction to resist the displacement shall not at any time fall to zero. Lateral control displacement shall not produce longitudinal control forces in excess of 40 % or pedal forces in excess of 100 percent of the associated lateral force. Pedal displacement shall not produce longitudinal control forces in excess of 8 percent or lateral control forces in excess of 6 percent of the associated pedal force. For helicopters employing power-boosted or power operated controls, there shall be no longitudinal control forces developed in conjunction with lateral or directional control displacement.

3.3.15 The response of the helicopter to lateral-control deflection, as indicated by the maximum rate of roll per inch of sudden control deflection from the trim setting, shall not be so high as to cause a tendency for the pilot to over control unintentionally. In any case, at all level flight speeds, including hovering the control effectiveness shall be considered excessive if the maximum rate of roll per inch of stick displacement greater than 20 degrees per second.

3.3.16 There shall be no objectionable or excessive delay in the development of angular velocity in response to lateral or directional control displacement. The angular acceleration shall be in the proper direction within 0.2 second after control displacement. This requirement shall apply for all flight conditions specified in3.2.1, including vertical autorotation.

3.3.17 The helicopter shall not exhibit excessive lateral trim changes with changes in power or collective pitch, or both. Specifically, when starting from trim at any combination of power and airspeed within the flight envelope of the helicopter, it shall be possible to maintain lateral trim with a control displacement amounting to no more than 2 inches from the initial trim position as the engine power or collective pitch, or both, are varied either slowly or rapidly in either direction throughout the available range.

3.3.18 Lateral control power shall be such that when the helicopter is hovering in still air at the maximum overload gross weight or at the rated power, a rapid 1-inch step displacement from trim of the lateral control shall produce an angular displacement at the end of one-half second of at least $27/3x(W+1000)**1/3$ degrees. When maximum available displacement from trim of the lateral control is rapidly applied at the conditions specified above, the resulting angular displacement at the end of one-half second shall be at least $81/3x(W+1000)**1/3$ degrees. In both expressions W represents the maximum overload gross weight of the helicopter in pounds.

3.3.19 To insure satisfactory initial response characteristics following either a lateral or directional control input and to minimize the effect of external disturbances, the helicopter, in hovering, shall exhibit roll angular velocity damping (that is, a moment tending to oppose the angular motion and proportional in magnitude to the rolling angular velocity) of at least $8(Ix)**0.7$ ft-lb/rad/sec, where Ix is the moment of inertia about roll axis expressed in slug-ft. The yaw angular velocity damping should preferably be at least $27(Iy)**0.7$ ft-lb/rad/sec., where Iy is the moment of inertia about the yaw axis expressed in slug-sq.ft.

3.4 Vertical characteristics.

3.4.1 It shall be possible to maintain positive control of altitude within +/- foot by use of the collective-pitch control while hovering at constant rotor rpm under conditions of 3,2.2. This shall be accomplished with a minimum amount of collective stick required, and in any case it shall be possible to accomplish this with less than +/- ½ inch movement of the collective stick. When a governor is employed, there shall be no objectionable vertical oscillation resulting from lag in governor response.

3.4.2 The collective-pitch control shall remain fixed at all times unless moved by the pilot and shall not tend to creep, whether or not cycled or directional controls moved. The maximum effort required for the collective control shall not exceed the values specified in table II. The breakout force (including friction) shall be within the acceptable limits as specified in table II.

3.4.3 Movement of the collective-pitch control shall not produce objectionable forces in the cyclic control; in no case shall these forces exceed 1 pound. In helicopters where power operated or power boosted controls are utilized, there shall be no control force coupling.

3.5 Autorotation, rotor characteristics, and miscellaneous requirements.

3.5.1 It shall be possible while on the ground to start and stop the rotor blades in winds up to at least 45 knots, For helicopters with a gross weight of less than 1000 pounds, this requirement shall be at least 35 knots. For all ship-based helicopters, this requirement shall be at least 60 knots while headed into the wind.

3.5.2 It shall be possible without the use of wheel chocks to maintain a fixed position on a level paved surface with takeoff rotor speed while power is being increased to takeoff power in winds specified in 3.5.4.1.

3.5.3 It shall be possible to perform all required maneuvers, including taxiing and pivoting, without damage to the coning stops and without contact between the blades and any part of the structure.

3.5.4 The helicopter shall be capable of making satisfactory landings and takeoffs. Specifically, the following conditions shall be met.

3.5.4.1 It shall be possible to make satisfactory, safe vertical takeoffs and landings in steady winds up to 45 knots and winds with gusts up to 45 knots. This shall apply to all helicopters, except those with a gross weight less than 1,000 pounds, which shall be capable of the foregoing in winds and gusts up to 35 knots.

3.5.4.2 From a level paved surface, it shall be possible to make satisfactory, safe running takeoffs with wheel-type gear, up to ground speeds of at least 35 knots.

3.5.4.3 For both power-on and autorotative conditions, it shall be possible to make satisfactory, safe landings on a level paved surface, with wheel and skid gear, up to ground speeds of at least 35 knots. This shall be construed to cover landings with 3-knot ground speed in any direction and up to a side drift of at least 6 knots when landing with a ground speed of 35 knots.

3.5.4.5 For all helicopters equipped with emergency floatation gear in both power-on and autorotative conditions, it shall be possible to make satisfactory, safe landings, on smooth water up to at least 15 knots surface speed. This shall be construed to cover landings with 3 knot surface speed in any direction and up to a side drift of at least 5 knots when landing with a surface speed of 15 knots.

3.5.5 The helicopter shall be capable of entering into power-off autorotation at all speeds from hover to maximum forward speed. The transition, from powered flight to autorotative flight shall be established smoothly, with adequate controllability and with minimum loss of altitude. It shall be possible to make this transition safely when initiation of the necessary manual collective-pitch control motion has been delayed for at least 2 seconds following loss of power. At no time during this maneuver shall the rotor speed fall below a safe minimum transient autorotative value (as distinct from power-on or steady-state autorotative values). This shall be construed to cover both single and multiengine helicopters.

3.5.5.1 Sudden power reduction, power application, or loss of power with collective control fixed, shall not produce pitch, roll, or yaw attitude changes in excess of 10 degrees in 2 seconds, except that, at speeds below that for best climb, a 20-degree yaw in 2 seconds will be accepted.

3.5.6 The control forces during the transition to autorotative flight under the conditions of 3.5.5 shall never exceed the values specified in table II.

3.5.7 It shall be possible, in still air at sea level, at the end of stabilized autorotative descents, to make repeatedly safe, power-off autorotative landing at speeds of 15 knots or less. Reduction of this autorotative landing speed to zero is highly desirable. This shall be construed to cover both single and multiengine helicopters.

3.5.8 For helicopters equipped with power boosted or power-operated controls, the following conditions should be met:

(a) In trimmed level flight at any speed, out-of-trim conditions resulting from abrupt power-operated control system failure shall be such that:

(1) With controls free for at least 3 seconds, the resulting rates of yaw, roll, and pitch shall not exceed 10 degrees per second, and the change in normal acceleration "shall not exceed +/- 1/2 g.

(2) It. shall be possible to continue level flight with zero sideslip with forces to operate the controls not exceeding 80 pounds for the directional control, 25 pounds for the collective and longitudinal controls, and 15 pounds for the lateral control.

(b) With power-operated control system off, it shall be possible to trim steady longitudinal, lateral, and directional control forces to zero under all the conditions and speeds specified in 3.2.1 and 3.3.2.

(c) With power-operated control system off, the collective-pitch control shall not tend to creep, whether or not cyclic or directional controls are moved.

(d) With the helicopter trimmed in steady level flight at 40 knots under power-operated control system failure conditions, it shall be possible without retrimming to make a normal landing approach and landing with control forces not exceeding the limits given in 3.5.8(a) (2).

(e) Engine failure or electrical system failure, or both, shall not result in primary power-operated control system failure.

(f) Power-operated control system failure shall not result in failure of the trim systems.

(g) For helicopters having two or more completely independent power-operated control systems, the requirements of 3.5.8 (a) shall be met upon failure of one of the complete systems during the period of transfer from one system to another. With the remaining system or systems 3.5.8(b) shall apply and the rates of control motion attainable shall be such that safe operation of the helicopter is in no way compromised, and shall in no case be less than 50 percent of the normal rates. In such operations, including the approach and landing specified in 3.5.8(d), the control forces stated in 3.5.8 (a) (2) shall be considered as absolute maximum, and it is desired that these forces be considerably lower.

3.5.9 Automatic stabilization and control or stability augmentation equipment, or both, may be employed to meet all of the above stated requirements of section 3, provided that suitable separate requirements for system reliability are met, if such equipment is employed, the following conditions shall be met.

(a) With the automatic stabilization and control or stability augmentation equipment or both engaged, and from steady level flight for a period greater than 30 seconds, out-of-trim conditions resulting from abrupt complete disengagement or from abrupt complete failure of the equipment shall be such that with controls free for 3 seconds following the disengagement or failure, the resulting rates of yaw, roll, and pitch shall not exceed 10 degrees per second and the change in normal acceleration shall not exceed +/- ½ g. When engaging the automatic stabilization and control or stability augmentation equipment, there shall be no apparent switching transients.

(b) For helicopters employing completely independent dual automatic stabilization and control or dual stability augmentation equipment, or a completely independent combination of both, the requirements of 3.5.9(a) shall be met upon the failure of one complete system during the period of transfer from one system to another, but need not be met for a simultaneous failure of both.

(c) It shall be possible on the ground, with the, automatic stabilization and control or stability augmentation equipment or both operating; to move the controls to all limits without exceeding the forces of table II. For helicopters with power-operated controls, this requirement shall apply also with rotor stopped.

(d) Helicopters employing automatic: stabilization and control or stability augmentation equipment or both shall possess a sufficient degree of stability and control with all the equipment disengaged to allow continuation of normal level flight, and the maneuvering necessary to permit a safe landing under visual flight conditions.

(e) In cases where automatic stabilization and control or stability augmentation devices, or both, are used to compensate for divergent tendencies of the basic airframe, a considerable margin of control power beyond that needed to overcome airframe instability under simple flight conditions shall be provided. For this purpose, sufficient control margin over the amount required to perform maneuvers and to accomplish stability augmentation shall be provided. Specifically for pitch, roll, and yaw control, the augmentation system in combination with pilot controlled inputs shall not utilize more than 50 percent of the available control moment in the unstable direction from the trim position for straight level flight at a given speed when performing the following maneuvers:

(1) Steady level-flight turn at cruising speed to maximum load force attainable in actual operation, or the design or placard load factor, whichever occurs first.

(2) Steady sideslips in both powered and autorotative flight at the combinations of speed and sideslip angle set forth in 3.3.9.

3.5.10 For all operating conditions, there shall be no dead spots in any of the control systems which permit more than +/- 0.2 inch of the cockpit control without corresponding motion of the rotor blades, control surfaces, etc.

3.5.11 For all operating conditions, longitudinal, lateral, directional, or vertical control motions shall not produce adverse response of the helicopter due to mechanical coupling in the control system.

3.5.11.1 If mechanical intermixing of longitudinal, lateral, directional, or vertical control motions is required to achieve the above requirements of section 3, no adverse limitations in control power shall exist with any possible combination of control inputs throughout the entire range each of the control motions.

3.6 Instrument flight characteristics.

3.6.1 For any helicopter required to operate under instrument flight conditions, the more stringent supplementary flying qualities requirements of 3.6 shall apply, in addition to the foregoing requirements of section 3. It shall be possible, without demanding undue pilot effort, to fly on instruments at all speeds, from hover to design cruise speed; for this purpose automatic stabilization and control or stability augmentation equipment, or both, may be employed, in addition to any required for compliance with visual flight criteria. The failure or disengagement of the equipment that provides the instrument flight characteristics shall not result in a degeneration of the stability and control characteristics of the helicopter below any of those specified in this specification for helicopters required to operate under visual flight conditions.

3.6.1.1. For any helicopter required to operate under instrument or all-weather conditions the following control power and angular velocity damping requirements shall apply in hovering:

Movement ~ Angular displacement at the end of 1 sec. for a rapid 1 inch control displacement-degrees

$\qquad$ ~ Angular velocity damping ft-lbs/rad/sec.

Longitudinal ~ 73 / (W+1000)**1/3 ~ 15 (Iy)**0.7

Directional   ~ 110 / (W+1000)**1/3 ~ 27 (Ix)**0.7

Lateral   ~ 110 / (W+1000)**1/3 ~ 25 (Ix)**0.7

3.6.1.2 Longitudinal- and lateral-directional oscillations with controls fixed following a single disturbance in smooth air shall exhibit the following characteristics:

(a) Any oscillation having a period of less than 5 seconds shall damp to one half amplitude in not more than one cycle. There shall be no tendency for undamped small amplitude oscillations to persist.

(b) Any oscillation having a period of less than 10 seconds shall damp to one half amplitude in not more than two cycles. There shall be no tendency for undamped small oscillations to persist.

(c) Any oscillation having a period greater than 10 seconds but less than 20 seconds shall be at least lightly damped.

(d) Any oscillation having a period greater than 20 seconds shall not achieve double amplitude in less than 20 seconds.

3.6.2 The requirements specified in 3.3.9 shall be extended to include control force stability, and the variations of pedal force and lateral control force with sideslip shall conform to the requirements specified in 3.3.9 for the corresponding control displacements. In addition, the requirements specifications in 3.3.9.1 shall apply with pedals free.

3.6.3 The helicopter shall, at all forward speeds and at all trim and power conditions specified in table I, possess positive, static longitudinal control force, and control position stability with respect to speed.

3.7 Vibration characteristics.

3.7.1 In general, throughout the design flight envelope, the helicopter shall he free of objectionable shake, vibration, or roughness. Specifically, the following vibration requirements shall be met:

(a) Vibration acceleration at all controls in any direction shall not exceed 0.4 g for frequencies up to 32 cps and a double amplitude of 0.008 inch for frequencies above 32 fps; this requirement shall apply to all steady speeds within the helicopter design flight envelope and in slow and rapid transitions from one speed to another and during transitions from one steady acceleration to another.

(b) Vibration acceleration at the pilot, crew, passenger, and litter stations at all steady speeds between 30 knots rearward and Vmax shall not exceed 0.15 g for frequencies up to 32 cps and a double amplitude of 0.003 inch for frequencies greater than 32 cps. From Vcruise to Vlimit the maximum vibratory acceleration shall not exceed 0.2 g up to 36 cps, and a double amplitude of 0.003 inch for frequencies greater than 36 cps. At all frequencies above 50 cps a constant velocity vibration of 0.039 cps shall not be exceeded.

(c) Vibration characteristics at the pilot, crew, passenger, and litter stations shall not exceed 0.3 g up to 44 cps and a double amplitude of 0.003 inch at frequencies greater than 44 cps during slow and rapid linear acceleration or deceleration from any speed to any other speed within the design flight envelope.

3.7.2 The magnitude of the vibratory force at the controls in any direction during rapid longitudinal or lateral stick deflections shall not exceed 2 pounds. Preferably, those vibratory forces shall be zero.

3.7.3 The helicopter shall be free from mechanical instability, including ground resonance, and from rotor weaving and flutter that influence helicopter handling qualities, during all operating conditions, such as landing, takeoff, and light.

4. QUALITY ASSURANCE. PROVISIONS Not applicable.

5. PREPARATION FOR DELIVERY

Not applicable.

6. NOTES

6.1 Intended use.

This specification establishes design requirements for flying and ground handling qualities of military helicopters.

Notice: When Government drawings, specifications, of other data are used for any purpose other than in connection with a definitely related Government procurement operation, the United States Government thereby incurs no responsibility nor any obligation whatsoever; and the fact that the Government may have formulated, furnished, or in any way supplied the said drawings, specifications, or other data is not to be regarded by implication or otherwise as in any manner licensing the holder or any other person or corporation, or conveying any rights or permission to manufacture, use, or sell any patented invention that may in any way he related thereto.

Custodians:

Army-TC   Navy- Wep

Air Force- AFSC   Preparing activity   Air Force- AFSC

Laminar Airflow Patent

United States Patent [19]

Millcroi et al.

[11] Patent Number: 4,988,471

[45] Date of Patent: Jan. 29, 1991

[54] APPARATUS AND METHOD OF FORMING A CONTINUOUS LAYER OF THERMOPLASTIC MATERIAL

[75] Inventors: Paul S. Brydel, Rockaway; Esgralo Millerol, Fort Lee, both of N.J.

[73] Assignee: Sano, Inc., Passaic, N.J.

[21] Appl. No.: 407,898

[22] Filed: Sep. 15, 1989

[51] Int. Cl.⁵ B29C 47/88
[52] U.S. Cl. 264/211.12; 264/216;
                                       425/224
[58] Field of Search 264/216, 212, 336, 355,
                          264/211.12; 425/224

[56]   References Cited

U.S. PATENT DOCUMENTS

2,946,927	12/1960	Crosby et al.	425/224
3,131,983	2/1964	Heller, Jr.	264/216
3,134,658	10/1919	Aronsa	264/101
3,347,962	10/1967	Dieck et al.	425/224
3,579,734	3/1971	Mehta	425/224
3,733,711	5/1973	Haydenthwaite	34/23
4,034,903	7/1977	Launckh	226/193
4,031,354	7/1977	Remmington et al.	264/216

FOREIGN PATENT DOCUMENTS

3505252	11/1912	France	264/212
48-2218	1/1973	Japan	264/216
53-125252	11/1978	Japan	264/216
2064272	4/1967	United Kingdom	264/216

Primary Examiner—Jeffery Thurlow
Attorney, Agent, or Firm—Lerner, David, Littenberg, Krumholz & Mentlik

[57]   ABSTRACT

In a cast film process, a thin sheet of semi-molten resin is extruded from a die and directed onto a rotating cylinder such that the transverse length of the sheet and the rotational axis of the cylinder are parallel to one another. During this stage of the process, it is desired to prevent air from entering and subsequently being trapped between the film and the casting surface of the rotating cylinder. Ultimately, entrapped air results in localized film distortion and non-uniform heat dissipation from the film. The cast film process employs an air deflector to deflect this laminar air flow layer which is entrained with the surface of the rotating cylinder before it can be forced between the casting surface and the film. The elimination or substantial reduction of the tangential velocity of the entrained air layer provides more uniform adherence of the extruded film to the casting surface.

36 Claims, 4 Drawing Sheets

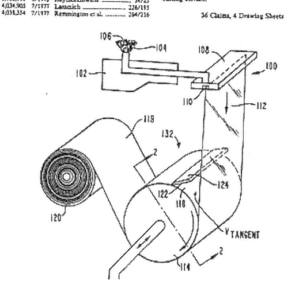

ABSTRACT

In a cast film process, a thin sheet of semi-molten resin is extruded from a die and directed onto a rotating cylinder such that the traverse length of the sheet and the rotational axis of the cylinder are parallel to one another. During this stage of the process, it is desired to prevent air from entering and subsequently being trapped between the film and the casting surface of the rotating cylinder. Ultimately, entrapped air results in localized film distortion and non-uniform heat dissipation from the film. The cast film process employs an air deflector to deflect this laminar layer air flow which is entrained with the surface of the rotating cylinder before it can be forced between the casting surface and the film. The elimination or substantial reduction of the tangential velocity of the entrained air layer provides more uniform adherence of the extruded film to the casting surface.

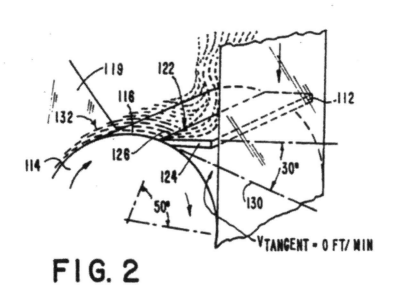

FIG. 2

TABLE I
DEFLECTOR POSITION~AIR VELOCITY, FT/MIN

Deflector Position	Air Velocity, FT/MIN
*NO DEFLECTOR	~ 210 to 230
* 1 (90 degrees)	~ 210 to 230
* 2 (80 degrees)	~ 210 to 230
* 3 (70 degrees)	~ 125
* 4 (60 degrees)	~ 75
* 5 (50 degrees)	~ zero

TABLE II
RADIAL DISTANCE, INCHES ~ AIR VELOCITY, FT/MIN

Radial Distance, Inches	Air Velocity, FT/MIN
0.020	~ 220 to 240
0.060	~ 220 to 240
0.100	~ 210 to 230
0.200	~ 180 to 200
0.300	~ 150 to 170
0.400	~ 140 to 160
0.500	~ 120 to 140
0.600	~ 100 to 120
0.800	~ 90 to 110
1.000	~ 60 to 80

Supplemental Information

The following is some further clarification of the important terms defined in section Basic Helicopter Theory:

Inviscid Flow
Euclidean Space
Vector
Vector field
Flux
Divergence
Gradient
Vortex

Let's begin by talking about the natural phenomenon that occurred to me this January while planning a ski trip. The weather forecast was predicting an unusually rare artic "polar vortex" wind to pass thru the West Virginia ski area I was going to. Sure enough the temperatures dropped into the minus numbers and the winds exceeded 20 plus mph.

Why did they classify this artic wind a "polar vortex?" By definition a polar vortex, also known as a circumpolar whirl, is a persistent, large-scale cyclone located near either of a planet's geographical poles. In the Northern Hemisphere the Artic vortex is elongated in shape, with two centers one near Baffin Island, Canada and the other over northeast Siberia. These polar vortices exist from the stratosphere downward into the mid-troposphere, have a counter clockwise spin, cover approximately 620 miles and their intensity is driven by the temperature differential between the equator and the poles. This cyclone type wind is an area of closed circular fluid motion rotating in the same direction as the earth. In the Northern Hemisphere they are usually characterized by inward spiraling winds that rotate anti-clockwise. The largest types are cold-core, low atmospheric centered polar cyclones. There are many types of cyclones but the following facts are common to all:

1) A cyclone, from a meteorological point of view covers any closed low-pressure circulation.

2) Its center is at the lowest atmospheric pressure and is referred to as the eye.

3) There are two forces at play that keep a cyclone from collapsing, imploding or flying apart. One is the outward force generated by the Coriolis effect created by the shear nature of the wind flow around the cyclone (cyclonic circulation). The other is the opposing balancing force due to the negative pressure gradient at the core and the area over which it acts.

At the time of my ski trip, this artic airflow path was amplified somehow and the center of one of the polar vortex axes shifted south. Due to this aberration the usual mid-latitude airflow pattern buckled and the artic vortex pushed into the Mid-Atlantic States. This ultimately generated the severe cold conditions I experienced.

Of course this particular vortex of Mother Nature was on the large scale but more typical is your tornado or twister. Those are also technically called a vortex if the violently rotating column of air is in contact with both the surface of the earth and the base of a cumulus cloud. As for your everyday experience of a vortex just look at the circular flow pattern the water makes as it drains out of your kitchen sink.

Returning to our to our "X" marks the spot example of the hill (pages 34 thru 36), let's say that instead of just going to "X" I am also required to build a 6-inch diameter water pipeline directly to it. Additionally a specific volume of water must flow each hour from the "source" to the "exit point". The building of the theoretically straight pipeline is easy but to measure the flow rate requires some thought. To achieve this goal we will insert two fine mesh grids into the pipe, perpendicular to the flow, one at the source the other at the exit point. At each intersection point on the grid we will attach a short string with a small sphere attached to its free end. If we have water flow then all these strings will position themselves in the direction of the flow path. Once these strings have magnitude (the force of the water pushing them forward) and direction (the string following the flow path) they can be considered vectors at each point on the grid. This can now be considered a Vector Field. In this string / sphere setup we know there is sufficient force on each one to point in the direction of the flow but we really don't know the exact magnitude of the force on each of the string / sphere vectors. If we add an infinitesimally small spring between each grid point and the string / sphere when we look at the vector field the string assemblies with the longest length will be the ones having the greatest flow force on them. Since we know from Hook's law the force exerted on the spring is proportional to its extended length we can determine exactly the force associated with each vector in the field. This Vector Field can represent the total force of the grid and with some mathematics, such as taking the line integral of the Vector Field; we can obtain the work done by the force as it moves along the path. Knowing the work done by the water can be used to determine its flow rate. As mentioned previously Vector fields can represent the velocity of a moving flow in space.

Next would be the concept of flux which depends on the size and strength of the Vector Field. Returning to our pipeline example of the 6-inch diameter pipeline to transfer water from one point to another. Now we are told that it must be able to handle the water transfer of 10,000 gallons of water to point "X". The water source will be a moveable "water cannon" used in strip mining. Intuition tells us in order to get the most water (most flux) passing thru our pipe entrance wire-mesh-grid the water cannon should be aligned with the centerline of the pipe and be as close as possible to the pipe entrance. In this setup the "water cannon" (flux source) delivers water at its maximum strength. Its straight-line flow (orientation angle) into the pipe is optimized, and since the wire-mesh-grid size remains the same any change in either increasing water cannon distance from the pipe or its orientation to it will result in less water (less flux) entering the pipe. In fact no water enters the pipe if the water cannon is perpendicular to the wire-mesh-grid. To sum things up "flux" depends on the magnitude of the source, the angle between, and the size of the surface.

Recalling our definition of divergence as the rate of change of volume of a flow the question becomes what is the volume when we are talking about a point.

Returning to our pipeline example looking closely at our wire-mesh-grid, we would see that every grid space is equal in the x and y direction. Lets define a very small volume at this grid location such that it's z direction (a direction parallel to the pipe centerline) is equal to either x or y to make it a uniform cube. Under steady-state conditions the amount of water entering (flux entering) the front cube face equals the amount of water exiting (flux leaving) the back cube face. Under normal operating conditions, water being virtually an incompressible fluid, the steady-state condition described exists for all but extreme flow dynamics.

What if this pipeline was used to transfer a gas to point "X"? Under steady-state conditions the amount of gas entering and leaving the cube would be in equilibrium, similar to the water flow case. In this particular example, since we are moving a gas we will use a transfer pump at both the entrance and the pipe exit. This means for steady-state flow the entrance pump will be forcing gas into the pipe at the same rate the exit pump is exhausting the gas out. Looking at our reference cube flux what if the operator mistakenly reverses one of the two pump's directions? If it is the entrance pump reversed then any gas in the pipeline will be evacuated. At the cube level "flux" will be leaving it from all sides and this is referred to as "positive divergence". If the exit pump direction is reversed then gas is being pumped into the pipe at both ends increasing the amount of gas in the pipe. At the cube level flux is increasing from all sides and this is called "negative divergence". Obviously, positive and negative divergences are the extreme flux transfer conditions. In reality it stands to reason the flux entering or leaving any particular face of the cube, more than likely, will be the same.

Turning our cube over to the math guys for further explanation, the first thing they would do is calculate the incremental average flux change in the x, y, and z face directions then sum those values to get a total flux change. In our steady-state example everything entering and leaving is equal so net flux crossing the entering and exiting cube face in the x direction is 0, net flux in the y direction cube faces is zero and net flux in the z direction cube faces is zero. In this case the "total net flux" = 0 +0 +0 = zero which means there is no net flux. Zero flux does not infer there is no flow. Actually the magnitude and direction (velocity vector) of the flow in the z "cube face" direction can be quite large depending how much force is being applied to transfer the gas, and yet in the x and y "cube face" direction there is no flow at all. Eventually, from the analytical sequence of things our cube will be investigated in incrementally smaller dimensions until what remains of "our cube in a grid" is just a "point in the grid plane with sides dy, dx and dz" with a vector attached to it aligned with the pipe centerline.

If we put some imbalance in the flux field at the point center of our small cube such that "total net flux" = 1 –3 +6, this tells us at that point, where x

=1, y =-3, and z = 6 there is divergence. This implies our velocity vector in our example is no longer traveling in the direction of the pipeline centerline. It has been diverted to a different direction or in math vernacular it has divergence. Divergence represents the rate of change of volume of a flow.

Next up is the concept of gradient meaning the rate of change of a function which was previously explained in our hill example. Since we started with the pipe example lets go back to it for a moment. At one point during our gas transfer process our pipe gets a hole in it resulting in automatic shut down of both pumps and sealing off both ends of the pipe. The equations developed in this case refer to the pressure drop due to the size hole. At any point we can get pressure and if we differentiate the equation to get the rate of change of that pressure, the gradient vector will point to the direction where the greatest change in the function is taking place, which indicates where the pressure is lower, that being the hole.

Curl was defined as circulation per unit area, rate of circulation, or better yet, the rotation of a flow. To explain this further we go back to our gas pipeline example. Looking back at our pipe entrance vector field we notice that the vectors are not pointing in the direction of the pipe centerline but are skewed at some angle. This indicates a rotation of the vector field. Stated another way the vector field is curling and the magnitude of the vectors in the field indicate the twisting force at each point.

What caused the vector field to rotate? The centrifugal pump that added a swirl to the flow as it forced the gas into the pipe probably caused it. The "vorticity" term now denotes the extent a fluid will rotate about itself (its "curl"). Depending on our pump size it is conceivable to see the resultant flow separate into a multitude of independent circulatory streams in the overall swirl of the downstream flow.

In the gas pipe example we have the multitude of independent circulatory streams in the downstream flow. Each one of these streams can be visualized as a thin vortex tube that has vorticity (rotational flow about itself). If you shrink the diameter to an infinitesimally small size (in the limit as the diameter is made small) but keep the circulation fixed, this region of vorticity is identified as the vortex filament. The surface of the vortex filament is developed as an infinite number of vortex lines drawn through each point of the closed curves defining the surface with each vortex line's tangent always parallel to its local vorticity vector. Simply said a vortex tube filament is a bundle of vortex lines.

Let's return to the tornado reference and look at it from a fluid dynamics standpoint. Upon inspection of the tornado you would observe that this vortex tube is three-dimensional and its spout shape dynamically erratic along with its core centerline. All tornado funnels I have seen have some spiral curvature to them so it follows its vortex line would also have spiral curvature. If you were to draw a line tangent to any point on this particular vortex line it would point in the direction of vorticity vector at that point. In something as violent as a tornado it is hard to imagine fluid flows not rotating about themselves and those vortex lines would have substantial vorticity (flow rotation or twist to them).

Bibliography
1. Theory of Wing Sections by Ira Abbott and Albert E. Von Doenhoff; Dover Publications, Inc. New York
2. Helicopter Theory by Wayne Johnson; Dover Publications, Inc. New York
3. Standard Handbook for Mechanical Engineers by Baumeister and Marks; McGraw-Hill Book Company, New York
4. Analytical Mechanics of Gears by Earle Buckingham; Dover Publications, Inc. New York
5. Fluid Dynamics by William F. Hughes and John A. Brighton; Schaum's Outline Series; McGraw-Hill Book Company, New York
6. Mechanical Engineering Reference Manual by Michael R. Lindeburg; Professional Publications, Inc. Belmont, Ca. 94002

Printed in Great Britain
by Amazon

70362981R00258